Engineering Rules

Hagley Library Studies in Business, Technology, and Politics
Richard R. John, Series Editor

Engineering Rules

Global Standard Setting since 1880

JoAnne Yates
Craig N. Murphy

Johns Hopkins University Press
BALTIMORE

This book was brought to publication through the generous
assistance of the Robert L. Warren Endowment.

Printed in the United States of America on acid-free paper

Johns Hopkins Paperback edition, 2021
2 4 6 8 9 7 5 3 1

Johns Hopkins University Press
2715 North Charles Street
Baltimore, Maryland 21218-4363
www.press.jhu.edu

The Library of Congress has cataloged the hardcover edition of this book as follows:

Names: Yates, JoAnne, 1951–, author. | Murphy, Craig, author.
Title: Engineering rules : global standard setting since 1880 / JoAnne Yates, Craig N. Murphy.
Description: Baltimore : Johns Hopkins University Press, 2019 | Series: Hagley Library studies in business, technology, and politics | Includes bibliographical references and index.
Identifiers: LCCN 2018036808 | ISBN 9781421428895 (hardcover : acid-free paper) | ISBN 9781421428901 (electronic) | ISBN 142142889X (hardcover : acid-free paper) | ISBN 1421428903 (electronic)
Subjects: LCSH: Standards, Engineering—History.
Classification: LCC TA368 .Y37 2019 | DDC 620.002/18—dc23
LC record available at https://lccn.loc.gov/2018036808

A catalog record for this book is available from the British Library.

ISBN 978-1-4214-4003-3

Special discounts are available for bulk purchases of this book. For more information, please contact Special Sales at specialsales@press.jhu.edu.

Johns Hopkins University Press uses environmentally friendly book materials, including recycled text paper that is composed of at least 30 percent post-consumer waste, whenever possible.

contents

acknowledgments

We began our research on voluntary standard setting more than a decade ago and started to focus on this book in 2012, after we had published a short volume on the International Organization for Standardization (ISO) and after JoAnne had completed a five-year administrative assignment. Over those years, we have been helped by dozens of people and many institutions—our funders, those who have provided us with information, our research assistants, and those who gave us feedback on the work in progress.

Both of us are deeply grateful to the Center for Advanced Study in the Behavioral Sciences (CASBS) at Stanford for its incomparable support of JoAnne as a fellow and Craig as a visiting scholar during the 2012–2013 academic year. Craig also thanks the Radcliffe Institute for Advanced Study at Harvard for a productive year (2008–2009) as a fellow, working on both the ISO book and the research for this book. Both of us thank our respective home institutions, Sloan School of Management at Massachusetts Institute of Technology, and Wellesley College, for their research support.

This book would have been impossible without the gracious help of the many people who gave us access to, and helped us understand, the many materials on which we relied. Tricia Soto and Amanda Thomas at the CASBS library helped us identify and obtain nineteenth- and early twentieth-century published sources through the Stanford University Libraries. Béatrice Frey graciously gave us access to the early records of the ISO and its predecessors and the files of the secretary-general's office. Stacy Leistner gave us similar help with the records of the American National Standards Institute (ANSI) and its predecessors. Many people at the International Electrotechnical Commission (IEC), including Gabriela Ehrlich, Guillaine Fournet, and especially Claire Marchand, helped us with IEC records. Pierre Sebellin at the IEC helped us with materials for Comité international spécial des perturbations radioélectriques (CISPR) that were held by the IEC. Heather Heywood provided expert help at the International Telecommunication Union (ITU) Archives. Wang Ping of the Chinese National Institute of Standardization was extremely helpful with materials about China. Peter Anthony, archivist at Deutsches Institut für Normung, and Alice Tepper-Marlin at Social Accountability International also went out of their way to help with documents and photographs.

We were able to provide an in-depth and detailed look at the development of the postwar global standard-setting system and at the work of particular standard-setting committees only because of the extraordinary help we received in gaining access to the files of individual standard setters. We are grateful to Lars and Lolo Sturén for letting us use their father's papers. Terry Snyder, librarian of Haverford College, alerted us to the existence of the Murray Freeman Papers and helped us gain early access to them.

Our access to the materials that made chapter 6 on EMC standardization possible started with Sheldon Hochheiser, then at the IEEE History Center, who referred us to Dan Hoolihan and Don Heirman, experts in electromagnetic compatibility (EMC) who were working to compile historical documents of the Institute of Electrical and Electronics Engineers' (IEEE) EMC Society. Without Don and Dan, we could not have written this case study. After he had scanned some of the documents for his EMC history project with IEEE, Dan sent JoAnne the Leonard Thomas Papers he had obtained from Thomas's family after his death. He also put us in touch with Ralph Showers, who died before we could interview him but whose daughters, Janet Showers Patterson and Virginia Showers White, generously gave us access to the Showers Papers, in an apartment in a senior living center and in the attic and basement of the family home, and found photos of him for illustrations. In the revision stage, Janet was enormously helpful in sharing personal family documents to provide a fuller picture of her father. Don Heirman gave JoAnne early electronic access to the documents he was scanning for the Heirman Papers at the Purdue Archives. Our colleague Leslie Berlin at CASBS and her colleague Henry Lowood, both of the Silicon Valley Archives, pointed us towards the MacQuivey Papers in the Stanford University Special Collections. We also thank Erik Rau of Hagley Museum and Library for providing a home for the Showers and Thomas Papers.

For chapters 7 and 8, we very much appreciate the access to the Web Cryptography Working Group that Jeff Jaffe, CEO of the World Wide Web Consortium (W3C), gave to JoAnne. In the summer of 2012, he responded to JoAnne's request by giving her full access as an invited expert to the recently started working group. He also read the draft chapter for us and provided corrections and comments (any remaining errors are, of course, our own). We also thank working group chair Virginie Galindo for welcoming JoAnne to the group and W3C staff contacts Harry Halpin and Wendy Seltzer for explaining technical and process issues to her throughout the project and Wendy for providing aid at the final stages of manuscript writing and preparation, including giving us permission to use her photograph of Virginie. We thank all the members of the working group who allowed JoAnne to interview them and who graciously provided permissions to quote from these interviews. Maria Auday of W3C also helped us obtain a photograph of Tim Berners-Lee.

In addition to those contributing to chapters 6, 7, and 8, we thank the many other standardizers we met during the course of our work, who were generous with their time, explanations, and contacts, including the following: Scott Cooper from ANSI, Trond Arne Undheim from Oracle, Pekka Isosomppi from Nokia, Russ Housley from the Internet Engineering Task Force (IETF), and Andy Updegrove, an attorney who established many standards consortia. In addition, scholars who have worked on standards were very supportive and encouraging of this project; JoAnne particularly appreciates Marc Levinson's willingness to respond to queries for detailed information on container standardization not spelled out in his book *The Box*.

We are also grateful to our research assistants over the years: Yulia Poltorak, who read and summarized articles on standardization, Veronica Jardon, who worked on the color television case, and Hilary Robinson, who did research in the ITU archives and interviewed ITU leaders, all from MIT; Naa Ammah-Tagoe, who helped us with French language sources, and Honor McGee, who worked on the social responsibility cases, both at the Radcliffe Institute; Maria Nassén at Wellesley College, who conducted interviews in Sweden and helped us with Swedish-language sources. Finally, we thank Michael Wahlen at MIT, for working with us during the summer of 2017 to prepare the first complete manuscript for submission.

We have presented papers on this work to a variety of audiences and received useful comments from them that have helped sharpen our ideas. We presented this work at an early stage to our very interdisciplinary set of colleagues at CASBS and received encouragement and feedback. In addition, our CASBS colleague Deborah Tannen suggested the working title we used for many years, *Standards Bearers*. We also thank our audiences at presentations in the following conferences and seminars: the Radcliffe Institute's seminar series; the American Political Science Association; the Business History Conference; the European Business History Association; the Sociology Department at Emory University; Harvard Business School's Business History Seminar; the International Studies Association; the Wellesley College Department of Political Science; the University of Massachusetts Boston's Seminar on Organizations and Change; the MIT Sloan School's Organization Studies seminar and Technological Innovation, Entrepreneurship and Strategy seminar; the Wharton Seminar on the Evolution of Organizations and Industries; the London School of Economics' Department of Accounting; the Davis Conference on Qualitative Research of the University of California, Davis; the Governing through Standards conference at the Royal Danish Institute for International Studies; the University of Southern California's Center for International Studies; University of California, Irvine; Stanford University's Seminar in Work, Technology, and Organization; and the European Academy for Standardisation.

We are indebted to Robert J. Brugger at Johns Hopkins University Press for his enthusiasm and encouragement in early stages of the project, to Matt McAdam for seeing it to completion at the Press, and to series editor Richard R. John for his close reading and developmental advice. We also appreciate Carrie Watterson's careful copyediting and cooperative approach. We deeply appreciate Deborah Fitzgerald's willingness to read the entire first-draft manuscript and to give us very useful comments calibrated to the project's late stage and looming deadline. We thank both the anonymous reviewers of the first full draft for their very helpful comments and suggestions, which helped shape the final draft. We also thank many colleagues for reading and commenting on specific chapters and responding to questions, including Meg Graham, Andrew Russell, Andrew Updegrove, and Marc Levinson. We thank our friends and colleagues for engaging in endless conversations about standardization. Particular thanks go to Fred Turner at Stanford; Christopher Candland and Beth DeSombre at Wellesley; Jinyoung Kang at the University of Massachusetts, Boston; and Deborah Fitzgerald and Harriet Ritvo at MIT.

We also thank our extended families for their support during the research and writing process. We gratefully acknowledge our two cats, Max and Minnie-the-Who, who spent most of their lives with us supporting (and sometimes getting in the way of) our work on this project. Sadly, they did not live to see the final product. Finally, we thank each other for this rare opportunity to work together on an intellectual project, and we are pleased to say that, contrary to some expectations, our marriage appears to have survived it!

acronyms

ABNT	Brazilian Association of Technical Standards
AESC	American Engineering Standards Committee
AFNOR	French Standards Association
AIEE	American Institute of Electrical Engineers
AIME	American Institute of Mining Engineers
ANAB	American Quality Society National Accreditation Board
ANSI	American National Standards Institute
API	application program interface
AREMWA	American Railroad Engineering and Maintenance of Way Association
ARPA	US Defense Department's Advanced Research Projects Agency
ASA	American Standards Association
ASCE	American Society of Civil Engineers
ASCII	American Standard Code for Information Interchange
ASME	American Society of Mechanical Engineers
ASQ	American Society for Quality
ASTM	American Society for Testing Materials
BBC	British Broadcasting Corporation
BESA	British Engineering Standards Association
BSI	British Standards Institution
BITA	British Iron Trades Association
CBS	Columbia Broadcasting System
CCIF	International Long-Distance Telephone Consultative Committee
CCIR	International Radiocommunication Consultative Committee

CCIT	International Telegraph Consultative Committee
CCITT	International Telephone and Telegraph Consultative Committee
CEN	European Committee for Standardization
CENELEC	European Committee for Electrotechnical Standardization
CERN	European Organization for Nuclear Research
CFT	French Television Company
CISPR	International Special Committee on Radio Interference
CR	candidate recommendation
CSA	Canadian Standards Association
CSR	corporate social responsibility
CTI	Color Television Incorporated
DIN	German Institute for Standardization
ECMA	European Computer Manufacturers Association
EEC	European Economic Community
EMC	electromagnetic compatibility
ESC	Engineering Standards Committee
ETSI	European Telecommunications Standards Institute
ETVs	German regional electrotechnical societies
EU	European Union
FCC	Federal Communication Commission
FPWD	first public working draft
GATT	General Agreement on Tariffs and Trade
HTML	hypertext markup language
IAB	Internet Architecture Board
IAF	International Accreditation Forum
IATM	International Association for Testing Materials
ICC	Interstate Commerce Commission
ICE	Institution of Civil Engineers
IEC	International Electrotechnical Commission
IEE	Institution of Electrical Engineers
IEEE	Institute of Electrical and Electronics Engineers
IETF	Internet Engineering Task Force
IFIP	International Federation for Information Processing
IFRB	International Frequency Registration Board
IISD	International Institute for Sustainable Development
ILO	International Labor Organization

IMechE	Institution of Mechanical Engineers
INWG	International Network Working Group
IPTO	Information Processing Techniques Office
IRE	Institute of Radio Engineers
ISA	International Federation of the National Standardizing Associations
ISEAL	International Social and Environmental Accreditation and Labeling
ISO	International Organization for Standardization
ITU	International Telecommunication Union
JOSE	JavaScript object signing and encryption
JTC1	ISO/IEC Joint Technical Committee 1
Marad	US Maritime Administration
MIT	Massachusetts Institute of Technology
NEDCO	Netherlands' statement concerning ISO liaisons and activities
NEMA	National Electrical Manufacturers Association
NPL	National Physics Laboratory
NTSC	National Television System Committee
NWG	Network Working Group
OECD	Organisation for Economic Co-operation and Development
OMB	Office of Management and Budget
OSI	open systems interconnection
PAL	phase alternating line
PAS	publicly available specification
PR	proposed recommendation
PTT	postal, telegraph, and telephone agencies
QMS	quality management system
RAND	reasonable and nondiscriminatory
RCA	Radio Corporation of America
RFC	request for comment
RFI	radio frequency interference
RKW	Reich Board for Economic Efficiency
RTU	Radiotelegraph Union
SAE	Society of Automotive Engineers
SAI	Social Accountability International
SECAM	sequential color with memory

SG	study group
SIS	Swedish Standards Institute
SNA	IBM's System Network Architecture
TC	technical committee
TCP/IP	transmission control protocol / internet protocol
UIR	International Radio Union
UHF	ultrahigh frequency
UNCTC	United Nations Centre on Transnational Corporations
UNESCO	United Nations Educational, Scientific, and Cultural Organization
UNSCC	United Nations Standards Coordinating Committee
URL	universal resource locators
USNC	US National Committee
VDE	Association of German Electrotechnicians
VDI	Association of German Engineers
VHF	very high frequency
W3C	World Wide Web Consortium
WEF	World Economic Forum
WG	working group
WHATWG	Web Hypertext Application Technology Working Group
WHO	World Health Organization

Engineering Rules

Introduction

If you are reading this text in book form, the page in front of you is likely to be of a standard size that was determined almost a century ago by a committee of German engineers within a private standard-setting organization, and it was probably shipped in containers standardized more than half a century ago. If you are reading this on a screen, the specific set of electrical impulses that creates the characters that appear before you was standardized by a more recent international committee of engineers, as were the software languages that make it possible for your e-reader to generate this text. Other such committees established the standards for the battery your device uses and for every switch and junction, transmission line and tower, between your battery's charger and the power plants that provide its electricity. Similar committees of engineers standardized the cement and steel that make up those towers, the rivets or nuts and bolts that hold them together, and all the machinery that erected those towers and strung those wires. The power plants where the wires begin rely on scores of other standards created by yet more committees and organizations. Even the two or three little screws that you might see on the back of your device conform to a standard. In fact, the problem of creating standards for the size, shape, and spacing of screw threads was one of the motivations for a precursor to this kind of standard setting in the nineteenth century. And the vast network from which you downloaded this text relies on the standard setting done by some relatively new private organizations, including the Internet Engineering Task Force (IETF) and the World Wide Web Consortium (W3C), that follow procedures similar to those developed more than a century ago.

There are similar standards for almost every other object we use and every built space we inhabit. These standards have shaped the course of industrial development by fixing the technological platforms on which further innovation occurs. (For example, all of the later innovations that gave us today's electronic readers depended on the first American Standard Code for Information Interchange, or ASCII, set in 1963.) Without such standards, most of what we buy would be more difficult to produce, and conflicts between merchants and customers likely would be more intense; moreover, without such standards the power grids, water and sewer systems, and networks for communication that we rely on every day would not exist in the same form they do now. This kind of private standardization has come to provide a critical infrastructure for the global economy.

Although the standards developed this way may not always be optimal technically, reflecting conflicts and compromises among engineers representing different firms and interests, much of what these private processes have achieved has been for the good. While there have been downsides to economic globalization, the standard container has certainly lowered the cost of consumer products everywhere in the world, and software standards have encouraged firms to compete by developing new features, products, and services that are of much greater interest to most of us than the platforms that have been standardized.

Standards Bearers

This book is a history of the engineers and organizations that develop and operate the vast yet inconspicuous global infrastructure of private, consensus-based standard setting, a process with an astonishingly pervasive, if rarely noticed, impact on all our lives. We might call this process, these people, and these organizations "standards bearers" because they have borne standards—"documents that provide requirements, specifications, guidelines or characteristics that can be used consistently to ensure that materials, products, processes, and services are fit for their purpose"[1]—in so many different ways. They gave birth to most of the private standards we use, they supported them, and they displayed them as marks of honor. They also turned them in different directions at different times, bearing them toward the differing national interests of the warring governments during two world wars, toward international markets in the 1960s through 1980s, and toward the global concerns of businesses since the 1990s. More often, they just steered them toward a vision of a world united through private, voluntary standardization.

The standards bearers' story starts in the late nineteenth century. Anxious to solidify their claim to professional status by offering some kind of social service, engineers on both sides of the North Atlantic invented a new process to create technically sound standards for industry. The engineers' process provided a timely way to set desirable standards that would have taken much longer to emerge from the market and that governments were rarely willing to set. The new process involved consensus-seeking committees of technical experts representing a range of stakeholders. Through iterative research, discussion, deliberation, and often voting, members of these committees attempted to reach a consensus that had the buy-in necessary for voluntary adoption by all parties. This process has regularly created standards that have then been widely adopted—for screw threads that held things together, for shipping containers that enlarged global markets, and for the World Wide Web.

This private, nongovernmental activity occurred within a host of not-for-profit standard-setting bodies at many levels, a new type of organization first established in the years immediately before and after 1900. These bodies quickly produced many national standards (and some international standards) that began to help regulate economies around the world. From the end of World War II through the mid-1980s, a new generation of engineers and related experts launched additional organizations that established and that still maintain many of the product standards that made an increasingly international economy possible. At the end of the twentieth century, a later generation of engineers—along with a much more diverse set of additional experts—began using the process in new, more global ways. They created one set of new organizations to maintain the internet and make it more widely usable, and another to help regulate some negative side effects of the new global economy. Today, most existing industrial standards, as well as many significant social, environmental, and service-sector standards, have been set through this process. Its effects are ubiquitous, critically important, and all but invisible.

Private Regulations

Given the importance of such standards, it may seem surprising that so many have been set by the relatively obscure committees of technical experts that are the subject of this book. In the real world, you might ask, aren't governments the ones responsible for assuring that "materials, products, processes and services are fit for their purpose" in those cases when the market fails to do so? Perhaps they should be, but, since the late eighteenth century, national governments have been slow to take on the task, even when they believed that standardization was essential to the success of industrialization, a goal that governments increasingly embraced, and even in fields with especially clear public interest in common standards, such as fundamental standards (weights and measures) and standards for safety and health. This reluctance originally stemmed from the fact that *local* political leaders or businesses often had a vested interest in maintaining traditional systems of multiple standards.

France and the United States are cases in point. Unsurprisingly, creating universal standards was a hope of the Enlightenment, which gave both countries their current forms of government. One of the demands of the French Revolution was for a single national set of weights and measures that would abolish the *seigneurage* reaped by the local nobility (the *seigneurs*) who maintained incompatible standards from one district to the next. Yet, even the great rationalist Denis Diderot doubted that the National Assembly would quickly agree upon and universally enforce a national system. He was correct; it took France until 1840 to fully adopt the metric system introduced in 1799. The US government proved even slower. Its 1787 Constitution granted Congress the power to set fundamental standards, and President George Washington's first State of the Union address urged the legislature to take up the task immediately. Nevertheless, more than a century passed before Congress established a weak National Bureau of Standards in 1901. At the end of the nineteenth century, 25 different fundamental units of length (e.g., inches, feet, and rods) were still used in different parts of the United States— "three had the same length but different names, the remainder different names and values that were unrelated to each other."[2] Many attempts to impose the widely used, Enlightenment-inspired metric system in the United States followed the 1901 breakthrough. The last ended with President Ronald Reagan's defunding of the office that was to enforce the 1975 Metric Conversion Act, an action he took in response to deeply committed constituents, some with a nationalist attachment to the American traditional systems of measurement, but most—especially smaller businesses—concerned about the cost of changing to a new system.[3]

More generally, when any governmental or intergovernmental body sets standards and makes them compulsory, the new standards tend to create costs for, or take some advantage away from, deeply committed groups with the power to resist or even block legislation or enforcement. For the same reason, governments tend not to oppose voluntary standards set by private actors, standards that are not officially enforced, though they may be referred to in regulations. Nevertheless, one political scientist who studied the long history of governmental standard setting in multiple countries argues that it is unsurprising that "public actors may impede standardization even in the face of high private demand and clear public welfare gains." For that reason, he says, what really needs to be explained is not governments' frequent inability to set and enforce standards but the infrequent situations when governments decide that they must do so.[4]

Those cases of governmental (or intergovernmental) standard setting are not the focus of this book, although government agencies regularly appear as promoters of private standard setting and as users of private standards. For example, the early history of the engineers' consensus-based process is intertwined with the growing recognition by governments that they should set safety standards to regulate some of the new and terrifying products of industrialization, especially the new steam boilers that seemed to explode so frequently. Late nineteenth-century governments also wanted standards to refer to when making their own purchases of steamships, steam locomotives, and the like. This interest in making sure that the things governments purchased were "fit for purpose" has continued through every generation of innovation since, making governments interested in standard setting but not necessarily in doing it themselves.

Many private actors—final consumers, companies that produce and use various goods and services, engineers and others who design those goods and services, and scientists whose work informs the engineers—have long had more consistent and compelling interests in setting common standards. For that reason, a wide range of nongovernmental organizations—professional societies, trade associations, and ultimately the dedicated standard-setting bodies that are the focus of this book—began to experiment with different modes of standard setting during the nineteenth century, eventually converging on a voluntary, consensus-based process after 1880. Economists call the result "standardization by committee" and argue that there are theoretical reasons for considering it superior to standardization by government or through the market (the latter of which has produced ubiquitous, if imperfect, standards such as the QWERTY keyboard and Microsoft Word). In an important 1988 article, Joseph Farrell and Garth Saloner compared formal models of the different kinds of standard setting.[5] Under the reasonable assumptions that they make, standardization through the committee process outperformed the market. This fits with common experience. Markets tend to maintain competing standards for a long time, leading to a great deal of unnecessary cost and frustration. Think of all the different electrical plugs and outlets that world travelers need to worry about or the strange fact that every mobile phone seems to have its own special connector cord. Both are the result of market failures that private standard setting has been unable to change.

Farrell and Saloner also point to a mechanism that is even more effective than committees acting alone: standard-setting by private committees in a world in which powerful actors (dominant firms, leading states, or formally organized trading regions such as the European Union) can ignore the process and set a standard that many others are likely to follow. This is a fairly accurate description of the world in which we actually live, a world of widespread voluntary standard setting by consensus-seeking committees of stakeholders, a system that, by its very nature, allows powerful companies and governments to have the option of setting their own standards before consensus has been reached.

How did the actual committee system of private standard setting arise and develop in our real world of interested and unequally powerful companies and governments? That is the question we set out to answer. Our focus is global (but with a US bias reflecting our linguistic limitations and the availability of archival material). We occasionally discuss the other ways standards get set, primarily to tease out why the private, consensus-based system ended up with its specific, if vast, focus. For example, governments and intergovernmental organizations have become the leaders in setting or enforcing standards around such issues as steam boiler safety in the nineteenth century, worker safety in the mid-twentieth century, and pollution control in the late twentieth century. Yet the private system was also involved in each of these fields, and the boundaries established between the competing systems help explain why both systems continue. We also discuss equally illuminating cases where governmental agencies, including the US Federal Communications Commission and the European Commission, chose to rely on the private national, European, or global standard-setting bodies, rather than to do the work of setting standards themselves. In addition, we examine a case in which a public-private hybrid committee modeled on private consensus-based standardization emerged within an intergovernmental body, the International Telecommunication Union.

The market's role in standardization similarly comes into our story in connection with standards wars and consortium standards. Standards wars arise when companies producing new technologies strive to establish their own product as a de facto standard through the market. (Older readers will remember the failure of the market to give us a timely and high-quality standard for video tape recordings from the 1970s through 2000, the period when Sony's Betamax competed with JVC's VHS.[6]) Standards wars have sometimes undermined, and sometimes spurred, the voluntary, consensus-based process. Standard setting by closed consortia of cooperating companies has become a significant practice in high-tech fields since the 1990s. These consortia set and maintain standards that consortium members hope will prove successful in the marketplace. Some champions of the older system of private standard setting see consortia as a major challenge to that process. Others note that standard setting by consortium is becoming more like the consensus-based system. Moreover, to avoid losing relevance, international standards bodies have developed new processes to allow them to bring such one-off consortium standards into their broader standardization framework.

The relative ambiguity of the boundaries between the governmental (or intergovernmental) and private systems of standard setting in some areas—even the typical confusion about where the boundaries do and should lie—is part of what has made the private system so powerful and so attractive to generations of public policy theorists and social activists across the political spectrum.[7] At the end of World War I, Sidney and Beatrice Webb, leaders of the Fabian Society, argued that further extension of the practices of the first national standards body, the British Engineering Standards Committee, should be an important part of the constitution of a socialist Britain; leading liberal internationalist theorist Mary Parker Follett saw the process as essential to an emerging progressive "world state"; and Republican US commerce secretary (and engineer) Herbert Hoover championed private voluntary standardization as part of his modern "associational" state.[8] A little less than a century later, another prominent liberal internationalist, UN secretary-general Kofi Annan, and a worthy ideological successor to Hoover, the World Economic Forum's Klaus Schwab, made similar claims.[9]

The Organizations, the Process, and the People

Organizations are a critical component of the private standardization system. In a recent history of a related topic—the modern idea of world government—Mark Mazower calls ISO "perhaps *the* most influential private organization in the contemporary world, with a vast and largely invisible influence over most aspects of how we live, from the shape of our household appliances to the colors and smells that surround us."[10] That seems a very strange thing to say of an organization with a few hundred paid employees in an unremarkable building near the Geneva airport overlooking a suburban train station and a highway cloverleaf. It is nothing like the cabinet ministries of any major government (or even the UN agencies in Geneva) with their magnificent headquarters filled with thousands of employees and tens or hundreds of thousands of well-paid field operatives working at distant sites across the country (or the world).

ISO is a different, confusing sort of beast, an organization made up of organizations, the national standard-setting bodies in more than 160 countries. Most of these organizations, in turn, are also made up of organizations (although some have individual members, as well), thousands of organizations that also sponsor committees to establish and maintain standards. Mazower's conclusions about ISO's power make sense when we think not of its few hundred employees but instead of the entire network of national standard-setting bodies (and their own member societies and organizations) that are members of ISO, the tens of thousands of committees and subcommittees that they sponsor, the hundreds of thousands of experts forming the community of standardizers who serve on the committees, and the consensus process that all of them follow. Mazower and many other commentators, understandably, even get ISO's name wrong. Its name in English is not the International Standards Organization (as he refers to it), but the International Organization for Standardization. ISO is a pseudo-acronym, chosen precisely because it was *not* an acronym in any of its original official languages. (The French name is Organisation international de normalization; Russian, of course, uses a different alphabet.) In addition, those who were there at the beginning say that the frequently repeated story that the founders chose ISO because *isos* was the ancient Greek word for "same" is a myth.[11]

Today, the realm of private, consensus-based standard setting is even larger than the entire international ISO network, which fifty years ago included most of the organizations that followed that process. Now this realm includes newer organizations such as W3C and IETF (both of which see themselves as global and have no formal systems for representing different countries or regions of the world), as well as all the new private organizations competing with ISO in environmental and social responsibility standard setting. Moreover, unlike in 1946 when ISO was founded, many potential standardizers in today's cutting-edge high-tech industries view all organizations as bureaucratic and stodgy, leading them to create bodies that look as little like organizations as possible; the IETF, for example, officially has *no* membership. Be that as it may, in the history of private standard setting, organizations have been critical in providing structure for the process and recruiting the people that make it work.

ISO organizes the work of creating standards that are meant to be adopted internationally. ISO's member national bodies organize the work of creating national standards. Other standard-setting organizations, such as the IETF and W3C, often focus on standards for particular sectors that end up being adopted globally. (As we have noted, the boundaries within this system of governance and regulation can be ambiguous.) Nevertheless, all organizations involved in private standard setting use variations of the same *process*.

From the beginning, at the heart of the process have been technical committees that worked to reach consensus on documents—the documents *are* the standards—that defined specific qualities of products, technical processes, or (more recently) organizational practices. At the national level, committees were typically designed to balance membership of engineers from producer companies, user or consumer companies, and those unaffiliated with either category. No category of participant was allowed to dominate. Committees typically worked on a given standard or set of standards over many years, by correspondence and in periodic meetings. They exchanged results of technical studies relevant to the standardization task, proposed potential specifications for the standard, discussed and deliberated over the proposals, tried to reach unanimous consensus, and often voted on them in the final stages. In ISO and similar private international standards organizations, representation and voting has typically been by country delegation, but those delegations generally have been made up of technical experts chosen by the country's standardization or technical bodies, not officials chosen by governments. In more recent global organizations such as W3C, no national delegations exist, although technical experts from multinational firms are well represented.

From the inside, the decision-making process in standard-setting committees has looked a lot like deliberative democracy—a careful process that allows all voices to be heard and all positions to be considered before working toward a consensus. From the outside, the system of voluntary standard setting certainly appears to be a technocracy rather than a democracy, since expert knowledge was usually a requirement for being included in the committees—and always a necessity for getting taken seriously. Moreover, most of the technical experts represented profit-making companies, the producers and the main consumers of most of what has been standardized. Requirements for balanced representation of stakeholders and for what we might call due process (e.g., considering every negative vote and responding to each objection in writing) were intended to prevent any committee member or set of members from railroading a particular standard over the reasonable objections of other stakeholders, but these requirements also made agreement on standards much more difficult. National or firm players with strong, self-interested agendas could certainly slow down a process and, by the rules of many standardization organizations, could force a stalemate. However, the consensus process almost always assured that such players could not force agreement on their desired standard.

In the late nineteenth and early twentieth centuries, the people who created and enacted this process were almost all engineers, with a few scientists and technically skilled executives. They made up a fairly small community, a few thousand men (and, as far as we can determine, no women), largely European and American. Today, there are hundreds of thousands of people around the world—many but not all engineers—who spend some part of each year helping create voluntary standards. We describe three waves of institutional innovation, each led by a few individuals who acted as Schumpeterian political entrepreneurs of standard-setting organizations, whom we call *standardization entrepreneurs*.[12] One of the first and most influential of these was British electrical engineer Charles le Maistre. He was instrumental in founding the first of the broad-based or general national standard-setting bodies, the British Engineering Standards Committee, in 1901; the first surviving domain-specific international body, the International Electrotechnical Commission, in 1906; and the broad-based ISO, in 1946. In the post–World War II era, Swedish engineer Olle Sturén helped shift the focus of standard setters from national to international standards as the longest-serving head of ISO in the second half of the twentieth century. British software engineer Tim Berners-Lee, inventor of the World Wide Web and founder of W3C, is representative of institutional standardization entrepreneurs of the most recent era, when software engineers have championed an egalitarian internet and non-engineers with a similar egalitarian intent have experimented with applying the process to create standards for social responsibility.

These men (and one woman, very late in our story) led the movement, but they could not have done so without the hundreds, thousands, and ultimately hundreds of thousands of people who did the hard work of sitting in technical committees and deliberating until they reached consensus on standards. For almost all of them, this work was not their primary job. For some it was one small part of their job; they worked for large companies affected by proposed standards (e.g., the Pennsylvania Railroad in the early period and Google today), and they were paid (for time and travel) by the companies to represent their interests. Other standard setters, such as independent consultants and academics, paid their own way for a combination of other reasons: because they viewed the work as contributing to their skills and visibility; because they identified with the community; and, in many cases, because they were committed to the standardization process itself, which they believed served the common good. In fact, even many representatives of large firms with clear agendas became, through the process, what today would be called "standards geeks"—standardizers who deeply believed in the process, kept the general interest in mind, and were willing to sacrifice time and comfort to help make it work. Although outsiders sometimes viewed the extensive travel involved in attending standardization meetings around the world as a perk, standardizers quickly learned the downside of deep engagement in the process, as well. At the end of his 35-year career with ISO, during which Olle Sturén traveled to more than 60 countries, he noted that one of the best parts of his job had been the opportunity to see so much of the world; standardizers became part of a worldwide community of engineers doing standardization work and enjoyed the social and educational aspects. Nevertheless, he complained, "Anyone who thinks that attendance at technical committee meetings is a comfortable, touristic experience is mistaken." He explained that the work involved wrangling with "the best brains in the relevant industry, and somebody who is not completely confident technically may hesitate before contributing the mildest comment."[13]

Yet, perhaps the most important thing that we learned in researching this global history of private standard setting is that, ultimately, the effectiveness of the community was not just a matter of the competence of the people involved; it was a matter of their commitment. The history of the standards bearers turns out to be the history of a social movement, a standardization movement that has ebbed and flowed since the late nineteenth century. That is perhaps the most surprising part of this book's story.

Where This Book Fits into the Current Literature

Since the turn of the twenty-first century, research on private standard setting and its consequences has surged. Business and economic historians have studied the role of private standards in specific arenas, such as railroads and digital computers.[14] Others have written accounts of specific national or international standard-setting bodies, often as official histories.[15] Some have emphasized the centrality of standard setting in the history of individual economies and of modern industry as a whole.[16] Similarly, recent work in the social sciences has investigated the role of voluntary standards as an alternative to intergovernmental agreements in specific fields as well as the impact of different variants of the process on the operation of global markets.[17] Some scholars have raised broader questions about the ways in which standard setting may shift, reinforce, and conceal social power.[18] Moreover, a recent reflection on the most significant research questions in the field of global governance identified the history and impact of voluntary standard setting as one of the most important topics for further research.[19]

This type of standardization is a complex phenomenon not easily explained by a single theory or discipline. The standardization literature that emphasizes the roles of strategy and power is useful because it shows how a firm's standardization policy can become a determinant of market success or failure[20] and how, as a given standard becomes dominant, the decision to adopt it becomes less and less a matter of free choice, even if the standard is not referenced in legislation, as many private standards, in fact, are.[21] But this is far from the whole story. In political scientists' terms, standard setting is not just about "power over," or who controls things. We (and most standard setters themselves) approach standard setting as more about "the power to do," the ability of people, through cooperation, to do more than they otherwise would have been able to.[22]

Some economists' work on standardization recognizes this by treating the voluntary consensus process as a means of lowering the transaction costs of coordination among firms,[23] but a strictly economic view of standard setting is also incomplete. This approach has trouble accounting for instances like one firm's decision to give up patent rights to what became the standard container corner fitting even though the decision would not be helpful to the company in the short term. That action fits better with the standardization movement's ideas about transcending interests and acting to support technical excellence and the public good, as well as with the general belief held by standardizers that, in the long run, any standard is better than no standard. Accounts based purely on economic interest also leave unexplained, for example, Charles le Maistre's and Olle Sturén's efforts to bring more and more countries into the international standardization network, lessening the influence of the developed countries that pioneered it. Neither do such explanations account for the standardizers who were unaffiliated with or had retired from a firm but continued to work in standardization, paying for their own travel (often extensive) to do so. The fellowship of the community of engineers may go further to explain that. Even today, as the traditional voluntary standardization movement has lost some of its steam, the moral commitment of social movements around internet and web technology and around environmentalism and corporate social responsibility seem to be important elements of successful standard setting.

We believe that a better way to approach private standardization is to see it as an entirely different realm with a very different logic from either commerce or politics, something that developed in response to the greater social complexity that accompanied the pressure toward the greater economic integration of industrial capitalism. The early standardization entrepreneurs would probably want to add that it was a social realm that developed because neither the market nor the state could successfully fill the standardization realm's function.[24]

What Lies Ahead: Three Waves of Private Standard Setting

Three waves of institutional innovation shape our history of private, voluntary standard setting. The first wave, which focused primarily but not exclusively on national standardization, is the subject of part 1 (chapters 1–3). As engineering spread and professionalized in industrial countries, engineers wanted to demonstrate their professional status by serving the public good. They began to establish standard-setting committees within national professional and industry associations. The first general national standard-setting body and the limited-domain International Electrotechnical Commission (IEC) both took form in the first decade of the twentieth century. Charles le Maistre became secretary of the British body and general-secretary of (and chief moral and operational force behind) the IEC. He encouraged other nations to form national organizations on the British model before, during, and after World War I. The first wave of the movement crested in the decade after the war, when national standard-setting bodies spread across the industrial world; then, with the Great Depression, it receded until World War II.

The second wave of innovation, the subject of part 2 (chapters 4–6), started in World War II and focused on international standardization. Standardizers of the victorious Allies launched a new international organization, ISO. It quickly moved to incorporate the defeated powers and to bridge the Cold War divide through the next several decades. As head of ISO, Sweden's Olle Sturén urged developing countries to create national standardization bodies and to join ISO in its push to set international standards and to work with governments and intergovernmental organizations to reference many international private standards in legislation. Successful international standards such as that for the shipping container helped internationalize commerce, while failures to achieve international standards, as in the case of color television, created significant barriers to trade. A close look at standard setting around radio frequency interference and electromagnetic compatibility from the end of the war through the 1980s demonstrates that, in spite of US military and economic dominance, its standardizers had to learn to participate effectively in the international standard-setting arena.

Both national politics and the lead industries of the industrialized world underwent fundamental change in the late 1980s, initiating a third wave of innovation in standard setting, discussed in part 3 (chapters 7–9). New types of private standard-setting organizations grew out of computer internetworking, including the IETF and W3C. In addition, bodies that seemed to violate many of the established principles of voluntary standard setting, corporate consortia, emerged as well, further challenging the traditional private standard-setting bodies. A close look at a W3C standard-setting committee shows points of similarity and difference between old and new types of standardization bodies. Finally, the third wave included another institutional innovation as well, this time from within traditional standard-setting processes as well as outside them—the application of private and voluntary standard-setting processes to a broader set of managerial and social issues, from ISO 9000 quality assurance standards to standards for environmental management systems and social responsibility, including the work of standardization entrepreneur Alice Tepper Marlin in establishing the SA8000 global labor standard and in helping ISEAL (International Social and Environmental Accreditation and Labeling), a group of non-ISO organizations, to define and adopt its own version of the voluntary consensus principles.

The book ends with a discussion of the most active threats to the survival of this mode of standard setting in anything approaching its current form, including the aging of the community of standard setters; its limited gender, racial, and ethnic diversity; and the shift of economic power toward companies and governments in Asia and other parts of the developing world. We consider some of the actions being taken by existing standardization organizations to respond to these challenges and discuss the increasingly important role of China and the developments there that may shape the next wave of voluntary standardization and the world that it affects.

part I

THE FIRST WAVE

The voluntary standardization movement arose out of the professionalization of engineering in industrial countries in the nineteenth century. With shared beliefs in scientific methods and efficiency, along with a desire to demonstrate their professional status, engineering societies began to establish standard-setting committees late in the century. Just after 1900, engineers in Great Britain formed the first broad-based national standard-setting body, and shortly after that some of the same individuals founded the first successful limited-domain international body in electrical engineering. The movement established principles including the voluntary adoption of the standards, the need for a balance of interested parties (engineers from producer and user firms, along with unaffiliated engineers); a respectful process that considered all points of view; and, in international bodies, the principle of one vote per country. Although an international strand was present from the start, the first wave, up through the interwar period, focused primarily on building national standard-setting bodies.

Chapter 1 examines early attempts at standardization, culminating in the 1880s and 1890s with the first (ultimately unsuccessful) attempt to establish an international standard-setting body. Chapter 2 traces the rise of the first national standard setting body in Great Britain and the first surviving international body, the International Electrotechnical Commission (still active today), and highlights the role of standardization entrepreneur Charles le Maistre in both. Chapter 3 takes the story to the crest of the first wave at the beginning of the Great Depression, analyzing the standardization movement and community.

1

Engineering Professionalization and Private Standard Setting for Industry before 1900

Nineteenth-century developments in engineering led to the first organized attempts at private standard setting within professional associations of engineers before the century's end. Initially, scientific communities in Europe and beyond met in various venues to discuss and agree on standards for scientific terminology and units in many domains. As networked communication and transportation—telegraph and railroad—spread across national boundaries, scientists were joined by engineers, industrialists, and diplomats in setting treaty-based administrative standards in those applied settings. Meanwhile, engineering was establishing itself as a profession, with its own communities centered on professional societies. Engineering societies grew and differentiated on a national level in industrializing Europe, the United States, and Japan. Engineers began to form committees within their engineering societies to establish voluntary national technical standards for industrial purposes, including standards for steam boilers, screw threads, and steel rails for railroads. By the last two decades of the century, many engineering societies were taking on standardizing to bolster their status as professionals.

The end of the nineteenth century also saw an early attempt at creating an international, limited-domain private standard-setting organization. Engineers in German-speaking universities, led by Professor Johann Bauschinger, initiated a series of increasingly international congresses to agree on methods for testing materials used to construct bridges, roads, and buildings. In the 1890s they established the International Association for Testing Materials (IATM) to set voluntary, nongovernmental technical standards in this domain. Although the association would not survive World War I as a standard-setting organization, Bauschinger introduced some early principles of voluntary standard setting. In addition, the IATM spawned an American body that broke off to become an important, limited-domain American standard-setting organization in 1902. Under the leadership of the Pennsylvania Railroad's Charles B. Dudley, it participated in the US standardization of steel rails that had begun two decades earlier.

In this chapter we briefly trace these developments, which laid the groundwork for the turn-of-the-twentieth-century founding of the first ongoing national and international organizations established solely to set private voluntary standards, discussed in chapter 2.

Nineteenth-Century International Scientific Communities

In the second half of the nineteenth century, many international scientific and technical congresses took place, often in conjunction with the increasingly popular international exhibitions and world's fairs, starting with London's Crystal Palace Exhibition. At these associated congresses, international scientific communities in particular fields often attempted to achieve agreement on terminology and metrological standards for their fields.

At the 1851 Crystal Palace Exhibition, the French government began a push to internationalize the metric measurement system it had legislated as a national standard by displaying a meter rod, a kilogram weight, and a liter container.[1] This display triggered much discussion about metrological standards among scientists and government representatives beyond France. In 1855 some participants at the Second International Statistical Congress in London created the Association for Obtaining a Uniform Decimal System of Measures, Weights, and Coins, with international banker Baron Jacques Rothschild as its first president, to promote universal adoption of the metric system.[2] Subsequent regular meetings helped increase the number of nations signing on to this system. By 1867, representatives of forty nations met at the Paris Exposition Universelle to talk about establishing an international metrological organization that would broaden the system's sponsorship from France to the international community; in an 1872 meeting, an intergovernmental organization was formally proposed, and in 1875 a treaty established this organization, initially called the International Metric Commission (later to become the International Bureau of Weights and Measures).[3] Despite their reservations about the metric system, the United States and Great Britain both sent representatives to the conferences, but only the United States signed the treaty agreeing to the formation of the commission—a signature that did not require acceptance of the metric system. Neither Great Britain nor the United States adopted the system, creating an ongoing source of contention in future international industrial standardization.

In addition to the metrologists and statisticians, scientists in seismology, mathematics, and cartography also held meetings at international expositions. The private, nongovernmental International Geodetic Association, which first met in 1864 and 1867, brought together scientists interested in collaborating to map the world; this group also supported refining the metric system and establishing it universally, to assure a single, well-defined, and precise measure for their cartography. The first International Geographical Congress met in 1871 in Antwerp and at intervals after that.[4]

Cartography raised political and diplomatic as well as scientific issues. In 1884, the US government invited 25 countries to send representatives to the First International Meridian Conference in Washington, to agree on a single prime meridian, or line of zero longitude, as well as a universal day, starting at midnight at the prime meridian.[5] Admiral C. R. P. Rodgers, president of the conference, noted in his opening address that "delegates renowned in diplomacy and science" attended the meeting—that is, members of the international scientific community as well as diplomats representing national governments.[6] Diplomatic posturing by France dominated the initial stages of the intergovernmental gathering, reflecting in part France's resentment that the United States and Great Britain planned to make a strong push for Greenwich, rather than Paris, as the meridian. The French delegate, A. Lefaivre, declared his opposition to a US proposal to allow participation of several eminent scientists who were not representing a country (a compromise allowed them to speak at the invitation of the president and with the approval of the delegates) and flatly rejected another US proposal that the meeting be open to the public. He argued (as summarized in the proceedings report) that the conference was

> an official and confidential body; scientific, it was true, but also diplomatic; that it was empowered to confer about matters about which the general public have now nothing to do; that to admit the public to the meetings would destroy their privacy and subject the Conference to the influence of an outside pressure which might prove very prejudicial to its proceedings, and he would object to this resolution absolutely.[7]

Thus, France asserted the primacy of diplomatic representatives of national governments over the international scientific community and the public. This opening interchange also highlighted the US delegation's stated view that scientific and technical expertise was more important in this matter than status conferred by the state. The eventual ratification of the vote to adopt the Greenwich meridian at that gathering (over one objection, from Santo Domingo, and abstentions from France and Brazil[8]), simplified and standardized both cartography and time keeping.

The international scientific community's shared belief in scientific accuracy and rationality drove demand for metrological standard setting. Industrialization and expanding networks of communication and transportation meanwhile created demand for different types of standards.

Electrical Units and Administrative Regulations for Networks of Communication and Transportation

Over the century, similar congresses, societies, and treaty organizations—increasingly including engineers and industrialists—also addressed applied standards issues emerging from the creation of a network of telegraphic communication and rail transportation that crossed national boundaries. During the nineteenth century, standardization of measurements for electrical resistance and other electrical units came out of cooperation among scientists (primarily physicists), engineers, and ultimately diplomats, driven by the pressing needs of telegraph engineers installing and maintaining long-distance and undersea telegraph cables. As early as 1851, German physicist Wilhelm Weber, building on Carl Friedrich Gauss's previous work on magnetism, had proposed an "absolute" system based on units of work and energy. This system, however, was hard to explain or to embody in material form; moreover the resistance unit, expressed in terms of velocity through a wire as a meter per second, was "ludicrously small (less than 1/100 millionth of the resistance of a mile of ordinary wire)."[9] Telegraph engineers constructed resistance coils to use in their work but had no way to calibrate them with those used at different telegraph companies. In 1860 German telegraph engineer and industrialist Werner von Siemens, founder of Siemens AG, proposed a measure of resistance based on a mercury column one meter long and one square millimeter in cross section at zero degrees Celsius—a measure that had a material embodiment and was of an appropriate scale.

By this time, however, broader efforts to establish a new system of electrical units, of which resistance was only one, were gaining momentum. In response to the 1858 failure of the first Atlantic cable, laid by the Atlantic Telegraph Company, the British government and the firm set up a joint investigative committee to assess causes of the failure and to propose a future course of action. In its 1861 final report, this committee recommended that future contracts specify standard resistances. Meanwhile, others—including most prominently physicist William Thomson (later Lord Kelvin), formulator of the second law of thermodynamics and consultant to the initial failed Atlantic cable project—suggested that the British Association for the Advancement of Science form a committee to standardize electrical units.[10] At an 1861 British Association meeting, the Committee on Electrical Standards was formed for this purpose. It originally included six members—five scientists (including Thomson) and one telegraph engineer—but added several other members during its work, including prominent scientist James Clerk Maxwell (who formulated Maxwell's equations), and several prominent telegraph electricians, including Werner von Siemens's brother, Carl Wilhelm Siemens, who headed the London office of the family company.[11]

This committee worked hard to satisfy the needs of telegraph engineers as well as laboratory scientists, ultimately basing the system of units on Weber's absolute system but setting units at a practical level (e.g., the resistance unit, later to be called the ohm, was based on the Weber meter per second but increased by several orders of magnitude). The committee issued preliminary resistance standards (embodied in resistance coils) in 1863 to fill the pressing needs of telegraph engineers and issued the official British Association standards in 1865.[12] Nevertheless, the battle between these standards and the mercury-based unit advocated by Werner von Siemens continued for years, as did attacks on the precision of the committee's measurements establishing its standard. An international compromise was finally achieved during the First International Electrical Congress in 1881. Hosted by France in Paris, the meeting turned into a debate between the Germans and British, centered on the units of resistance. By the end of the meeting, the 200 scientists and engineers attending from throughout Europe had agreed on a system of the ohm, volt, ampere, coulomb, and farad based on Weber's absolute system and expressed in centimeters, grams, and seconds, but they had also agreed on a measurement of resistance calibrated to a mercury column. Scientists and electrical engineers would continue to convene such electrical congresses over the next several decades, addressing further issues of electromagnetic measures.[13]

The telegraph raised issues of international diplomacy as well as science and technology. The International Telegraph Union (ITU), which was created at an 1865 diplomatic conference in Paris, is usually identified by historians of intergovernmental organizations as the first of several universal public international unions established in the nineteenth and early twentieth centuries.[14] In Europe, telegraph systems were run by national governments, and crossing borders posed potential national security problems from the start. Indeed, before the establishment of the ITU, at each border crossing, transnational telegrams were written out on paper, physically carried across the frontier, and then retransmitted on the other side.[15] The right of transit across borders, tariffs for telegrams, and the use of telegraphic codes—all essentially administrative rather than technical issues—needed to be negotiated. Beginning as early as 1849, bilateral and then regional treaties were established between and among some states, but problems in sending a telegram across Europe remained.[16]

In 1865 the French government invited representatives of 20 European countries to convene in Paris to establish a uniform European telegraph system. The delegations were headed by diplomats, but to negotiate the details of the agreement they established a special committee composed of high officials from each country's national telegraph administration. This committee successfully negotiated the First International Telegraph Convention (the treaty itself) and a set of Telegraph Regulations establishing the ITU as a permanent organization to maintain, facilitate, and update these agreements. At the 1868 Vienna Telegraph Conference, the ITU established an International Bureau to handle details between conferences, and at the 1875 St. Petersburg conference it divided meetings into two types: plenipotentiary conferences, attended mostly by diplomats and focused on diplomatic issues concerning the convention only (subsequently held very rarely—indeed, none occurred between that year and 1932); and administrative conferences, attended mostly by experts from telegraph administrations and focused on updating the tariffs and administrative regulations.[17] Changes to the International Telegraph Convention required a majority vote of national telegraph administrations at the administrative conferences, followed by ratifications by governments. Nations voting against a decision could exercise a "reservation," essentially refusing to be bound by that decision. Changes to the Telegraph Regulations did not require this additional level of governmental ratification; attendees of the administrative conferences could agree on them by vote.

By 1911 the ITU's membership had grown to 48 nations, with additional observers permitted to attend the conferences in an advisory capacity. This practice allowed the United States, in which the telegraph system was under private rather than state control, to participate (though not to vote) via representatives of the private telegraph companies (e.g., Western Union). Observer countries such as the United States generally voluntarily adopted the ITU administrative protocols for cross-border communication.

Other such primarily European conventions, unions, and agreements followed the ITU's lead. National governments and their postal services established the Universal Postal Union on the ITU model when they agreed on a convention at an 1874 conference in Berne, then ratified it in 1875.[18] On a slightly different model, European railroads coordinated themselves around tariffs and right of transit for freight traffic with a series of meetings leading in 1890 to the establishment of the Convention international sur le transport de merchandise par chemin de fer. They then addressed scheduling of passenger services, with discussions culminating in 1891 in a regularized biannual European Time-Table Conference (but no treaty, central office, or union). Finally, European railroads negotiated a separate 1886 railroad convention (revised in 1907) that covered such matters as railway gauge, loading of cars, and maintenance of rolling stock.[19] In response to the emergence of wireless telegraphy and the need for ships to be able to communicate no matter how far from their home nations, the Radiotelegraph Union (RTU) was established in 1906 (following the ITU model) to establish administrative protocols for such communication.[20]

During the nineteenth century, treaty-based public international unions set primarily administrative rules, rather than technical standards, that extended communication and transportation networks across national boundaries. National delegations to these bodies represented governments or governmental bodies (e.g., national telegraph administrations), not technical societies. In the 1920s the ITU and RTU would establish separate technical committees (see chapter 3), creating hybrid public-private bodies composed primarily of engineers and focusing on technical standards that would interact with and influence private standard-setting bodies.

The next section turns to the main protagonists of the private standardization story—engineers.

The Professionalization of Engineering

During the nineteenth century the industrializing countries—including Great Britain, continental European states, the United States, and Japan—experienced the rise of engineering and the effort to professionalize it.

In 1818, the Institution of Civil Engineers (ICE) was founded in London to form the core of a British engineering profession in the civil (as opposed to military) sector.[21] The demand for roads and bridges in Great Britain and throughout the British Empire encouraged the British engineers' early move toward professionalization. In the wake of the revolution of February 1848, French engineers founded the Société des ingénieurs civils de France, and the French state officially recognized the society's public utility in 1860.[22] An important predecessor of US engineering associations, the Franklin Institute, was founded in 1824 as a mechanics institute, one of many launched in the 1820s to teach workers to apply science to practical uses.[23] This Philadelphia-based institution ultimately transcended its fellow mechanics institutes to become the nation's leading scientific and technical institution, with its own prestigious publication, the *Journal of the Franklin Institute*, in which scientists, engineers, and industrialists came together. In 1852, engineers in New York founded the American Society of Civil Engineers (ASCE), a regional association that soon went dormant; in 1867 they revived and expanded it into a national association, the first professional engineering society in the United States.[24] Meanwhile, in 1856 German engineers created the Verein Deutscher Ingenieure (VDI, the Association of German Engineers) to champion German unification and industrialization, as well as to promote professionalization and the middle-class aspirations of engineers.[25] The Engineering Society of Japan was formed in 1879, the year in which the first class graduated from the country's first college of engineering.[26] Engineers were organizing and working to professionalize themselves in large countries and small, those with empires or without them, as they industrialized.

Although engineers in all these countries defined the initial engineering societies broadly, soon those working in specific engineering disciplines and industrial domains founded additional, more specialized societies, with the British again leading. In 1847 the Institution of Mechanical Engineers (IMechE) was founded in Birmingham as a regional organization; by 1877 it had moved to London and become a national organization. British engineers and industrialists also created associations in particular industrial domains during this period, including, for example, the Institute of Naval Architects in 1860, and the Iron and Steel Institute in 1869.[27] In 1871 British engineers interested in developing what they termed "telegraphic science" founded the Society of Telegraph Engineers.[28] In Germany, the Verein Deutscher Eisenhüttenleute, an association for engineers from the mining and metals industries, broke off from the VDI in 1880.[29]

In the US, the American Institute for Mining Engineers (AIME) was established in 1871, and the American Society of Mechanical Engineers (ASME) in 1880.[30] Railroad associations that served as trade associations as well as quasi-professional societies, such as the Master Car-Builders Association, founded in 1867, also proliferated.[31] As historian Edwin Layton has noted, during this period American engineers were torn between allegiance to the engineering profession and allegiance to business, in the form of their firms and industries.[32] Those most devoted to making engineering a recognized profession like law and medicine tended to form or join engineering societies that had higher membership qualifications (years working as an engineer at a particular level, examinations, etc.) and that organized around engineering disciplines. Those more oriented toward business tended to form or join associations for engineers and managers within specific industries, such as railroads or mining, and to have very low membership qualifications (e.g., simply working in the industry, whether as engineer or manager).

The 1881 International Electrical Congress in Paris (at which attendees agreed on the units of electrical resistance) spurred the professionalization of electrical engineering in many countries. In 1883, French electrical engineers founded the Société internationale des électriciens, and Austrians formed an electrotechnical association in Vienna; in 1884 Belgians founded the Société royale belge des electriciens.[33] In anticipation of the 1884 International Electrical Exhibition in Philadelphia (sponsored by the Franklin Institute), a group of American electrical engineers formed the American Institute of Electrical Engineers (AIEE) earlier the same year, so the United States would have a professional society to host the luminaries expected to attend the exhibition.[34] From the beginning, this society had a wide theoretical and practical scope and a roster of officers and members that included inventors (e.g., Alexander Graham Bell and Thomas Edison), professors of electrical engineering, telegraphic engineers and electricians, and managers from electrical industries (e.g., the president of Western Union, Norvin Green).[35] Meanwhile, Britain's Society of Telegraph Engineers expanded its scope beyond the telegraph to include other aspects of electricity, and, by 1888, it became the Institution of Electrical Engineers (IEE), reflecting its new identity as a true electrical engineering society.[36] In 1888, Japanese electrical engineers founded the Institute of Electrical Engineers of Japan, and in 1889 Swiss electrical engineers founded the Association suisse des électriciens.[37] In still incompletely centralized Germany, regional electrotechnical societies (ETVs) were founded in Berlin (by Werner von Siemens) in 1879 and in Frankfurt in 1881; in 1893 members of these ETVs came together to form the VDE, the Verband Deutscher Elektrotechniker (Association of German Electrotechnicians), as a German umbrella association for electrotechnology.[38] In 1897, the Associazione Elettrotecnica Italiana was founded in Italy.[39]

During the nineteenth century multiple engineering disciplines sought to raise their status by professionalizing across the industrialized world, creating organizations of technical experts. Because professionalization was often associated with public service, they also sought ways to serve the public to bolster their professional claims. The next section traces their first forays into nongovernmental voluntary standard setting in the service of society.

Early Engineering Standardization: Steam Boilers, Screw Threads, and Steel Rails

Three early examples of industrial standardization through engineering and technical societies occurred during the nineteenth century—around steam boilers, screw threads, and steel rails—showing both engineers' desire to supply privately set standards and industry's, governments', and the public's demand for such standards.

Engineers' and technical experts' earliest and most publicly visible efforts to set standards through their societies concerned the safety of steam boilers. In the United States, the Franklin Institute took the lead, building its excellent professional reputation in part on its research into steam boiler explosions in the 1830s. When steam boilers on riverboats blew up, they caused many widely publicized and horrific deaths among crew and passengers. Some members of the Franklin Institute, motivated by "the urge for useful service," decided to launch an investigation into the causes of such explosions.[40] At the same time the US government, in response to public pressure, gave funds to Samuel D. Ingham, the secretary of the treasury, to undertake such an investigation. After some initial, relatively unsuccessful efforts to gather relevant information on the problem, Ingham funded the Franklin Institute's study, allowing it to undertake experimental work as well as gather existing information. The Franklin Institute had already organized itself to undertake special investigations by establishing a Committee on Inventions, which could itself establish further special committees when needed. The special committee that undertook the study of steam boiler explosions included a mix of scientists, engineers, and industrialists. Based on the study, its 1836 report recommended standards and regulatory legislation on the design, materials, construction, and maintenance for steam boilers to improve the safety of steamships; the US government took another 14 years to adopt legislation based on it, with failures and half measures preceding its full enactment into law in 1852, greatly improving riverboat safety.[41]

Several decades later, in 1884, the German VDI followed the US example and published its first "guideline" (*Richtlinie*) or standard for the thickness of steel plates to be used in steam boilers.[42] Both steam boiler standards, though designed by private organizations of technical experts, were intended to, and ultimately did, become legally mandatory safety standards in their respective countries.

The next, and iconic, example of industrial standard setting, the screw thread, was a voluntary standard, never intended to be legislated. Screw threads, which were the focus of the first technical committee of many twentieth-century standard-setting organizations, also provided a very early and important example of standardizing supported through, though not created by, nineteenth-century engineering and technical societies. Having a standard system of screw threads enabled interoperability, so screws made in one machine shop could be used as replacements for those lost from machinery created in another. In both Great Britain and the United States, an engineering or technical association built support that led to considerable voluntary adoption of screw thread standards by machinists.

A standard set of screw threads was proposed first in Great Britain. In 1841, Joseph Whitworth, one of the leading British mechanical engineers of the Victorian era, presented a paper, "On a Uniform System of Screw Threads," to a meeting of ICE.[43] In it, he proposed a system based on examining existing threads from bolts he had collected from many British workshops and designing one as close to average as possible in both cross section and number of screw threads per inch. This approach suggests that he sought not to find the best technical solution but to find a solution that would be most acceptable to all, since he saw shop owners and customers as benefiting from a standard. His own shop and a few others had already adopted his proposed system. ICE did not develop or officially endorse the standard system, but Whitworth's ICE presentation launched it more broadly, and by 1858 it was widely enough used that he was able to claim, although with some exaggeration, that it had been "universally adopted" in Great Britain.[44]

A few years later in the United States, another celebrated mechanical engineer and machine tool manufacturer, William Sellers, developed a different model for screw threads, with a cross section that was more easily manufactured by less-skilled workers than was the Whitworth screw thread.

After *Scientific American* called for standardization of screw threads in American machine shops in 1863, Sellers, then president of the Franklin Institute, gave a talk there on his recommended American standard.[45] Because of the Franklin Institute's reputation for setting steam boiler standards, its public support of Sellers's screw thread standard (later often referred to as the Franklin Institute standard) carried great weight, persuading many American engineers, including those of the Pennsylvania Railroad, two railroad industry associations (the Master Car-Builders' Association and the Master Mechanics Association) and the US Navy to adopt his system as their standard. Although changing to this new standard involved an initial cost for shops that had made other types of screws, and it could increase price competition in the making of replacement screws, it was clearly beneficial for customers (especially large ones like railroads), and it broadened potential replacement markets for suppliers.[46] Its adoption in the United States paralleled that of the Whitworth screw in Great Britain—broad enough to allow an exaggerated claim to universality but not broad enough to eliminate the interoperability problem. The problem of screw thread compatibility would be a perennial one for standard setters, reemerging between allies (such as the United States and Great Britain) during the world wars of the twentieth century, as shown in chapter 4.

A third, and more complicated, example of standardizing through engineering societies centered on performance standards for steel rails used on railroad tracks in the United States. In this case, unlike in the previous two, the voluntary standards created by committees of engineers were more openly established, and in some cases endorsed, by engineering societies as standards. Moreover, multiple engineering societies played a role in the drawn-out process that extended through the 1870s, 1880s, and 1890s.

The first stage stopped short of achieving standardization. During the late 1860s and early 1870s, railroads first substituted steel for iron rails, a substitution expected to greatly increase their life span. By the mid-1870s, however, railroads began to question their longevity and performance. Many railroad engineers belonged to the ASCE, now established as the first and most professionally oriented American engineering society. They dominated an ASCE committee on the "form, weight, manufacture, and life of rails" established to survey railroad managers about how steel rails were performing under various loads.[47] Although this committee neither performed tests on steel rails nor attempted to establish standards for them, it issued a series of reports revealing that railroad managers now believed that steel rails were only marginally superior in durability to iron rails. The committee also suggested that railroads demand rails better shaped for durability, rather than taking the easily manufactured but less durable shapes that steel manufacturers offered.

Meanwhile, the steel producers had turned to a different, more industry-oriented engineering association, the American Institute of Mining Engineers (AIME), to discuss how they could assure that steel was more durable than iron in order to justify its higher price.[48] In addressing hardness, metallurgical engineers and steel manufacturers wanted physical tests (e.g., force required to bend) and chemical specifications for rails with long-term durability.[49] So both railroad consumers and their steel manufacturing suppliers were turning to the still-young engineering societies for help with this problem of durability in rails, but they were generally working independently and with different societies, rather than collaborating to find solutions to the problem.[50]

One exception to this alignment of steel producers with AIME and railroads with ASCE was Charles B. Dudley, who became the chief chemist of the Pennsylvania Railroad in 1875, and who, the next section will show, also played a key role in establishing one of the oldest bodies for private standardization, the American Society for Testing Materials, or ASTM (figure 1.1).[51] Dudley served in the Civil War as a young man, suffering a severe leg wound that almost resulted in amputation and that left him with a serious limp for the rest of his life. After the war he went back to school, earning his way through the Sheffield Scientific School of Yale College to a PhD in chemistry by working as night editor for the *New Haven Palladium*.[52] He established the first chemical testing lab at a railroad company and researched chemical aspects of materials used by the railroad, from the steel rails to soaps used to scrub railroad cars. Over his career, he joined a long list of scientific and engineering associations; his active memberships in AIME and other engineering societies (he belonged to ASCE, ASME, and AIEE), suggest that he identified with engineers as well as with scientists.[53]

Dudley was a strong believer in standardizing methods of chemical analysis to use in creating standard specifications for materials being purchased by railroads. Beginning in 1878 he published a series of papers on the relationship of the chemical composition of steel rails to their wear in the *Transactions of the AIME*, recommending a chemical formulation of steel to make the hardest rails.[54] These papers generated considerable controversy, and the steelmakers in AIME vociferously challenged his findings and the notion of

Figure 1.1. Charles B. Dudley, chemist for the Pennsylvania Railroad and the first president of the ASTM, was respected by manufacturers and purchasers of steel rails, and would advocate balancing the views of both constituencies in standardization. Photograph courtesy of Hagley Museum and Library.

the railroads imposing chemical standards rather than simple performance criteria on steelmakers. Nevertheless, William R. Jones, the steelmaster of Carnegie Steel, noted, "Although the Doctor may be wrong, and I believe he is only partially correct, yet he was the first to endeavor to establish a formula of this kind, and is therefore entitled to the thanks of steel makers."[55] While further research led Dudley to abandon this specific formulation a few years later, he clearly retained the respect and goodwill of most members of AIME (the professional engineering association dominated by steel producers), as he was made vice president of that organization from 1880 to 1882. And he will reappear in our story around the turn of the century. Still, in the early 1880s, getting producers and consumers to agree on standards for rails did not seem likely.

Despite the diverging allegiances of steel producers and railroads, by the end of the 1880s both suppliers and purchasers of rails were ready to acknowledge the potential value of standards for steel rails; indeed, Jones of Carnegie Steel told an ASCE audience at its 1889 meeting that focusing on a few shapes for rails would also help steel company operations.[56] Consulting engineers and rail inspectors (who were hired by railroads to go into the steel mills to assure that rails were being made to specification and thus were familiar with the needs of railroads and the processes of steel mills) began to mediate between steel companies and railroads on this issue. In 1889, one such individual, Robert Hunt, agreed to serve as secretary of an ASCE committee on standard rail sections.[57] A civil engineer who had served as consulting engineer in building various railroad suspension bridges chaired the committee, which included as members both railroad engineers and consulting civil engineers, with many of the latter having had railroad experience at some point in their careers. In 1893 this committee issued a report proposing rail sections for various weights and providing advice on manufacturing processes. The report stated, "In deciding upon a series of sections, your Committee have given consideration to the manufacturing details of rail making, while seeking designs whose forms would be best adapted to meet the various requirements of traffic," reflecting the perspectives of both manufacturers and users.[58]

In 1895, the committee added chemical guidelines and a physical test to the specifications, further reflecting steel producers' desires. In that same year Secretary Hunt addressed members of the (steel-dominated) AIME about the ASCE committee process, describing "the honest and careful efforts made to obtain the views of the leading railroad engineers of the country, to harmonize differences and to design a series of rail-sections which would be in accord with their experiences and generally acceptable." Moreover, as he pointed out, "at the same time steel-makers were consulted, so that the proposed sections might not present special difficulties in manufacture."[59] He went on to extoll the successful outcome of the process: widely adopted, though not official, voluntary standards.

> While the Constitution of the American Society of Civil Engineers prevented that society from officially adopting as its own [standards] the rail-sections recommended by the committee, they have been popularly so regarded and called; and, what is better, they have been largely adopted already by the railroads of the country, and promise soon to be absolutely the standard American sections.

Although engineers who were currently working, or at some point had worked, for the railroads dominated the committee, many of them also understood the steel companies' processes and needs, and the report's conclusions considered the needs of both producer and consumer.[60] This focus on the needs of producer and consumer would become a hallmark of private standard setting.

These three early industrial standardization efforts through technical and professional societies addressed safety (steam boilers), interoperability (screw threads), and performance (steel rails) standards, and they reflect in part engineers' belief in standardization based on scientific principles and in part their desire to demonstrate "social beneficence" by contributing to the public good and thus reinforcing their professional status.[61]

Late Nineteenth-Century Standardization in Mechanical and Electrical Engineering

As the railroad case suggests, national engineering societies became increasingly involved in standard setting in the last two decades of the nineteenth century. In the United States, in addition to the ASCE, the American Society of Mechanical Engineers was also drawn into standard setting, despite its initial reluctance.[62] During the final decades of the nineteenth century and initial decade of the twentieth, some prominent members of ASME were leaders in the systematic and scientific management movements, which supported standardization of materials and processes *within* firms.[63] Initially, however, standard setting *across* firms was a source of contention in ASME; when a member urged the new organization to endorse the Sellers screw thread standard, for example, the leadership refused, choosing to leave it to the marketplace. During its first decade, the 1880s, members repeatedly discussed whether ASME should set standards, with sharply divided opinions. Some members considered it inappropriate for the association to take a position on commercial matters, while others believed that taking authoritative stands on technical matters was a very appropriate and public-spirited activity for a professional society of engineers. When the Pratt & Whitney Company developed a gauge that the firm proposed as the standard for measuring screw threads in the early 1880s, ASME established a Committee on Standards and Gauges to examine and report on it but not to set a standard for screw thread gauges. Although Frederick Hutton's proposal to have the society create a standard for rating steam boiler capability was turned down around the same time, a year later William Kent's proposal to develop a standard method of testing steam boilers was seen as worthy of establishing a committee to study and report on it. When that committee reported in 1885, many ASME members wanted to endorse the proposed testing method as a standard, but opponents of endorsing any standards won the ensuing debate.

Gradually the leaders warmed to standard setting. In 1887 the society established a committee to look at standardizing pipes and pipe threads—a committee that included representatives of pipe makers and pipe users and that consulted with a number of similar committees from industry-oriented organizations such as the Pipe Manufacturers Association. Another committee was set up to standardize pipe flanges for pumps and valves, and its report in 1888 moved "that the committee's proposals for flange diameters 'be adopted and recommended as the ASME standards.'"[64] These recommendations were referred to as ASME standards, even though the society had not yet formally decided to take on the standardizing role. In 1895, when the boiler testing code that was never made an official standard needed updating, the society established a committee to revise it. By that point, they seem to have settled the internal argument, referring to both the original and the revised recommendations as ASME standards. Thus ASME, like ASCE, came to embrace standard setting beyond firm boundaries as appropriate and desirable for the engineering profession.

Germany's VDI also embraced standard setting around the same time, publishing its boilerplate standard in 1884, but it followed a somewhat different path into standard setting than had engineering associations in the United States. Rather than addressing machine interoperability standards (e.g., screw threads) and performance standards (e.g., steel rail sections), as well as safety standards (e.g., a steam boiler code), VDI started later and initially focused solely on safety standards. German national machine standards, including those for screw threads, came much later, something that one of the first scholarly analysts of German standard setting, American economist Robert A. Brady, considered significant to the comparative history of industrialization across countries because "the most important impetus given to the whole modern elaborate [global] standardization movement came from [the] machine industry."[65]

At the time of VDI's establishment in 1856, many German engineers believed that the country's fragmentation into more than 30 separate sovereign states had slowed German industrialization and the privileged position that German governments granted to "architects" (architects / civil engineers)—who were civil servants overseeing the design of state-sponsored buildings, bridges, and roads—but not to other engineers impeded the professionalization of engineering.[66] Hence, the German association emerged as a united front of less-advantaged engineers promoting national unity and higher status for all engineers.[67] By the 1870s, both of those initial goals had been achieved, and German engineers had begun to experience some of the conflicts between allegiance to the engineering profession and allegiance to industry that troubled their American colleagues. The Germans also faced a larger problem that united their professional and commercial interests: the realization that, despite the high level of German science, many German "products were generally of low quality, imitative, and awkward" compared to those displayed at world's fairs and international exhibitions by the Americans, the British, or the French.[68] For everyone involved in German manufacturing, the invidious comparisons made by visitors to the 1876 Philadelphia Centennial Exposition were particularly disturbing. "Philadelphia," Friedrich Engels wrote, was "the industrial Jena," a reference to the battle in which Napoleon defeated the Germans.[69]

Many German engineers and industrialists committed themselves, instead, to creating their foreign competitors' Waterloo. The machine industry, especially, pledged to suffer no rivals in quality, originality, and design. By the time VDI published its steam boiler standard in 1884, according to a German machine-tool manufacturer, its "machine shops that had followed the 'shining example' of the United States in the use and specialized manufacture of specialized machine tools had no lack of orders or profits." Indeed, he cited two products, sewing machines and what he termed "locomobiles," as having "become just as good as American or British products."[70] Clearly, German industry believed that following these American methods led to quality in line with that of American and British products. Yet, these Germans did not initially follow one part of the American formula for industrial goods—standardization of machines and their parts. "The main objection brought against early attempts at machine standardization," Brady notes, "was that standards would throttle technical development."[71]

By the beginning of the twentieth century, VDI included many engineers who embraced such standardization *within* the firm as part of a new commitment to economic efficiency and to Frederick Taylor's scientific management, but "the Taylor system proved to be incapable of carrying a majority" of VDI members.[72] Only the rare German enterprise such as electrical giant Siemens even promoted standardization at a factory level, with its standards office founded in 1908. State pressure to produce more efficiently during World War I ultimately turned German businesses and engineers to Germany's rationalization movement (a parallel to the systematic and scientific management movements within US firms) and to "general basic standards" (*Grundnormen*) such as screw threads.[73]

The new field of electrical engineering, with its proliferation of national electrical engineering societies beginning in the 1880s, was another arena where interest in standardizing arose. In 1882 the British IEE established a committee to look at fire risks associated with electrical lighting; this committee rapidly reported back to IEE's council with a set of rules and regulations for the prevention of fire risks arising from electric lighting, which the council accepted and published.[74] The opening of the document noted that the rules were created "not only for the guidance and instruction of those who have Electric Lighting Apparatus installed on their premises, but for the reduction to a minimum of those risks of fire which are inherent to every system of artificial illumination." These safety standards would eventually become embedded in legal requirements.[75]

Other electrical engineering societies, such as the German VDE, also showed interest in standardizing. As the official chroniclers of the VDE have noted:

> While the mission of the [earlier] regional ETVs was to foster the technical applications of electricity and technical sciences and assemble the stakeholders of electrotechnology, the motivation for founding an umbrella organization went further. The mission of the [national] VDE was to create a central organization for technical, standardization and professional issues on the legislative and especial[ly] industry policy basis.[76]

In Germany, as in Great Britain and the United States, industrialists played an important role in standardization. Before his death and the founding of the VDE shortly thereafter, Werner von Siemens, cofounder and chairman of the Berlin ETV, had also lobbied for and helped found the Physikalisch Technische Reichsanstalt, a government lab intended to standardize basic physical units in Germany. Within two months of the 1893 creation of the VDE, at its first general assembly in Cologne, it created a permanent VDE commission to develop standards for electrical installations. By 1895, that commission had also issued "safety rules for high voltage installations."[77]

The US AIEE was interested in standardizing, as well, though it focused less on safety and more on measurement, compatibility, and rating standards for industry. The historian of that association noted that from its founding, the AIEE declared that it would "settle 'disputed electrical questions' within the industry—a sign of the importance given to uniform industrial standards from the beginning."[78] Nevertheless, some within the AIEE, as in ASME, initially resisted taking on standardization, and proponents had to overcome this resistance. During the early 1890s, the AIEE appointed a Committee on Units and Standards, which played an active role in achieving standards for the naming of electrical units in the United States. More importantly for the future of voluntary standardization for industry, but also more contentious, were the AIEE's attempts during the late 1880s and early 1890s to create standards for wire gauge and electrical apparatus. The attempt to standardize wire gauge failed, stalled by lack of agreement on what role AIEE should play in standardization.

Its attempt to standardize electrical apparatus in response to the New York Electrical Society's request was more successful. In 1898 AIEE president Francis Crocker—a Columbia University professor of electrical engineering and cofounder of Crocker-Wheeler Company, a small manufacturer of dynamos and other electrical apparatus—chaired a meeting titled "The Standardizing of Generators, Motors, and Transformers (A Topical Discussion)."[79] To avoid the earlier stalemate on wire gauge standards, Crocker framed the discussion carefully, citing the members' qualifications to set standards and precedents in other engineering societies, American and British:

> The general desirability of standardizing electrical apparatus has often been spoken of and seems to be very generally believed in. The feasibility of it, or the policy of it, is another matter. If such standardizing is to be effected, this body, THE AMERICAN INSTITUTE OF ELECTRICAL ENGINEERS, is undoubtedly the only one which is competent to decide such a question. As to precedents, I think we have many excellent ones.

He cited the ASME and ASCE forays into standardizing, as well as ones by "our sister engineering societies in England." Based on that, he asserted that "we should be doing nothing radical or unusual in establishing such standards provided they are found desirable." Finally, he turned to the voluntary nature of the standards recommended, noting the lack of any legal or moral requirement on anyone to follow such standards: "It is simply left to any person to follow it, as he sees fit, whether he is a member or not, and it is merely for the convenience of the members and the public generally that any such standards would be recommended." His framing of standard setting as common among engineering professional societies (perhaps a slightly exaggerated claim at this time) and the standards to be set as entirely voluntary no doubt helped shape the discussion that followed, which focused on *what* and *how* AIEE should attempt to standardize, rather than *whether* they should attempt to standardize anything.

The point of greatest contention was whether manufacturers should serve on the committee that would develop the standards. A Massachusetts Institute of Technology (MIT) professor, Cary Hutchinson, argued that they should not be included because they would naturally favor their own firm's interests. Crocker, however, took the more pragmatic view that "there are three sides to the question, as Mr. Rice has pointed out in his paper—the manufacturer, the purchaser and the consulting engineer, and leaving out any one of them you do not necessarily produce any better result."[80] A consulting engineer, Arthur Kennelly, agreed, pointing out that "having a committee to recommend how manufacturers should make apparatus, without having any manufacturers on that committee, was something like playing Hamlet with Hamlet left out."[81] Charles P. Steinmetz, the famous General Electric engineer, also supported including manufacturers, disputing Hutchinson's assumption that they would act only in their firm's interests:

> If the INSTITUTE intends to produce something of lasting value, which will be accepted and adopted by the whole continent, then the committee doing the work must be composed of men of such standing and reputation that, regardless of whether they are connected with manufacturing concerns or not, there can be no question that they will be impartial and not influenced by the fact that they are connected with this or that company.

A combination of the reputation argument and the pragmatic argument apparently defeated Hutchinson's objections. After some discussion, AIEE appointed a Committee on Standardization composed of men from all three of the constituencies: two professors (Hutchinson and Crocker), two famous General Electric engineers and managers (Steinmetz and Elihu Thomson), and two engineers from power companies using the equipment (John Lieb and Lewis B. Stillwell).[82]

In 1899 the committee issued its first "Report of the Committee on Standardization," which focused primarily on defining various terms used in describing apparatus (e.g., efficiency, rise of temperature, rating) and recommending how to measure them for various types of apparatus, rather than attempting to standardize the apparatus itself.[83] The accompanying transmittal document, however, suggested that the committee saw itself as having just begun its standardization work, noting that "while it is the opinion of the committee that many other matters might advantageously have been considered, as, for example, standard methods of testing; yet it has been deemed inexpedient to attempt to cover in a single report more than is here submitted."[84] AIEE had clearly taken on standardization as part of its professional mandate. It would add to and revise this set of standards multiple times over the next several years, making the committee a standing committee, creating multiple subcommittees in specific technical areas, and greatly expanding the content of the initial report, with the 1916 version occupying more than 100 pages of the *AIEE Transactions*, compared to just over 10 pages in the 1899 *Transactions.*[85]

By making these moves toward standardization, national engineering associations attempted to raise the profession's status by serving the public and industry. At the same time, in Europe a group of academic materials testing engineers reached across national borders with a similar purpose but without lasting success.

International Standardization in Materials Testing—an Aborted Attempt

The IATM, the first international association created solely for setting voluntary, limited-domain standards for industry, did not emerge from existing national engineering societies; indeed, no national engineering societies in the domain of materials testing existed at this time. Rather, it emerged from the collaboration of directors of elite academic testing labs in several German-speaking countries. Although the international association officially formed in 1895 did not survive World War I as a standard-setting body, an important national standard-setting association in this domain, the American Society for Testing Materials, of which Charles Dudley was elected the first president, emerged from it at the turn of the century. One early scholar of standardization noted that ASTM (and, by extension, the IATM) was not really a professional engineering association but a clearinghouse for standards in the area of testing.[86] That is, it did not perform many of the functions of a typical professional engineering association of the time (e.g., holding meetings where engineers presented papers, publishing proceedings); it focused solely on agreeing on standard methods for testing materials. As such, it was a new type of association that established an organizational form, goals, and membership that reflected this different mandate.

In the later nineteenth century, with engineers using materials such as steel to make bridges and rails, and concrete to make buildings, both practical and academic engineers found the lack of uniform methods for testing the quality and safety of such construction materials problematic.[87] In Germany, Austria, and Switzerland, elite engineering schools obtained government support to establish materials-testing laboratories for academic and industrial use. In 1871 mechanical engineering professor Johann Bauschinger established the first such laboratory at the Polytechnic Institute in Munich; in the same year another laboratory was established at the Berlin Polytechnicum. In 1873 a similar lab was established in Vienna, and in 1879 two more in Zurich and Stuttgart.[88] In 1884 Professor Bauschinger convened a group of these lab directors and engineering professors in Munich to develop and promote uniform methods of testing materials.[89] They organized a series of formal conferences (in Dresden in 1886, Berlin in 1890, and Vienna in 1893) that included engineers who worked for both manufacturers and users of materials, as well as professors and lab directors.

Bauschinger stood out among the growing group of lab directors. Born in 1834, he had excelled in math and science as a boy, and he had also learned diplomacy as part of a large family of modest means. After studying at the University of Munich and then briefly teaching in nearby Augsburg and Fürth, he accepted the chair in mechanics and statics at Munich's polytechnic university, where he developed what were the most accurate means of testing the properties of dozens of materials from building stones to steel beams. Although Bauschinger held these initial conferences in German-speaking cities, delegates from many countries attended. As Bauschinger noted when he published the resolutions adopted at these first four conferences, "Not only has there been an increase in the number of delegates from countries already represented (Germany, Austria-Hungary, Switzerland, Russia), but delegates have come from other countries (France, America, Norway, Holland, Italy, Spain), so that the conventions have assumed a truly international character."[90] Professor Mansfield Merriman, the first president of the American Section of the IATM, explained this spread as follows: "The reports of the proceedings of these conferences . . . attracted wide attention, and the great value and importance of the discussions became universally recognized in engineering circles. In short, the movement assumed an international character."[91] This effort resembled the earlier nineteenth-century scientific congresses in which scientific communities came together across national boundaries to discuss important scientific matters. Indeed, the "engineering circles" Merriman mentions may initially have been national engineering communities, but these congresses recruited them into an international engineering community as the scientific congresses did for scientific communities.

From the beginning, Bauschinger and the testing engineers established and worked through committees of engineers to create voluntary standards. They created "permanent committees" in various areas (e.g., Standard Specifications for Iron and Steel) to seek agreement on testing methods; these committees communicated by correspondence between one conference and the next and created reports and resolutions to present for vote at each subsequent conference.[92] In Bauschinger's introduction to the "Resolutions of the Conventions Held at Munich, Dresden, Berlin and Vienna," before the official formation of IATM, he asserted that determining suitable standards for testing required balancing representatives of both the producers and consumers of materials to avoid either set of interests dominating.[93] Like standardizing committees in engineering societies, these committees sought to reach consensus on recommendations, then forwarded them to the broader body to be voted on and eventually published. As Bauschinger explained, "Votes and resolutions have no other aim than to bring out the methods of testing which the majority of the members prefer. In conformity with the first resolution of the first conference, 'deliberations are to be free and resolutions not obligatory.'"[94] Thus, these international resolutions about agreed-upon testing methods were voluntary standards for which no enforcement mechanism existed or was desired. Unlike the ITU, the IATM did not seek to inscribe key standards into intergovernmental treaties or national legislation. The resulting resolutions were published in existing technical journals and as pamphlets, attracting considerable attention from the engineering community.

Up to the 1890s, this community of testing engineers had met at periodic conferences and established permanent committees, but it had established no ongoing organization nor any publication mechanism. At the 1893 Vienna conference, under Bauschinger's leadership, the engineers decided to establish a more permanent organization. Bauschinger himself did not live to see the decision fulfilled—he died suddenly later in that year. His obituary in the *Transactions of the American Society of Mechanical Engineers* (ASME) in 1894 noted that

> it was mainly due to Bauschinger's energy, tact, and indefatigable labors and uniform amiability, that the now famous "Conferences on the Unification of Tests and Methods of Testing Structural Materials" have been so eminently successful. Had it not been for his guiding hand, constant vigilance to avoid dangerous discussions and to straighten out tangled arguments, it is doubtful whether so much could even have been achieved, with a voluntary assemblage of manufacturers on the one hand and users of materials on the other, with a third party of investigators or directors of laboratories between them.[95]

That they were so influential, the obituary continued, "is mainly due to Bauschinger's honesty, integrity, and character as a man" (figure 1.2).[96]

Delegates at the 1895 Zurich conference, which included representatives from the United States and all the European countries except Turkey, formally organized the International Association for Testing Materials with the stated purpose of "the development and unification of standard methods of testing for the determination of the properties of the materials of construction and other materials, and also the perfection of apparatus for that purpose."[97] The Zurich conference thus became known as the First Congress of the IATM, and Professor Ludwig von Tetmajer, the civil engineer who ran the laboratory at the Zurich Polytechnic Institute, served as its first president.[98] At that meeting an International Council was established to handle the affairs of the organization between meetings. By the IATM's second congress, in Stockholm in 1897, international membership exceeded 1,000, with the largest number from Germany.[99] When the council met early in 1898, it named an official series of technical committees, with designated chairs and committee members. It also established a policy, based on Bauschinger's belief in the importance of balance between producers and consumers of materials, "that its Technical Committees should be nearly equally divided between producers and consumers."[100] The council also recommended that IATM members in each country form national sections to organize their work and committee appointments.

Membership in the IATM differed from membership in typical professional engineering societies in at least three ways.[101] First, membership was not by technical qualifications, as in typical engineering professional societies, but by agreement to forward the IATM's objectives (and presumably willingness to pay dues). One reason for this lay in a second unusual factor: professional societies, firms, and government departments, as

Figure 1.2. Johann Bauschinger, pictured on his gravesite memorial, held the conferences that led to the founding of the IATM. Photograph taken March 10, 2000, by Evergreen68, Creative Commons Attribution-Share Alike 3.0, accessed August 13, 2017, https://commons.wikimedia.org/wiki/File:Bauschinger_Johann_1.JPG.

well as individuals, could be members. Organizations then designated individuals to represent them. The American members included the Franklin Institute, ASME, engineering journals, and firms such as steel companies; German members included many urban public works departments, boards of railroads, Berlin's police department, and technical societies. The third difference in membership was the presence of engineers from multiple engineering disciplines (e.g., mechanical, civil, and chemical engineering) and even from other professions (e.g., architects, scientists, and industrialists), as needed to address complex industrial problems such as determining methods and equipment to test steel rails. All three of these differences in membership from professional engineering societies reflect practical aspects of the IATM's sole focus on standard setting, and similar membership policies would be adopted by other standardizing bodies subsequently, as would the policies around balance of producers and consumers of materials.

The American response to the early 1898 IATM council's call for national sections to be formed was immediate and, based on the subsequent long history and important role of ASTM, consequential. By June of that year Gus C. Henning, temporary American representative on the International Council and a consulting mechanical engineer, called American IATM members together to organize the American Section of the IATM, and in late August they conducted the new American Section's first annual meeting.[102] At that meeting they elected Mansfield Merriman, professor of mechanical engineering at Lehigh University, as chairman; Henry M. Howe, professor of metallurgy at Columbia University, as vice chairman; and two practicing engineers in testing departments (one at a railroad and one at Philadelphia's town hall) as treasurer and secretary. The By-Laws of the American Section of the IATM, adopted at the August meeting, defined its purpose, like its name, with respect to IATM: "The objects of the American Section shall be to unite more closely those members of the International Association living on the Western Continent, and to co-operate in the promotion of the purposes of this Association."[103] This apparent subservience to IATM would not last long. Members of the American group quickly chaffed at what they perceived to be poor organization and communication as well as excessive control on the part of the international association, and a longer-than-expected gap between IATM congresses exacerbated the problems. Organizers of the 1900 Paris exposition refused to allow IATM to meet in Paris that year, as planned, since the French organizers had appointed their own officers to conduct a (non-IATM) congress on testing materials instead. Thus, IATM moved its next meeting to Budapest a year later.

Correspondence between the IATM council and officers of the American Section between the 1897 second congress in Stockholm and the 1901 third congress in Budapest revealed strong disagreement over American representation on the IATM council during this gap.[104] At the second congress in Stockholm, before the American Section had been organized, IATM leadership appointed an American military officer the US representative on the IATM council.[105] When the officer failed to respond to the council's correspondence, it appointed Gus C. Henning, who had represented ASME at the congresses, as temporary American representative until the (planned) Paris congress.[106] In June 1899, IATM president Tetmajer wrote to Chairman Merriman, asking him "how the American members of the Association would like to settle the representation of America in the Council until the Paris Congress."[107] Because the planned Paris congress was less than a year away, Merriman initially responded that Henning should remain until then. But when the American Section held its second annual meeting two months later members had learned that the Paris congress would not occur.[108] At this point, while Henning requested that the council allow him to stay in the position until the next congress of the IATM, whenever it occurred, the American Section declared its desire to name Professor Henry M. Howe, vice chairman of the American Section, as its representative. The council ignored this declaration and the American Section's subsequent votes, petitions, and resolutions in favor of Howe and decided to leave Henning in place.[109] The exchange with Tetmajer became quite heated, with the American Section stating that Henning was persona non grata (for reasons now unclear). This issue was resolved only in 1901 when the secretary of the IATM finally convinced the council to accept Howe as the American representative.[110]

A second area of disagreement, this one solved more rapidly and with less acrimony, concerned the American Section's membership on IATM technical committees. The IATM council had originally assigned Committee 1, Standard Specifications for Iron and Steel, 40 members, of which 5 were allocated to the United States. By September 1899 the American Section, believing that it had been given authority to expand its numbers, had increased the American subcommittee to more than 20 members, with most of the added members from steel companies requesting representation.[111] Tetmajer responded that the council refused the request to increase the number of Americans on the IATM committee because it would "disturb the balance [among representatives of producers and consumers of materials] and in voting matters the due proportion of the committee and will give other countries the right of demanding the same number of representatives," but he softened the refusal by saying that the five official American committee members could simply represent the larger American subcommittee at the International Committee meetings.[112] Finally, the American Section was unhappy with the IATM's lack of transparency and communication. The official journal the IATM had launched, *Baumaterialienkunde*, initially published papers from the congresses in French and German only. Moreover, neither proceedings of the council, nor financial statements, nor work of the technical committees were published.[113]

In his chairman's address to the American Section in October of 1900, Merriman summed up these problems and stated a position: "After the experience of over two years in conducting business with the International Authorities, I have been forced to the conclusion that a reorganization is necessary in order to enable it to effectively conduct its affairs and successfully carry on its scientific work."[114] He recommended making the national sections the primary units, responsible for the proceedings, elections to the International Council, dues, and so forth, and having the national sections meet together periodically for congresses in which they "unify" their positions. The IATM did not make the changes needed to satisfy Merriman's vision, and in 1902 the American Section took matters into its own hands, officially incorporating itself as the American Society for Testing Materials and electing Pennsylvania Railroad chief chemist Charles B. Dudley president. From then on ASTM operated on its own, though it nominally affiliated with the IATM for almost two more decades. Members of ASTM were not required to be members of the IATM, as members of the American Section of the IATM had been.

The IATM continued to issue resolutions about testing methods on the international level until World War I, which triggered its demise as a standard-setting organization. In March 1915, the general secretary of the IATM informed members that "the war has interrupted the activity of the Association" and no more dues needed to be sent until it resumed activity.[115] When the war was over, the 1919 *Proceedings of the ASTM* replaced its identification of ASTM as "Affiliated with the International Association for Testing Materials" with "Organized in 1898 / Incorporated in 1902."[116] In his address published in that same volume, the ASTM president at the time reviewed some inadequacies of the IATM from the perspective of ASTM, including its refusal to set standards for materials themselves, as well as for testing methods, and its dislike of strong national associations; he then went on to say, "It is questionable, if it is practicable or doable, to revive the old International Association."[117] He reported that ASTM's Executive Committee had passed a resolution saying that the society would go to Great Britain and France to support the formation of a new international association, and that it would support "admitting the neutral countries into this organization," but not, by implication, enemy countries. That statement suggests an additional political reason for not reviving the IATM—its German center of gravity was no longer desirable to the victors of the war.

The proposed new association never seems to have gotten started. Members from two neutral countries, Switzerland and the Netherlands, attempted a short-lived revival of the old IATM in the late 1920s and 1930s, holding an International Congress for Testing Materials in Amsterdam in 1927.[118] At the meeting, representatives of 17 countries debated whether an ongoing association (as opposed to just a series of congresses) was needed, and whether any standardization should be part of the mandate. The constitution that was adopted at the end of this debate stated the goals of "The New International Association for the Testing of Materials" as securing "international cooperation, exchange of views, experience and knowledge in regard to all matters connected with the Testing of Materials" and that this should be done principally through periodic congresses coordinated by a permanent committee.[119] The constitution, however, explicitly ruled standardization out of scope.[120] At the 1937 London congress, it returned to being just a series of congresses, each independently funded by its host country, and soon it disappeared even in that form.

Crippled by nationalism during and after World War I and by postwar disagreements about the association's goals, the IATM was a fairly unsuccessful first attempt at establishing an international standard-setting association in one domain. Nevertheless, it demonstrated engineering interest in such international standardization. The other major legacy of the IATM was the formation of ASTM, a domain-specific national standard-setting association that would play a role in creating the first American general standard-setting organization at the national level and would itself take on an international scope in 2001, renaming itself ASTM International.

Steel Rails Redux

More immediately, however, ASTM figured in the next stage of US steel rail standardization, discussed earlier in this chapter. Around the turn of the century, frequent rail breakage led railroads and the ASCE committee to look at the steel rolling and finishing process, in particular at firing temperatures. In trying to make their steel processing more efficient, steel mills had raised their firing temperatures and moved the rails from one stage of processing to the next without allowing them to cool down, weakening the steel.[121] Railroads wanted to convince steelmakers to change this process without raising their prices. This time the efforts to achieve standards that were agreed to by railroads and steelmakers, undertaken by intermediaries including Hunt, were complicated by the addition of two more associations to the mix: the American Section of the IATM, soon to become ASTM, and the American Railroad Engineering and Maintenance of Way Association (AREMWA), founded in 1900.

As with the AIME and ASCE, these two new organizations had industry alignments: initially AREMWA, a railroad industry association, got involved because its members wanted railroads to come together to exert more pressure on the steel producers around this issue. Hunt and William Webster (another consulting engineer and rail inspector who also felt further standards were needed) joined AREMWA's newly forming committee on rails (of which Webster was made chair). From that position, they tempered the group's initial one-sided approach and encouraged the members to work more as a community of engineers and less as advocates for the railroads. Webster then convinced the ASCE to revise its own guidelines and also became a member of the ASCE committee on rail standards.[122]

Meanwhile, the rail subcommittee of the American Section of the IATM (part of the IATM's committee on Standard Specifications for Iron and Steel) was initially perceived as slanted toward the steel manufacturers. With the IATM's policy of roughly balancing its technical committees between representatives of producers and consumers, the IATM council had initially allotted the American Section of this subcommittee a total of five representatives: three engineers from steel companies and two unaffiliated or railroad engineers (William Webster being among them and designated chair of the American subcommittee).[123] Recall that the American Section then added more members (the subject of one of their disagreements with the IATM), including many more from steel companies who had requested membership. In an October 1900 meeting, in response to criticisms of the committee's apparent slant toward steel producers, Chairman Webster defended himself, perhaps disingenuously, explaining that he had added the steel manufacturers to the committee to help with getting the specifications implemented, and that they had attended meetings and contributed to the effort: "This has all been misunderstood, and the proposed specifications have been referred to as 'Manufacturers' Specification,' and as being 'too lenient,' etc. This has come about by the engineers on this Committee being too busy to attend its meetings."[124]

By 1902, however, the body had transcended this criticism, in great part by its election of the popular and respected Charles B. Dudley as president of the newly independent and renamed ASTM, a role he maintained until his death in 1909. Not only did Dudley represent a consumer (railroad) company, though active in the producer-dominated AIME, but he had also long advocated balancing producers and consumers in setting standards. Indeed, for the first meeting of the Executive Committee of the newly formed American Section of IATM in June 1898, he had suggested a discussion session titled "The Relation of the International Association to Producers and Consumers."[125] In his "Personal Tribute" to Dudley upon his untimely death in 1909, Hunt would note how important Dudley's beliefs in balance were to the success of ASTM:

> When the Society was first formed, it was generally feared that it would become a manufacturers' organization, and therefore it was with hesitancy that others gave to its deliberations and recommendations unprejudiced consideration. I doubt if there was another man under whose leadership the Society could have so quickly been made successful. Dr. Dudley's whole business career has been as representative of the consumer, yet he had always made manifest his endeavor to be fair to the maker. Hence the confidence that both sides had in him. There might be, and often were, differences of opinion, but never doubt of intelligence or honesty of purpose.[126]

With the change in leadership and the transformation into ASTM, that society was no longer seen as slanted toward steel manufacturers. At this point Hunt and Webster managed to guide proceedings toward a convergence of standards among the ASCE, AIME, AREMWA, and ASTM that reflected both sets of stakeholders' concerns, adding a shrinkage clause to address the finishing temperature.[127] Dudley's work with standards in the materials testing domain would continue up to his death in 1909. In addition to his work for ASTM, in 1904 Dudley was elected president of the IATM (still active internationally before World War I).[128] But he made his biggest contribution in ASTM, for which one obituary named him "the father of a movement which is of as much philanthropic as commercial importance"—a movement that chapter 3 will take up.[129]

The practical impact of the nineteenth-century standards discussed in this chapter was profound. A 1900 study reported that one year before the Franklin Institute steam boiler standard was enacted into law in 1852, one in every 55,714 steamboat passengers in the Mississippi Valley and Gulf of Mexico lost their lives. Thirty years later, the number was one in 1,726,287. Of course there were improvements outside the boiler room, especially in navigation—in the 1840s about 1 in 10 steamboat accidents were "collisions," not "fire" or "explosions"—and while these standards were created by a private organization, they certainly were not voluntary as they were written into law.[130] The wide adoption of screw thread standards by leading companies and government offices in the United States and in Great Britain demonstrates the value of these private, voluntary standards. The nearly universal adoption of steel, rather than iron, rails after the development of standards confirming their strength, according to a leading economic geographer, significantly "reduced the long-distance rail transport costs for heavy and bulky raw materials," and was as important as "the development of the internal combustion engine for trucks and personal transport" in the development of the twentieth-century North American economy.[131] The same is certainly true of other economies that span vast territories. Finally, the early materials standards established by ASTM (for steel girders, concrete, etc.) and the early electrical systems standards created by the IEE, AIIE, and VDE, which would provide the basis for many of the early twentieth-century national standards, facilitated the building of the turn-of-the-century physical infrastructure that transformed urban life—the towering buildings, electrical systems, underground railroads, telegraph networks, and modern water and sewer systems that all of us take for granted.[132]

At the turn of the twentieth century, as the next chapter relates, professional engineers established the first still-surviving organizations focused solely on standard setting and in them continued to develop a process for achieving consensus on voluntary standards.

2

Organizing Private Standard Setting within and across Borders, 1900 to World War I

With the arrival of the twentieth century, the growing national and international communities of industry-focused engineers launched the first wave of private standardization organizations. Engineers sought to serve the public by broadening and coordinating standard setting across multiple engineering fields and industry communities as a way to improve the efficiency, interoperability, safety, and output quality of manufacturing industries at home (and, in some cases, within sprawling empires), and to improve their economic competitiveness internationally. Although this wave focused primarily on national standard setting, only a few years after the formation of the first national body, a group of electrical engineers reached across national boundaries to form the first, still-surviving, private international standard-setting body, albeit in a restricted domain.

This chapter traces the rapid rise of the first wave, starting with two significant organizational developments in private standardization that occurred just after 1900. In 1901, the first successful and still-surviving general national standards organization emerged in Great Britain—the Engineering Standards Committee (ESC), forerunner of today's British Standards Institution. Then, just five years later, electrical engineers formed the first successful and still-functioning domain-specific international standardizing organization, the International Electrotechnical Commission. The engineers who launched these organizations, apparently influenced by principles followed by the International Association for Testing Materials, solidified principles and processes that would inform private and voluntary standardization over most of the twentieth century. Joining Bauschinger and Dudley as pioneering standardization entrepreneurs are civil engineer Sir John Wolfe-Barry and electrical engineer and industrialist Colonel Rookes Evelyn Bell Crompton, both of the UK, further articulating principles and processes for industrial standard setting by committees of engineers. Charles le Maistre, the British electrical engineer who served as secretary of the British national body and general secretary of the IEC, spread and amplified these principles and processes from his base in London, acting as an ambassador for this model of voluntary, consensus-based standardization and became the most significant standardization entrepreneur of the first half of the twentieth century. The chapter ends with the spread of national standard-setting organizations around World War I, in some cases directly shaped by le Maistre.

Britain's National Standardizing Organization

Although civil engineering societies initially saw themselves as representing all nonmilitary engineers, the previous chapter showed that new societies representing both different fields of engineering and engineers in particular industries proliferated during the last two decades of the nineteenth century. Some of them established committees to set standards, but if any one engineering society wanted to set a nongovernmental voluntary industrial standard that would be widely adopted, it needed to coordinate with other relevant engineering societies. Some associations had already coordinated in ad hoc efforts such as the standardizing of steel rails in the United States discussed in chapter 1. At the turn of the twentieth century, British engineering leaders decided that they needed to become more involved in setting standards and that they needed a more formal coordination mechanism for setting standards for Great Britain and the British Empire. The national standardizing body they established would set a pattern followed by other countries during and after World War I.

At the end of the nineteenth century the Institution of Civil Engineers was the world's—as well as Great Britain's—oldest, largest, and most prestigious professional engineering society.[1] Indeed, most leaders of the other British engineering societies were also members of ICE, making it the core of the British engineering community. ICE had already played an indirect role in standard setting by launching the Whitworth screw thread, but it had not wanted to coordinate standard setting with other institutions (or countries). In the final years of the nineteenth century ICE's Governing Council responded unenthusiastically to invitations to participate internationally in the IATM and nationally in a committee to standardize brick size, proposed by the Royal Institute of British Architects.[2] In 1900, ICE refused another request from the latter body to collaborate on structural designs (saying that "it was 'the practice of the Institution of acting independently in matters coming within its scope'"), as well as a request from the British Iron Trades Association (BITA) to consider sizes and standards for structural materials. A change of heart by two of ICE's leaders, combined with a compelling problem, moved the Governing Council away from this position in 1901.

In 1897 a seed was planted for this change of heart when John Wolfe-Barry, ICE president that year, along with other representatives of scientific and technical societies, met with the prime minister, Robert Gascoyne-Cecil (Lord Salisbury), to discuss the need for a national laboratory. Ultimately, this meeting would result in Great Britain's establishment of the National Physical Laboratory, run by the Royal Society, to set fundamental scientific standards. The far-ranging conversation at that meeting, which included discussion of American standardization initiatives such as those around screw threads, steam boilers, and structural steel, may have begun to open Wolfe-Barry's mind to the notion of ICE participation in standardization for industry.[3]

Wolfe-Barry, born in 1836, two years after Bauschinger and six years before Dudley, was of the same generation, had an education similar to those of his German and American counterparts, and played a similar role as an institutional innovator, but Wolfe-Barry's own engineering work was better known to his own countrymen, both before and long after his death. In the early twenty-first century, Germans could find a plaque commemorating a Bauschinger testing formula in one Munich subway station, but the *entire* London Underground seemed to be a memorial to Wolfe-Barry, who designed the early key route along the River Thames. He also designed four of the bridges that span the river, including the iconic Tower Bridge, completed in 1894, with its sky-high pedestrian crossing and neo-Gothic exterior hiding a state-of-art (for the day) mechanism that opened the street-level bridge to maritime traffic. The bridge secured Wolfe-Barry's public reputation as one of Britain's greatest engineers, as well as his 1897 knighthood. For the next two decades, until his death in 1918, he used that reputation to promote standardization, first in the National Physical Laboratory project, then in the new committee soon to be formed by ICE. As his ICE obituary reported, "It was the subject of the Forrest Lecture delivered by Sir John in the year 1917, the title of which was 'Standardization and its Influence on the Prosperity of the Country.' No one can fully estimate the debt which the country and the Empire owe to his influence in this matter."[4]

Nevertheless, it was one of Wolfe-Barry's successors as ICE president, Sir Douglas Fox, who first promoted ICE's role in standardization in his 1899 presidential address. Fox surveyed the accomplishments of civil engineering during the nineteenth century, paying special attention to railroads and electricity. He also noted areas in which Great Britain had fallen behind other countries, lamenting that "it is discouraging to compare the present position of electrical industries with that of similar undertakings in the United States, Germany, Switzerland, Japan, and even in our own Colonies," ultimately identifying the real competition as coming from engineers in Germany and the United States.[5] In particular, he noted the progress the United States had made in standardizing shapes of structural steel for buildings, as well as steel rails for railroad lines. The US standardization of structural steel shapes for buildings was a case of de facto market standardizing, achieved by the Carnegie Steel Company.[6] The steel rails for railroads, however, had been standardized through the efforts of the various engineering and industry associations discussed in the previous chapter, including the American Section of the IATM (soon the American Society for Testing Materials). He complained that arcane "vested rights are difficult to overcome" in Britain and that the Board of Trade and parliamentary acts imposed constraints not faced in other countries, and he noted that "German and other continental engineers are greatly assisted in many ways by paternal Governments, whose officers they generally are, and who lay down valuable regulations, and in many instances establish standards of quality and design."

Fox went on to argue in favor of wider standardization not by the British government but potentially by ICE, wondering "whether competition in the world's race could be facilitated by the establishment, upon sufficient authority, of standard specifications for such materials as steel and cement, and the introduction of standard types for bridgework, roofing, and other structures, . . . and for locomotives and rolling stock." He warned against the potential danger of preventing improvements but used the example of "the experience with the Atbara Bridge" to argue "how important it is, when early delivery is a necessity, to be prepared with type designs, based upon ordinary merchant sections of steel." This example reveals that Fox was clearly thinking not just of Great Britain but also of its empire. The government had awarded a contract to build the bridge, a critical railway link in British Sudan, to an American firm because it could build it more rapidly and at a lower cost than any British firm by using standard American steel sections.[7] Fox extolled the contribution "to engineering successes of late years [of the] introduction of cheap steel of good quality and of high tensile strength, both for rails and for plates and rolled sections" but noted that engineers were worried about signs of metal fatigue after extensive use: "With reference to this matter and to other questions involving scientific research, the resources of this Institution might, I suggest, be advantageously employed." Any one firm might not be able to afford to perform the research necessary to solve such big problems, but ICE could do so and benefit all of its members, Fox observed. This plea for ICE to become more involved with testing and standardization sounded a new note from an ICE leader.

Fox had cited as one area ripe for standardization the compelling problem that would nudge ICE's Governing Council into recognizing both the need for standards and the need to cooperate with other engineering societies in setting them. Ironically, in light of ICE's recent refusal to collaborate with the British Iron Trade Association on standardizing structural materials, the problem was the vast array of sizes and shapes of rolled steel girders ordered in Britain. H. J. Skelton, a London iron merchant and BITA member, complained in an 1895 London *Times* letter:

> Rolled steel girders are imported into Britain . . . from Belgium and Germany, because we have too much individualism in this country in quarters and upon questions where collective action would be economically advantageous. As a result, architects and engineers generally specify such unnecessarily diverse types of sectional material for given work that anything like economical and continuous manufacture becomes impossible. In this country no two professional men are agreed upon the size and weight of girder to employ for given work, and the British manufacturer is everlastingly changing his rolls or appliances, at greatly increased cost of manufacture, to meet the irregular, unscientific requirements of professional architects and engineers.[8]

Although this letter had no immediate effect, in 1900 Skelton presented a paper advocating standardization of steel girders at a BITA meeting attended by Wolfe-Barry, who several years later would refer to Skelton's paper as having impressed him at the time.[9] This paper, combined with his earlier participation in the conversation with Lord Salisbury and Fox's presidential address, probably contributed to Wolfe-Barry's change of heart about standardization at the end of 1900.

In the six weeks between the last ICE council meeting of 1900 and the first one of 1901, perhaps Wolfe-Barry and Fox discussed the value of industrial standardization for ICE and for Britain, for in the January 22 meeting Wolfe-Barry moved and Fox seconded a motion "that a Committee of six members be appointed to consider the advisability of standardising the various kinds of iron and steel sections; and if found advisable then to consider and report as to what steps should be taken to carry such standardisation into practice."[10] The motion passed, and a prestigious committee of ICE leaders, including Fox (1899 president), Wolfe-Barry (1897 president), and James Mansergh (1901 president) was formed. The committee quickly added to their number by inviting representatives from the Institution of Mechanical Engineers, the Institution of Naval Architects, and the Iron and Steel Institute, since this industrial matter (like IATM's testing and the American efforts to standardize steel rails) required combining efforts of multiple engineering disciplines. Soon afterward, the "ICE Committee on Standard Sections supported by the IMechE, INA and the ISI," as it was initially designated, met to start its work, now with a chairman (Mansergh), an acting secretary (Leslie S. Robertson), and a representative of each of the four engineering societies.[11]

Robert C. McWilliam, historian of British standardization, has noted that the committee could easily have completed its assigned task and gone out of existence or become a semiautonomous, loosely organized body, as similar committees in the United States had during the last quarter of the nineteenth century and would continue to do through World War I. He attributes the fact that it developed into the first national standard-setting organization primarily to the peculiar nature of the British market for standards.[12] Two enormous customers dominated that market: the British military (the War Office and the Admiralty) and the civilian India Office. These governmental purchasing organizations desired standardization in service of the British Empire, though the government itself was not, evidently, willing or able to set standards itself.

Although initially the new committee had some financial support from ICE and the coordinating engineering societies, its leaders successfully turned to the government for additional help. A writer for the periodical *Public Works* explained two years later "that the movement was of such national importance that, in addition to the active and financial support of the five leading engineering institutions, the Government ought to furnish assistance in view of the great advantages anticipated . . . to the general trade of the country."[13] Indeed, when Chairman Mansergh, Wolfe-Barry, and Secretary Robertson approached leaders in the War Office and Admiralty, they gained immediate financial support through the Board of Trade. They also approached the India Office, which was eager to have standard locomotives that could be transferred among the various lines in the subcontinent. Thus, when Wolfe-Barry approached that office, its leaders became enthusiastic supporters, encouraging the committee to take on standardizing not just steel sections but also railway engines. By the end of 1901 the ICE council had enlarged the committee's mandate to include standardizing locomotives and tests for materials, and it had changed its name to the Engineering Standards Committee, a name it would retain until it became independent of ICE in 1918. This government support and partial funding undoubtedly helped convert a one-time committee to an ongoing organization.

The committee, acting more or less autonomously, developed its staff, structure, and membership to account for its enlarged scope. It added Charles le Maistre, a young electrical engineer and member of the Institution of Electrical Engineers (IEE), to the staff as assistant to Secretary Robertson—a fateful appointment that would start le Maistre's long career as a standards entrepreneur, the early stages of which will be discussed later in this chapter. Structurally, the ESC supplemented its main committee with four sectional committees: (1) Bridges and Building Construction, (2) Materials for Shipbuilding, (3) Locomotives and Railway Rolling Stock (chaired by Fox), and (4) Railway and Tramway Rails (chaired by Wolfe-Barry).[14] These sectional committees (and their subcommittees) accomplished the work of standardizing, in a pattern that resembled and probably followed that of standardizing committees that emerged in the late nineteenth century in the IATM as well as in German and American engineering societies. In 1902, at the request of the engineer in chief of the General Post Office, Sir William Preece, electrical plants were added to the committee's mandate.[15] In response, ICE encouraged the ESC to invite the IEE to be represented on the committee and thus to become the fifth and final of what were designated the Founder Institutions (an official designation retained in the constitution and committee structure of the current British standardizing body, the British Standards Institution, when it was reorganized and renamed in 1931).[16] Colonel Rookes Evelyn Bell Crompton, an electrical engineer and head of a British manufacturer of electrical generating and lighting systems, who will become a leading character in the next section, was named the IEE representative to the main committee of the ESC.[17] This addition necessitated a fifth sectional committee, Electrical, as well. The ESC added four additional sectional committees during 1903 (Machine Parts, Pipe Flanges, Cement, and Cast Iron Pipes), and three more in 1911 and 1912 (Vitrified Ware Pipes, Road Material, and Automobile Parts).[18] Before it became independent of ICE, a final sectional committee (Aircraft Parts) emerged in 1917, necessitated by wartime needs for aircraft standards. In later years, as a general (rather than domain-specific) body, it would continue to broaden its scope.

Although officially it had only five founding bodies, the ESC formed the sectional committees so they also included representatives from other domain-relevant associations (e.g., the Institution of Engineers and Shipbuilders of Scotland, and the IATM), trade organizations representing firms and industries (e.g., South Staffordshire Ironmakers' Association), and government departments (e.g., the Admiralty and Board of Trade), as well as individuals with knowledge in an appropriate domain (e.g., H. J. Skelton, the iron merchant who in 1895 advocated standardizing steel girders).[19] With this heterogeneous membership, the ESC, like the IATM (and surely influenced by it, given the latter's representation on sectional committees), sought broad and balanced input from what it saw as all interested groups—producers, purchasers, and independent engineers. For example, the sectional committee on ships and their machinery included steelmakers, shipbuilders, representatives of the Admiralty, and academic engineers.[20]

In late 1902, ICE gave the ESC a final power necessary to its new role: the right to publish its findings as British Standards.[21] Starting in early 1903, the committee published British Standard #1 (BS1), British Standard Sections for iron and steel, thereby accomplishing the original goal of the committee. The ESC published four more standards in 1903: BS2, Tramway Rails and Fish-Plates; BS3, Influence of Gauge Length and Section of Test Bar on Percentage of Elongation; BS4, Geometrical Properties of Standard Beams; and BS5, Standard Locomotives for Indian Railways. BS5 reflected the influence of the India Office; and BS2 and BS4 revealed the ongoing importance of steel sections, rails, and beams; BS3, a report of a technical investigation, exemplified a small category of informative standards.[22] During the first period of British standardization, from the ESC's founding in 1901 to its independence from ICE in 1918, 972 individuals served on 14 committees, and it issued 81 standards.[23] The standards "were generally written from the buyer's perspective" so those buyers could easily incorporate the specifications into their orders.[24]

Contemporaries saw the ESC and its British Standards as supporting both British national manufacturers and the British Empire. In 1903 one observer, echoing Skelton's earlier complaint, wrote, "In the past there has been an excess of originality in engineering work," with manufacturers and engineers all attempting to leave their own original mark on machinery, bridges, and other industrial items; "these and other impediments to national progress will be swept away in due time by the work of the Engineering Standards Committee, which is worthy of a foremost place among the great national institutions of the present century."[25] Wolfe-Barry, "the guiding spirit in the early days of the [standardization] movement"[26] (figure 2.1), reflected this mercantile nationalism in a 1908 speech:

> There are markets in which we cannot hope to compete successfully on equal terms, by reason of imposts and protective duties, but the work which has been effected by the Committee should very materially assist in keeping the trade of our Colonial Empire in the hands of British manufacturers. If the Colonial Governments will insist upon all materials bought by them being to British standards, they will know that they are receiving good designs and high-class material. . . . Standardisation affects the whole engineering trade of the Kingdom and Empire.[27]

He argued that British standards were important not only to British manufacturers and colonial governments but also to British working people, who depended on British manufacturers for jobs, a statement that may have indicated more concern about labor strife than about the working classes themselves. The attendance at a 1911 banquet in his honor (the 10th anniversary of the founding of the ESC) suggested that the India Office, shipbuilders, steelmakers, railway manufacturers, and the Admiralty all held Wolfe-Barry in high esteem.[28] Moreover, as a 1913 article in an American engineering journal noted, Great Britain was now recognized as "in advance of anything accomplished in the United States" in its cooperative work among engineering societies to create national standards.[29]

In his 1917 James Forrest lecture to ICE a year before the ESC became the independent British Engineering Standards Association (BESA), Wolfe-Barry spelled out eight principles for effective standardization that he had followed in his leadership of the ESC.[30] In his first principle, "participants should represent all interested parties, both producers and purchasers," he stated that all stakeholders (as we would now put it) should be included in the process. This was a principle that the Bauschinger congresses and the IATM, as well as engineering societies such as the American Institute of Electrical Engineers, had developed in the late nineteenth century and that the ESC had

Figure 2.1. Sir John Wolfe-Barry, builder of the Tower Bridge and other famous structures, guided the Engineering Standards Committee in its formative years. 1905 *Vanity Fair* caricature.

adopted as well, whether from direct imitation of the IATM or from its own discovery. Commercial entities were seen as essential to standard setting in this period. Wolfe-Barry stated the reason *private* standardization would be needed in his second principle, that "subjects [for standardization should be] introduced where participants recognised need for self-regulation to introduce order into a condition of things which had become more or less chaotic." This notion of self-regulation against chaos in areas where governmental regulation did not exist (and perhaps also where businesses preferred not to have governments regulate them) was important to private standard setting. A third principle, "efforts should be voluntary with no payment to participants," articulates another meaning of "voluntary" in this context: not only did firms adopt standards voluntarily, but engineers participated in the process voluntarily. While funding from the Founder Institutions and the British government paid for tests and the publication of standards, the time and travel of all participating engineers except for the full-time secretary were given freely or supported by the engineer's employer. At this time engineers, as new professionals, all aspired to be gentlemen. Still, this principle of voluntary service, which made it possible to carry out such work without enormous amounts of central business or governmental funding, would make it difficult for an engineer working for a small firm without adequate resources, or an independent engineer without personal resources, to participate in standard setting.

Principles 4 and 5 suggested that standard setting should be initiated to respond to clear demand, not for its own sake, thus putting limits on when standard setting was appropriate. Principle 6 highlighted the practical, engineering, and commercial orientation of standard setting, though it should be based on current scientific knowledge. Principle 7 said that the ESC should only develop and set standards, not become a testing authority, a principle that would be abandoned by today's successor to the ESC in the third wave of innovation in standard setting (see chapter 9). Finally, in principle 8 he emphasized the need for regular revision of standards to avoid having them become an impediment to progress.

Although we have no evidence of the direct influence of Bauschinger and Dudley or the IATM on Wolfe-Barry and the ESC, principle 1 as well as the fact that ESC committees were created with a heterogeneous mix of associations, firms, and government departments represented, suggests such an influence. Moreover, indirect evidence shows that Wolfe-Barry was clearly in the same international community of engineers, where it was extremely likely they at least knew of each other. In 1896, before the birth of the ESC, Wolfe-Barry (then ICE vice president) chaired and introduced a talk entitled "Physical Experiment in Relation to Engineering" that cited Bauschinger and his successor Tetmajer.[31] As chair of the ESC's Rails Committee, Wolfe-Barry would surely have looked at the results of the steel rail standardization in the United States discussed in the previous chapter, in which the IATM's and ASTM's principles of balance played a role. Moreover, the IATM was represented on some sectional committees of ESC (e.g., Professor Unwin, representing the IATM, served on a subcommittee of an ESC committee on steel sections for shipbuilding), providing Wolfe-Barry another source of exposure to IATM principles.[32] So whether Wolfe-Barry met and learned personally from Bauschinger before his death in the 1890s or Dudley in the first decade of the twentieth century, their ideas of the importance of standardization by engineering committees representing balanced interests of manufacturer and user organizations were part of the same general conversation among engineers of this period, as we will discuss in chapter 3.

In the year of his Forrest Lecture, "Sir John Wolfe-Barry himself led the panel which drafted the Memorandum and Articles of Association of the successor body" (BESA), though he died a few months before BESA's incorporation in May 1918.[33] In a later section of this chapter, we will discuss the influence of the ESC, primarily through Charles le Maistre, on the formation of other private national standard-setting bodies.

First, however, we look at another key development in private and voluntary standardization immediately after the formation of the ESC. Because so many of the new industries of the late nineteenth and early twentieth centuries were international, it is not surprising that, only three years after the first national standard-setting body was established, the first still-surviving private international standard-setting body was formed, though in a restricted technical domain.

The International Electrotechnical Commission

Although the IATM's attempt to create nongovernmental international standardization in the area of testing materials would not survive World War I, a much more successful international standard-setting body arose just a few years after the formation of the ESC, in the domain of electrical engineering: the International Electrotechnical Commission, an organization still active today.[34] Chapter 1 showed that nineteenth-century scientific congresses had standardized the system of electrical units, and that the intergovernmental ITU had standardized administrative protocols for telegraph systems across national boundaries of treaty signatories. In addition, the 1881 International Electrical Congress in Paris had encouraged the development of electrical engineering societies in several industrialized countries, and some of these national professional societies (e.g., the Association of German Electrotechnicians [VDE], the British IEE, and the American AIEE) also became involved in industrial standardizing around electricity. Early in the twentieth century, the work of these electrical engineers intersected with the ongoing scientific work of the International Electrical Congresses, resulting in the formation of the IEC, which extended electrical standard setting for industry into the international arena.

The IEC was conceived at the 1904 St. Louis International Electrical Congress, which met in conjunction with the Louisiana Purchase Exposition. Before the congress, a planning committee had met at Niagara Falls and elected Elihu Thomson—famed British-born but US-raised inventor, industrialist, cofounder of the Thomson-Houston Electric Company, and (since its merger with Edison General Electric Company in 1892) chief engineer of General Electric—to be president of the 1904 congress.[35] Alongside this scientific congress, the planning committee also invited a Chamber of Delegates appointed by and representing governments, to address matters of standardization in nomenclature not just among scientists but on a governmental and intergovernmental level as well. When the Chamber of Delegates (made up of 29 members representing 15 countries) met in St. Louis, it elected Elihu Thomson president of that body, as well. The chamber established a committee on international standardization, which came back the following day with a report that would have momentous consequences for private international standardization for industry. Along with recommendations concerning standardization of scientific nomenclature, the report recommended "that steps should be taken to secure the co-operation of the technical societies of the world, by the appointment of a representative Commission to consider the question of the standardization of the nomenclature and ratings of electrical apparatus and machinery." The committee expressed hope that "if the recommendation of the Chamber of Delegates be adopted, the Commission may eventually become a permanent one."[36]

The effort recommended here was momentous because it differed from scientific congresses' standardization of scientific nomenclature; it advocated standardization of electrical apparatus and machinery used by industry. Moreover, although representatives of governments made up the Chamber of Delegates, it called for the technical societies, not governments, to form the new, and potentially permanent, commission. The Chamber of Delegates promptly adopted the recommendation, asking delegates to inform their countries' technical societies (i.e., electrical engineering societies) and report their responses to two individuals who were tapped to lead the effort: the next president of the AIEE (as yet unnamed, and never very much involved), and Colonel Rookes Evelyn Bell Crompton, the British electrical engineer and

Figure 2.2. Colonel Rookes Evelyn Bell Crompton led the effort to establish the IEC, still an important standard-setting body today. https://upload.wikimedia.org/wikipedia/commons/8/8c/Rookes_Evelyn_Bell_Crompton_Vanity_Fair_30_August_1911.jpg.

manufacturer who represented the IEE on the ESC. Crompton had just given a well-received paper on the need for standardization at the Electrical Congress and had put forward the resolution in favor of such a body.[37] Figure 2.2 shows Colonel Crompton as depicted in *Vanity Fair* in 1911.

Born in 1845, Colonel Crompton had a varied career including military service, transportation engineering (early steam-powered road engines and later internal combustion engines), commercial leadership in electrical power and lighting systems, and standardization. At the age of 11, he had gone with his family to Crimea at the end of the Crimean War, where he joined his mother's cousin onboard the HMS *Dragon* and reportedly saw action, later often claiming (incorrectly) to have been awarded the Crimean War Medal and Sebastopol clasp at that young age. After he returned to school at Harrow, he indulged his second love by building a steam-powered road engine, the Blue Bell, which a machinist further developed for manufacturing. Then, when he was serving in the military in India, he combined the two passions by convincing the British to order one of the vehicles to try out as a replacement for the bullock trains then used to move army supplies and supervising its running himself. After these early military and transportation adventures he turned to the electrical work that would be his primary career for many decades, founding his company, Crompton and Co., in 1878.[38] He would later return to the military, as leader of a volunteer corps of electrical engineers in the Boer War, and to road transportation, as a founder of the Automobile Club of Great Britain and Ireland in the late 1890s and an engineer for the British government's Road Board, established in 1910.[39] His work in standardization was primarily, but not exclusively, linked to his identity as an electrical engineer. Indeed, he served on a sectional committee of the ESC that addressed screw threads and claimed later in life to have designed and tested the new form of screw threads developed by the ESC to replace the Whitworth standard thread "for all cases where the latter thread is considered too coarse for certain modern purposes, such as for automobile vehicles and many electrical purposes," a standard that came to be called British Standard Fine.[40]

When the 1904 Electrical Congress's Chamber of Delegates recommended creating a commission, Crompton, coordinating with the British IEE and the American AIEE, organized a 1906 meeting in London to form the International Electrical Commission (renamed during the meeting the International Electrotechnical Commission, to distinguish it from more scientifically focused organizations).[41] By 1905, nine countries' electrical engineering societies had responded positively to Colonel Crompton's correspondence about the 1906 meeting, indicating that an international community of interest existed in this relatively new field of engineering. In preparation for the 1906 meeting, the British IEE appointed a prestigious committee to draft a set of proposed rules and circulate it to electrical engineering societies of interested countries.[42] That committee included, among others, Alexander Siemens (cousin of Werner von Siemens and current head of the British arm of the

Siemens company), Lord Kelvin (William Thomson, the famous physicist and electrical engineer introduced in chapter 1), and Colonel Crompton himself. Participants at the 1906 meeting, chaired by Alexander Siemens, elected Lord Kelvin the first IEC president, and Colonel Crompton the first honorary secretary.[43] Crompton and Siemens were electrical engineers and industrialists, while Lord Kelvin gave the new association scientific luster. Crompton asked Charles le Maistre, a member of the IEE and assistant to the secretary of Britain's ESC, to serve as acting secretary.[44] Born in 1874 on Jersey, one of the Channel Islands between England and France, and completely fluent in French as well as English, he was the perfect choice for this international effort.[45]

Attending the 1906 meeting were 33 individuals representing 13 countries: the United States, Austria, Belgium, Canada, Spain, Japan, Great Britain, France, Germany, Holland, Hungary, Italy, and Switzerland.[46] Norway, Sweden, and Denmark accepted the invitation but had not yet appointed delegates. Initially Siemens asked the 33 delegates to focus discussion on the draft rules that had been circulated, but it quickly became clear that the group was too large to discuss the rules efficiently. Consequently, he established a subcommittee with one member from each of the 13 national delegations to hash out the details of the constitution.[47] The ensuing subcommittee discussion introduced many principles and processes and established a model that directly influenced subsequent voluntary international standard-setting institutions. Indeed, Colonel Crompton referred to what they were doing as "the preliminary stages of the movement."[48]

They adopted a basic principle that any participating country, no matter how small or large, would have one vote. As Alexander Siemens explained to the larger body of delegates, the original IEE committee draft "had endeavoured to place every country joining the Commission on an absolutely equal footing."[49] This practice was similar to that followed by the intergovernmental public international unions such as the International Telegraph Union in which every country's national—or, in the case of countries such as India, colonial—telegraph administration originally had a single vote. In the public international unions, the goal was to treat every organization that administered the public service equally.[50] In the IEC, the point was to treat every major technical body involved in electrical standardization equally. Although subcommittee members debated other aspects of representation and voting in the draft, they never questioned the notion of one country / one vote. Indeed, according to the report of the meeting, "The fact of each country contributing an equal share and having an equal vote would, [Siemens] thought, very materially help towards the success of the [international standardization] movement."[51] All the major private international standardization bodies created over the next 80 years would also follow this principle. The subcommittee considered how delegations would convey their votes and how they would handle internal disputes (local committees did not have to agree unanimously on how their single votes would be cast). They also considered how they would declare standards; they required unanimous agreement of the national members to publish a decision as an IEC standard, noting, "All decisions resulting from a divided vote may be published only when the names of the countries voting for and against are given."[52] When the body met again in 1908, it would reduce the requirement to declare an IEC standard from unanimity to a four-fifths majority but would not alter the principle of one vote per nation.[53]

In determining who would appoint each country's delegation or local committee to the IEC, they found it necessary to balance two different issues. Some delegations, such as the Germans from the VDE, "did not appear to think it advisable that the government of each country should be brought into the matter" and wanted each country's appropriate technical society to appoint that nation's local committee with no government involvement whatsoever.[54] The tension between the vast majority of German engineers and the state, with its privileged cadre of state-supported architects and civil engineers, left a long legacy of distrust; when the IEC was formed, it was still impossible for German engineers in recently developed fields—including, of course, electrical engineering—to move into the higher civil service.[55] Against this distrust, the subcommittee weighed the reality that many nations did not yet have any national engineering society, let alone a society of electrical engineers. In a private discussion with the lead VDE delegate, Siemens persuaded him that it was acceptable to have technical societies be the primary vehicle for representation within the IEC but to have the rule allow that, "in a country having no Technical Society dealing with electrical matters, the Government should appoint the Committee."[56] Diplomacy and the ability to compromise were important qualities of leaders in standard setting. The subcommittee also agreed, at the suggestion of Cyprien O'Dillon Mailloux, AIEE president in 1906 and leader of the American delegates, that "the Technical Society of any country desiring to appoint a Local Committee could do so provided such Society had been at least three years in existence previously," to avoid depending on a technical society before it was adequately established.[57] That they minimized governments' role in the IEC was particularly notable in light of both the precedent set by intergovernmental organizations such as the ITU and the IEC's origin in the government-appointed Chamber of Delegates at the 1904 International Electrical Congress. The significant involvement of Colonel Crompton, with his ESC experience, and of Mailloux, with his experience with the AIEE's standardization committee, no doubt played a role. Whatever the influences, the founders who came together in 1906 clearly envisioned the new organization as a nongovernmental association of national electrical engineering societies.

The subcommittee also discussed membership around manufacturing interests—and this time the issue was not whether to *allow* representatives of manufacturers to participate (as it had been almost a decade earlier in the AIEE discussions related in chapter 1) but rather whether to *require* such involvement. Several representatives of business interests attended the 1906 preliminary meeting as leaders and delegation members, including Chairman Alexander Siemens, Colonel Crompton from the British delegation, and Dr. Ichisuke Fujioka, Toshiba's founder (known as the "father of electricity in Japan") and the sole member of the Japanese delegation.[58] The Belgian representative raised the issue of manufacturing representation again in the final, full-group discussion, arguing that the rules should explicitly allow for the appointment of manufacturers on local committees. Delegates from the United States and Great Britain pointed out that the rules certainly did not prohibit such membership on local committees and that manufacturing interests were always represented in their electrical engineering societies. A Canadian delegate worried that if the technical societies were solely in charge, they might not appoint manufacturers, but others pointed out the dangers of divided responsibility. Ultimately, they compromised on this wording: "The persons appointed on the Committees need not be Members of Technical Societies."[59] Thus manufacturing interests were allowed but not required to be represented.

This issue arose again at the very end of the conference, after those present had approved the rules, subject to the ratification of the technical societies that had appointed them. At that point, Colonel Crompton offered his own services and those of Acting Secretary Charles le Maistre to show other delegates how the British technical society for electrical engineers (IEE) had organized its own standardization work. As summarized in the meeting report, he said, "The work in England had been carried on with very great difficulty, and it had been absolutely necessary to have the Manufacturers on their side[;] in fact it had been found by experience that no standardization could be successful which did not have the co-operation of the Manufacturers."[60] The IEC's founding leaders thus articulated this principle of representing manufacturing interests as well as more academic engineering interests from the organization's conception, setting a pattern for international standard-setting organizations, in line with the pattern already set in standardization committees in some national engineering associations and in the ESC.

At the end of the meeting, Colonel Crompton took a step back from organizational details to note what they had achieved, as summarized in the meeting report: "Colonel Crompton said it was one of the proudest moments of his life to hear a decision arrived at to continue the work of standardization," a startling claim from someone with many accomplishments he viewed with pride. He then referred to the speech Richard Haldane, the British secretary of state for war, made at the opening of a new building at the National Physical Laboratory, in which he "had remarked that this movement was an omen of the peaceful management of the world in the future. It was, by patiently working together and by smoothing down difficulties which were likely to arise in such peaceful matters as standardisation, that the world would gradually arrive at universal peace."[61] Crompton stated a theme that would appear frequently around private international standardization going forward—that it would promote world peace. Two world wars would certainly bring that notion into question, and it was only after the second of these that a general and lasting International Organization for Standardization was formed. Still, this rhetoric around international standardization was present even early in the first wave of innovation in voluntary standardization.

In 1908, after the required ratifications by national technical societies had been obtained, the IEC met for the first time as an official body, with 15 countries represented. Elihu Thomson was elected to replace Lord Kelvin as president, and Charles le Maistre was appointed general secretary, a central position he retained until his death in 1953.[62] Delegates to the meeting were welcomed by Arthur Balfour (who had served as British prime minister from 1902 to 1905) with the following words:

> As I comprehend it, the chief desire of the Commission is to so arrange, by international agreement, the tests which are to be applied to different kinds of electrical machinery, so to describe the qualities of different machines that,

> whilst the man who buys and the man who sells will know exactly what each, respectively, is doing, there will yet be the freest initiative left to both; to the man who desires machinery to be constructed to carry out some particular design of his own, and to constructors of machinery on their side, who desire to take advantage of the ever-increasing changes which growth in knowledge and increasing invention enable mankind to turn into account.[63]

In this welcome, Balfour referred to two key principles of the emerging system of private standard setting at all levels: the involvement of both manufacturers and customers in the process and the voluntary adoption of the standards created.

The IEC grew fairly rapidly up until World War I. As early as December 1910, le Maistre noted how quickly the community was becoming global, even while members remained attached to their own nations, sometimes parochially so. He reported to IEC members on the more than 2,000 letters that the IEC's Central Office had dealt with in the previous year.[64] The letters were primarily of two types: reports of the standardization work of national members (which were then Belgium, Brazil, Canada, Denmark, France, Germany, Great Britain, Hungary, Italy, Japan, Mexico, Spain, Sweden, the United States, and Uruguay) and requests for information about the organizations from engineers in nonmember countries (most recently, le Maistre reported, "Bulgaria, Turkey, Siam, and Switzerland"), often accompanied by reports of the work of the local electrical standard setters.[65] Le Maistre wished that such reports would come in one, or preferably both, official languages of the commission (English and French) so that "these Reports, of undoubted interest, would be read and digested by all Members, including those of the Anglo-Saxon race, which, it must be admitted, are not remarkable for linguistic ability."[66]

In 1911, 19 countries attended the meeting, and, in 1913, 24 were represented. By 1914, the IEC had established four technical committees and had issued its first standards (including standard nomenclature, a standard for copper resistance, and "definitions and recommendations relating to rotating machines and transformers").[67] The organization was off to a healthy start. The IEC also played an important role in the international standardization movement by providing Charles le Maistre (pictured in figure 2.3 at the 1911 IEC meeting with other standardization luminaries) with a platform from which he served as a roving ambassador of standardization, a role that will be elaborated in subsequent sections and chapters.

Figure 2.3. Many of the leaders of the early standardization movement pose for the official photograph of the 1911 IEC conference in Turin: Elihu Thomson is on the far right of the seated men; Alexander Siemens is the large man with white hair standing directly to his right; Col. Crompton is in the center (seventh from right in the first row of those standing); and, standing on the right in the front row, the dapper young man in the lighter three-piece suit with the large mustache is Charles le Maistre. IEC Photo File, courtesy of IEC.

Le Maistre's correspondence with Elihu Thomson, IEC president from 1908 to 1911, conveys the challenges of private international standardization and the important talents of the general secretary in forwarding it.[68] For example, the nature of the IEC and its mandate was not immediately clear to departments of national governments. He responded to a letter from Dr. Samuel Wesley Stratton, director of the US Bureau of Standards, in which Stratton evidently questioned the remit or "reference" of the IEC in relation to a British proposal for establishing an international unit of light. His response was diplomatic, thanking Stratton for writing "so fully to me on the subject," referring to a "misunderstanding," and sharing with him where the current proposal now stood (the national committees to the IEC from Hungary, Mexico, Sweden, the United States, and, unofficially, France had all agreed to the British proposal, and Germany was "very much against it"). Le Maistre explained, "The 'reference' given to the Electrotechnical Commission is not as confined as you would appear to believe," but he continued to say, "That the realisation and maintenance of an International Candle [the unit of light] is a matter for a standardising laboratory goes without saying," reinforcing the role of national standards laboratories such as the Bureau of Standards.[69] In his short cover letter to Thomson conveying a copy of his letter to Stratton, le Maistre said, "I have tried to be as diplomatic as possible; I think the least said the soonest mended. The influence of the Commission will be more and more felt and we need not insist upon its position which it will gradually make itself." Le Maistre would become widely known for his diplomacy in handling delicate situations and subjects, smoothing the way toward standardization.

He also carried on a phenomenal amount of correspondence with and travel to other countries to encourage their participation in the IEC and consequently their formation of national bodies in electrical engineering, if they lacked one. He reached well beyond the European countries and the United States to other parts of the world. While in Paris on another matter in 1910, he met with "the Chilian [*sic*] Engineering Representative in Europe" and as a result said, "I am sure we shall have no difficulty in obtaining the adhesion of Chile to the Commission."[70] A week later he informed Thomson that "the Japanese Government have now decided to cooperate in the work of the Commission. They will pay the subscription to the Central Office and are actively cooperating in the formation of the Japanese Electrotechnical Committee" to represent them on the IEC. Just weeks after that he announced to Thomson that Uruguay had joined and formed a national committee, also noting, "It is a small Republic and it is very gratifying to feel that the work of the Commission is finding such cordial support right through the whole of South America."

Le Maistre commented to Thomson in 1910 that, earlier, he had unknowingly laid the groundwork for this internationalizing effort through his actions as assistant secretary for the British ESC. In starting the process to get international agreement on standards for electrical machinery for the IEC, he began compiling a report showing the Standard Rules for Electrical Machinery in as many countries as had such standard regulations. In doing so he was initially surprised to find how many countries had them but then realized his own role: "I had forgotten that some four years ago I sent the Standard Rules of the British Engineering Standards Committee (which I assisted to draw up) to friends in different countries." He went on to note, "This must have helped them for I see in the Italian and in the Swedish Rules the effect of the English Rules."[71] This earlier sharing of national materials was helpful in internationalizing electrical standardization around this issue. Indeed, in 1911 when he transmitted to Thomson the printer's proof of this report, *Extract on Rating of Machinery, as Shown in Various Reports Issued by Various Countries*, he asked Thomson to write an introduction to it that emphasized that the similarities in these bode well for the international standardizing effort: "I should much like you to draw the attention of all to the fact that the examination of this Report will shew how very little divergence there really is and consequently how much hope of international agreement the study of this report should give." He noted that even his friend Colonel Crompton was "under the impression that the consideration of Rating from an international point of view is beset with many difficulties and this report is a surprise to him. It may be so to others."[72] Although his internationalizing activities were somewhat circumscribed during World War I, they would continue then and in the interwar years.

Le Maistre frequently communicated a belief that the value of international standards should take precedence over narrower interests, especially his own. When Thomson congratulated him on the success of the 1910 Brussels meeting, he deflected the personal congratulations to a broader, less personal reason:

> The success of the Brussels reunion was, I think, more due to the fact that everyone began to realize that they had come to do their best to obtain a working agreement on the different subjects put forward for discussion. Everyone being animated by this desire, they were more ready to give way on personal points which nearly always are the difficulties in the way of joint agreement.[73]

He asked of others what he asked of himself—putting the achievement of a standard above personal, national, or firm payoff—and by setting an example he often succeeded in encouraging such behavior. Similarly, in asking Thomson to write the introduction to the Ratings of Machinery report, he said, "Please do not mention me in any way. I want this to help the work forward and if it does that I shall be well repaid." This self-effacement would characterize le Maistre as he worked in the standardization world over the next 40 years.

In the conclusion of his annual report to the last general meeting of the IEC before World War I, a meeting held in Berlin in August 1913, le Maistre thanked electrical engineers for "the willing and gratuitous services rendered . . . often at much inconvenience to themselves . . . to further this international Movement." Then he returned to the theme of international standardization and world peace voiced by Crompton at the close of the 1906 meeting that established the IEC: "These regular international gatherings, at which many lasting friendships are made between electricians of different nationalities, must undoubtedly be a not unimportant factor in furthering the peace of the World."[74] Despite the sad irony of the message at this time, it was part of the vision that would continue to animate le Maistre's standardization entrepreneurship over subsequent decades and encourage the many engineers sitting on committees to perform their own laborious role in the standard-setting process.

Meanwhile, le Maistre also carried this entrepreneurship into his activities in support of national standardization, in Great Britain and elsewhere, up to and even during World War I.

The Spread of National Standard-Setting Organizations

Although Great Britain's ESC technically remained a committee of ICE until 1918, when it separated and became BESA, it operated relatively independently as a nongovernmental national standardizing body with its own organization and processes and involving a broad set of engineering societies, government departments, industry associations, and individuals. World War I would spur engineers in other countries to follow its lead in setting up private and general national standardizing bodies. Le Maistre and the ESC very directly influenced the formation and structure of the American Engineering Standards Committee (AESC), while any influence on its counterpart in Germany was less direct.

In an account of the AESC's origins written on the 25th anniversary of its official founding in 1918, its first acting secretary, Clifford B. Le Page, stated that a 1910–1911 visit to England by an engineer and member of the Society of Automotive Engineers, Henry Hess, provided the impulse that would ultimately lead to the formation of the AESC.[75] While he was there, Hess met with several British engineers involved in founding the ESC (as well as the IEC), including Colonel Crompton and Charles le Maistre. They inspired Hess to return to the United States and urge the organization of a joint committee like the ESC to perform voluntary national standardization for the United States as well as to coordinate the US role in the IEC. Over the next several years, representatives of four American engineering societies for mechanical, electrical, mining, and civil engineers—ASME, AIEE, AIME, and ASCE—met several times attempting to start a joint standardizing committee like the ESC.

Little progress was made until early 1916, when Professor Arthur Kennelly (of Harvard and MIT), the new head of the AIEE Committee on Standardization, took the lead in inviting representatives of the other three engineering societies as well as ASTM (the domain-specific American standard-setting organization that emerged from the IATM) to meet as an organizing committee to form a national standardizing committee.[76] The AIEE was an appropriate lead organization for the US body. As seen in chapter 1, it had already developed its own standard-setting capability and committee structure, and published its initial standards for terminology and testing of characteristics of electrical apparatus in the 1899 *AIEE Transactions*, updating them several times between then and 1916. It also was the US technical society involved in the formation of the IEC and in determining its US delegation. Its standardization process had long included a balance of engineers representing manufacturers, purchasers, and consulting engineers, similar to those of the IATM, British ESC, and the IEC. In June 1916, after several preliminary meetings among representatives of the five associations (AIEE, ASCE, ASME, AIME, and ASTM), they formed a committee to write a constitution for an American standard-setting organization; that committee held its first meeting in late December 1916.[77]

At this stage, le Maistre's influence was again felt. Significantly, despite the dangers of wartime overseas travel, le Maistre ventured to the United States in the early summer of 1916, in his role as IEC general secretary, to help the AIEE Committee on Standardization revise its rules. Undoubtedly he and Kennelly also discussed the concurrent attempt to organize a national standardizing committee in the United States, especially as le Maistre had become secretary (rather than assistant secretary) of the ESC that year, after Secretary Leslie Robertson, who had also taken on wartime duties as the director of production at the Ministry of Munitions, died when he was traveling to Russia with British secretary of war Horatio Herbert Kitchener, and their ship, the HMS *Hampshire*, was sunk off the Orkney Islands.[78] After two years of work on a constitution, the five societies, resembling the five Founder Institutions of the ESC, officially formed the American Engineering Standards Committee in 1918.[79]

The AESC followed several aspects of the British model established in the ESC. In its first meeting, after considerable discussion, the members agreed to invite the US Departments of Commerce, War, and the Navy to send representatives, as the ESC had done with British government departments.[80] The AESC Constitution also created a process by which a proposed standard was considered and approved at various levels, ending as an American Standard, similar to the British Standards the ESC established. The process depended on sectional committees modeled on those of the ESC, which "shall be made up of representatives of producers, consumers and general interests, no one of these interests to form a majority," a more precise formulation of the balance principle stated by Wolfe-Barry. In the second meeting of this body, Comfort A. Adams, an electrical engineering professor at Harvard, was elected chairman, and early in 1919 Paul Gough Agnew, who had a physics doctorate and had worked for the National Bureau of Standards, was elected secretary, a position he would retain until 1946.[81] Shortly thereafter, both Agnew and Adams referred to the importance of the British model for forming the American association. In particular, Agnew asserted, "All of the other national standardizing bodies have drawn heavily upon the experience and the methods of the British Engineering Standards Association. The British developed the method of co-operation which has come to be technically known as a 'sectional committee.'"[82]

Charles le Maistre continued to play an advisory role in getting the AESC organized and functioning. As secretary of the now renamed and independent BESA, he came to New York for the second AESC committee meeting (in August 1919), to counsel the new body. Secretary Agnew would state just three years later, "The American Engineering Standards Committee had the advantage, during the period of organization, of having the counsel and advice of Mr. le Maistre, the Secretary of the British Engineering Standards Association." He also noted that le Maistre had visited the United States twice during that period, "the last time being upon the specific request from the Americans for such assistance."[83] The first visit was clearly his 1916 meeting with the AIEE and the last his 1919 visit to the AESC. Almost half a century later, Comfort Adams would remark on the "invaluable" influence of le Maistre and the British model of a national standard-setting organization on the launch of the American organization:

> Mr. le Maistre's twenty years' experience in standards work proved invaluable to the organizers of the American national standards agency who consulted him freely. The structure of our technical committees followed closely that used by the British; it is now common for all national standards organizations.[84]

At the time of le Maistre's second visit, the ESC had just left ICE and established itself as BESA; this change no doubt also influenced the AESC, which in 1919 was already discussing the possibility of expanding its structure and mandate from a committee coordinating the standard-setting activity of the five founding societies and a few other bodies, to "invit[ing] all organizations and Government Departments interested in these standards to join in the formation of an American Standards Association," an independent standard-setting organization.[85]

World War I created pressure toward standardization in countries on both sides of the struggle. Around the same time that the AESC was taking shape, other countries were also beginning to organize their industrial standard setting on a national level. At the end of 1917 German engineers established their own national standard-setting body, initially the Standards Committee for the General Machine Industry and then its successor, the German Institute for Standardization (since 1975, called the Deutsches Institut für Normung, or DIN, which is the acronym we will use throughout the book[86]). Unlike the AESC, DIN was not actively shaped by le Maistre and the ESC, though the IATM, which arose in German-speaking countries, may have been a common influence on both.

DIN took off much more rapidly than the AESC, perhaps in large part because war-related government pressure had convinced the Association of German Engineers to embrace industrial standards. Waldemar Hellmich, a VDI officer from the group that followed Taylor's scientific management principles and supported the view that engineers needed to take on standardization as a public service (see chapter 1), was made head of the national body. He led it beyond setting safety standards to setting General Basic Standards (Grundnormen) for industry, including screw threads.[87] Germany's DIN also followed principles and processes similar to those of the ESC from the beginning. DIN's charter was complicated, but the American economist and scholar of German standardization, Robert Brady, remarked in the early 1930s, "The structure of the Deutscher Normenausschuss as set up in its charter is without significance as regards its standardization practice, for in actual practice the Committee is an exceedingly flexible organization, operating without fixed rules or procedure, constitutional limitations or formal bylaws." Indeed, Brady refers to it as a "flexible organization, largely the product of its guiding genius, Dr. H. [*sic*] Hellmich, [which] has been exceedingly successful in introducing standards into German industry."[88]

Brady summarizes five principles that Hellmich embedded into the organization's committee work: (1) initiating standardization only when rapid technical change and improvement of a product, part, or process had ended; (2) allowing change and revision; (3) creating standards with the needs of industry in mind; (4) making sure that all standards were consistent with one another; all of which requires that (5) "the process of standardization be made as scientific as possible."[89] The last principle, according to Brady, demanded "that all interested parties, not of choice but of necessity, be brought to bear upon the problem." He goes on to explain:

> "It is a fundamental principle in German standards work," Hellmich writes, "that all norms established shall be the product of the freely pooled labor of producer, consumer, and commercial interests, acting with the aid and cooperation of government and of science." The first three must see that the standard is technologically feasible and efficient, pecuniarily profitable (in capitalist industries), and, generally speaking, useful. The interest of the government lies in the application of standards to industrial processes owned, operated, and regulated by it, and in the relation of the norm to the general prosperity, security, and well-being of the entire community for which it is the general spokesman. To science and industry is delegated the task of preventing all standardization which might cramp future technical development, while at the same time they must see that standards proposed shall be scientifically the best possible, all factors included.[90]

The principle of involving all *Interressengruppen* (a word that now gets translated as "stakeholders") was not just a matter of making standardization as scientific as possible; it was also what made the resulting standards legitimate. In Hellmich's words, "A viable standard which is to be used by the public must be the result of balancing technical and economic viability. If one of the factors is equal to zero, the product also has the value zero."[91] Having all stakeholders at the table and demanding that they work toward the "scientifically best possible" standard helped limit control by any one economic interest. Unsurprisingly, although no direct influence of the ESC or the IEC or le Maistre is apparent in the German case, Hellmich followed principles that overlapped with those articulated by Wolfe-Barry for the ESC. The IATM's principles on balance, formulated by Bauschinger, may have been a common influence on both strands. The key issues of involving all stakeholders, keeping industry (not just technical issues) in mind and providing a system for revising standards, as well as the scientific approach to which all engineers had some loyalty, created many similarities in the processes followed.

Obviously, DIN, promoted by a government at war, arose in great part to serve specific national interests (indeed, the ESC and the AESC similarly emerged to serve national interests), but Hellmich and his colleagues recognized that standard setting needed to occur at multiple levels. From the beginning, according to Brady, the Germans believed in the need for international standards (as indicated also by the emergence of the IATM in Germany and German-speaking countries):

> If the field of application of the standard proposed is entirely localized and industrially limited, then only a limited number of persons need be consulted in the world. If the field is international (as with weights and measures and their conversion ratios, and nomenclature, fits and gauges, etc.), then the legislating body must be international.[92]

Nevertheless, the cause of private international standardization was not helped by the Great War, and another decade would pass before any international efforts beyond that of the domain-specific IEC would occur.

By World War I, private industrial standard-setting organizations and processes had emerged at the national and international levels. The ESC, building on the IATM and the work of some national engineering societies, established principles of voluntary standards and of balancing participation of producers, purchasers, and independent engineers. The IEC (directly linked to the ESC by Charles le Maistre's roles in both) focused on voluntary international standards and encouraged the involvement of manufacturing as well as purchasing and independent engineers in national delegations chosen by electrotechnical societies. It introduced principles unique to the international setting, including a belief in the positive benefits of internationalism and the principle of one nation / one vote. The next chapter will move into the interwar period and explore subsequent developments on the national and international levels during the 1920s.

3

A Community and a Movement, World War I to the Great Depression

From the outset in the 1890s and early 1900s, the individuals who created the first national and international private standard-setting bodies and who participated in their technical work formed a transnational community of engineers, businessmen, and government leaders from all parts of Europe, the overseas British dominions (Australia, New Zealand, Canada, India, and South Africa), the United States, Latin America, Japan, and a few other East Asian countries. These men shared an identity, rituals, and values, and their leaders regularly met face-to-face despite the distances that separated them and the expense and inconvenience of travel by steamship and rail. Members of this group considered themselves more than just a community. They saw themselves as a socially progressive movement similar to other internationally oriented movements of the day: the peace movement, the free trade movement, the humanitarian movement, and even the labor movement.

The launch of the standard-setting organizations described in the previous chapter initiated the first wave of such institutional innovation, which peaked around 1930 and then fell off through the Great Depression and World War II. This chapter analyzes the rise of that initial wave to its peak. We discuss how the early industrial standard setters formed an international community and understood themselves as a progressive movement that created a network of standard-setting organizations worldwide. Many standardizers even considered themselves the leading edge of a global movement of engineers that was making a more prosperous, peaceful, and humane world. Despite disagreement over cooperation with German engineers after World War I, long-standing commercial rivalries, the tenuous financial position of some national standard-setting organizations, and the personality conflicts that plague any movement, in the 1920s they established most of the private national standards bodies throughout the industrialized world that are still with us today, as well as the first international general standard-setting organization, which did not survive World War II. Other hybrid, public-private international standard-setting organizations also emerged. Many engineers, notably the organizers of the 1929 Tokyo World Engineering Congress, the largest international gathering of the profession in the first half of the twentieth century, treated the success of the standardization movement as the most persuasive evidence of the leading role that engineers could play in creating an ever-better world.

A Community

It is striking how often the same names appeared in the documents of the national and international voluntary standard-setting bodies founded before World War II, even in accounts of events separated by 40 years and more than 10,000 miles. The political champions of such standardization, including Britain's Arthur J. Balfour and America's Herbert Hoover, provided governmental support without government oversight. The scientists, such as Lord Kelvin, gave standard-setting organizations luster and scientific credibility. Industrialists—among them members of the transnational Siemens family, the British-born American Elihu Thomson, and Japan's Ichisuke Fujioka, all of whom were trained as electrical engineers—provided legitimacy and proof that a larger commercial community thought standardization was important. Most significantly, prominent standardization entrepreneurs such as Charles le Maistre and Waldemar Hellmich, who created the organizations, communicated constantly and met frequently.

At the center of the entire community was Charles le Maistre. His colleagues would later refer to him as "the father of international standardization" and its "*deus ex machina*."[1] As described in chapter 2, le Maistre became involved in the nascent community when he was hired in 1901 as assistant to the secretary of the new Engineering Standards Committee (renamed the British Engineering Standards Association in 1918); he served as that organization's secretary (and later director) from 1916 to 1942. He also served as general secretary of the International Electrotechnical Commission from 1906 until he died in 1953. Le Maistre played a critical role in founding the first short-lived international general standardization organization, the International Federation of the National Standardizing Associations (ISA), in the 1920s, and the enduring International Organization for Standardization after World War II, and he remained "married to standardization" until the end.[2] In this all-consuming focus, le Maistre was rare among the early standardizers.[3]

Most standardizers were part of a broad but not all-consuming community. In the 1920s there were perhaps 12,000 men regularly involved in industrial standardization around the world.[4] Most of them undoubtedly considered standardization an important purpose in their lives, but not the only one. "Standardizer" was only one of the identities embraced by the likes of the Siemenses, Elihu Thompson, Fujioka, or Hellmich, as well as the thousands of engineers who served on early technical committees. Most of the early standardizers thus formed one of the new kinds of professional communities that had become increasingly common since the mid-nineteenth century—a transnational professional group such as those of doctors, lawyers, and academics.[5] All of these communities can be characterized by (1) mutual orientation of members, (2) articulation around a common identity or common project, (3) a sense of reciprocal engagement, and (4) distinctive forms of active engagement and involvement, all leading to (5) a sustained sense of belonging.[6]

The standardizers oriented themselves toward each other as producers of standards, both as engineers writing the standards and as stakeholder representatives of manufacturers or purchasers of standard products. The engineers' common projects were the promotion of private standardization (including the creation of national standardization bodies and mechanisms for cooperation among them) and agreement upon specific industrial standards. Their sense of reciprocal engagement across industries, engineering specialties, and nationalities was linked to their regular interaction and to an engineering conviction that, at any point in time and in any field, there were likely to be "best practices," better ways of doing things, and that the best way to discover those practices was through the widespread engagement among experts in setting standards and the real-world testing of those standards through their widespread, voluntary adoption. Working on technical committees—the forums where standards were actually set—as well as establishing standard-setting bodies and managing them became distinctive forms of mutual engagement in the community of standardizers. By the early 1920s, standardizers throughout the industrialized world had come to think of themselves as belonging to a community and a broadly progressive "standardization movement," the collective identity that we explore in the next section. Le Maistre himself summarized the early standardizers' mutual orientation, common project, and sense of reciprocal engagement in a 1916 article about the value of standardization that was widely reprinted and quoted by contemporaries:

> Satisfactory results [standards that have been widely adopted] have been arrived at not by one section of the community imposing its opinions on the other, but rather as the result of co-operative action, mutual concession and ultimate agreement between all the interests concerned. The adoption of standards agreed on in this way undoubtedly promotes uniformity of practice, avoidance of waste, elimination of harsh and unnecessary conditions, reduction of manufacturing costs, and last, but by no means least, engenders a feeling of mutual confidence between user and producer such as could not be secured by isolated action on the part of either. Experience, moreover, has shown clearly that such procedure does not lower the standard of quality, but tends rather to raise it, for standardisation carried out along these lines reflects, in effect, the consensus of opinion as to what constitutes best modern practice.[7]

Although the community was transnational, it initially emphasized establishing national standardization bodies. In 1922 another devoted standardizer, Paul Agnew, secretary of the American Engineering Standards Committee, explained that "there are important considerations and powerful forces tending toward international standardization," including the existence of global scientific units and measurements, the growth of international trade that benefited from standard specifications, "the increasing interdependence of national industries," increasing knowledge of the world, and "the fact that industrial leaders are taking a larger and larger perspective in planning for the future."[8] Nonetheless, despite this expectation and despite the frequent interaction of leading early standardizers with collaborators abroad throughout the interwar years, the primary focus of most members of the transnational community of standard setters was on national standards. Like members of most transnational communities, the standardizers were, in the language of today's sociologists, "rooted cosmopolitans," involved in a project that was inherently global in scope yet "at the very same time embedded and rooted in other, often national or more local communities."[9]

The transnational community focused on transferring lessons from one country to the next. The early standardizers were, in that sense, also a "community of practice," a group of like-minded practitioners who participated in a common practice and who were interested in learning more deeply from other practitioners.[10] However, they did not constitute a new profession.[11] Rather than being *professional* standardizers, the community was largely made up of men who volunteered to work on technical committees because that was something they believed professional *engineers* should do to serve the public good. Their professional identities remained that of engineers in their particular fields: mechanical engineers, chemical engineers, electrical engineers, and the like. They were members of different engineering communities—experts in given fields—who were, through their common work, becoming members of a new community that overlapped all the others, the transnational community of standard setters.

World War I disrupted, but did not destroy, this community. Active engagement and mutual learning among former enemies restarted as soon as the war ended. For example, AESC chair A. A. Stevenson said of a 1921 London meeting, "Our secretary's trip to Europe to establish personal contact with similar organizations in other countries and learn of their methods has been of exceptional value. . . . I consider his report on Standardization in Germany of such value to the industries of America as to more than cover the cost of his whole trip."[12]

The German innovation that generated the most enthusiastic response from the national secretaries, and that was copied worldwide in the 1920s, was the practice of publishing each DIN standard as a separate numbered document, often with separate subsections, that could be sold in whole or in part to interested firms.[13] Today, this form of standards publication is a global genre. In fact, such documents (now often electronic) are what most people involved with private standard setting consider standards to be.[14]

Unsurprisingly, World War I encouraged national, rather than international, standard setting. Demand for rapid production of the machinery of war in the combatant countries and increasing concern for efficiency due to the war's global economic impact contributed to the establishment of national standardization bodies, and fears of a postwar recession led to further concern about economic rationalization. The wartime German and American bodies discussed in chapter 2 were joined by the Netherlands in 1916; Switzerland in 1918; Belgium and Canada in 1919; Austria and Japan in 1920; Hungary and Italy in 1921; Australia, Czechoslovakia, and Sweden in 1922; Norway and Poland in 1923; Finland and Spain in 1924; the Soviet Union in 1925; Denmark and France in 1926; Romania in 1928; China in 1931; and New Zealand in 1932.[15] Yet this rapid increase in the number of private national standard-setting bodies, from small countries to large, including those with and without imperial aspirations, engendered relatively little chauvinism or rivalry. All the new organizations followed procedures modeled on those of BESA, the AESC, and DIN. Moreover, American economist Robert Brady reported intense cooperation among national organizations throughout the 1920s, with Germany acting as "the leader in the standardization movement on the continent, as England has been in her own dominions"; indeed, he noted that "many of the German standards have been adopted in whole, or with certain modifications, by the Czechoslovakian, Swiss, Dutch, Austrian, Hungarian, Scandinavian, Italian, Russian, Polish and Belgian committees." In addition, he continued, "ten [Soviet] engineers are maintained in Berlin to translate German standards and others of use to the Russian Committee."[16]

Through the work of le Maistre, the British played a similar role. He was a consummate standardization entrepreneur. We have discussed le Maistre's aid to US standardizers in the formation of AESC. The historian of AFNOR (Association française de normalisation, the French Standards Association), Alain Durand, also credits him with assuring that French speakers would play a central role in international standardization efforts and with playing a critical role in the 1926 establishment of the French national standardization body.[17] In 1921, le Maistre, bilingual in French and English, had encouraged the first cooperation among Francophone national bodies (the Belgian and the Swiss), which helped assure that there would be a cohesive international community of French-speaking standardizers. France itself had lagged, but in 1926 le Maistre explained to his French colleagues that they would have no influence over the nature of the soon-to-be-created international standardization body unless they formed a private national body separate from the government's bureau of standardization with its historical focus on fundamental, metrological standards and little connection to industrial enterprises.[18] Durand concludes his tribute with this homage: "Amidst the clashes of the early 20th century, le Maistre's skills can only perplex the standardizers of the 21st century. . . . Without the Internet, fax, bullet train, or plane, it was his radiance and moral authority that allowed him to proselytize to the rest of the world, especially France." Moreover, as Durand also points out, his influence went far beyond France. "In his 50 years of work he attended almost all the meetings of all the major existing standardization groups; meeting, mobilizing, and advising all the national experts and representatives"; indeed, he adds, "the Canadian and Australian organizations particularly honor him for his encouragement during their beginnings."[19]

The United States, France, Canada, and Australia were not the only countries whose standardizers credit le Maistre. In Sweden, le Maistre was made a Knight Commander of the Order of Vasa for his work.[20] In New Zealand, le Maistre is given pride of place in local accounts of the founding of the national body.[21] A devastating 1931 earthquake convinced many New Zealanders of the need for a more coherent, locally relevant set of building construction standards. On a visit shortly after, le Maistre argued that a national standards body on the British model was needed. He wrote a British confidante that this was likely to happen because, at a public dinner in le Maistre's honor, the head of government "showed an interest and a knowledge of industrial standards, which I think few Prime Ministers have."[22] Five months later, the New Zealand Standards Institution was a reality.[23] Le Maistre may also have played a role in encouraging the Soviet Union's entry into national and international standardization. In 1923, as the IEC general secretary, he took an "extended tour" there, ostensibly to support the reanimation of that country's Electrotechnical Committee, which had been dormant since the October Revolution. He traveled with US senator Burton K. Wheeler of Montana, a leading advocate of normal relations with the revolutionary government. On his return to London, le Maistre offered to set up a voluntary network of Western engineering societies, firms, and individuals willing to provide technical literature to their Soviet counterparts, and he actively encouraged Soviet colleagues to take part in international standardization meetings in upcoming years, something they began to do after the formation of the Soviet national standards body in 1925.[24]

In his role as BESA secretary, le Maistre also initiated the first meetings among heads of the various national standardization bodies, helping to create a network of organizations within the community of standardizers. In September 1921 he hosted an Unofficial Conference of the Secretaries of the National Standardising Bodies in his London office with secretaries from the national bodies of Belgium, Canada, the Netherlands, Great Britain, Norway, Switzerland, and the United States. Paul Agnew's report on this meeting to the AESC suggested that a German representative was also present, but he is not mentioned in the minutes of the conference. The conference's stated purpose was to exchange information, "not to organize an international organization,"[25] but it began the process that resulted in the first international general (rather than domain-specific) standardization body. The secretaries followed up with a Second Unofficial Conference of the National Standards Bodies in 1923 (where the German secretary, Waldemar Hellmich, was definitely present), and a further informal conference in 1925.[26]

The minutes of these first three of many meetings and le Maistre's voluminous correspondence demonstrate the tact with which he pursued his agenda.[27] An issue of sartorial etiquette, for example, illustrated le Maistre's willingness to defer when necessary. In February 1926, representatives of 14 of the 19 national standard-setting bodies were preparing for the first official conference (after the two unofficial and one informal conferences). It would be part of a historic pair of April 1926 meetings of engineers in New York and Philadelphia. The IEC would hold its regular meeting there, and the national standard-setting bodies would meet to consider establishing an international general standardizing organization. These meetings were the first standards assemblies involving all the major industrialized countries that had taken place outside Europe since the 1904 gathering at the St. Louis World's Fair that had led to the creation of the IEC. In February, le Maistre informed his American co-organizer about the packing instructions he planned to send the many "European Engineers of world wide fame" who soon would be "embarking by the Cunarder 'Andania'" to attend it.[28] In light of the comparative informality of most American engineers and the conflicting demands of customs elsewhere, le Maistre wrote:

> I am advising [the Europeans] to bring:
> a. Evening dress
> b. Smoking dress
> c. Morning coat for official opening and that sort of affair. Foreigners as you know, use a tailcoat a great deal, which corresponds to the French.
> *d. No high hat*
> & I am telling the ladies [accompanying wives of standardizers] no furs are necessary.[29]

Alas, a photo of the European engineers disembarking from the Andania in New York (figure 3.1), shows le Maistre, top hat on head and cane in hand, at the head of a group of top-hatted men descending the gangway. Whether le Maistre's suggestion not to take high hats was rejected by the Americans (unlikely) or by his fellow Europeans, the significance and dignity of the occasion seems to have convinced "the father of international standardization" that, in fashion, at least, it was more diplomatic to follow than to lead.

Figure 3.1. Charles le Maistre, front left, with the group of top-hatted men representing European national standard-setting organizations, arriving in the United States for a conference to consider the establishment of a global general standardizing organization. IEC Photo File, courtesy of IEC.

A Movement

The pomp and ritual of this and subsequent international meetings throughout the 1920s served the larger goals of the transnational community of standardizers and contributed to make them, in their own minds, a movement. They had called themselves that since at least 1899 when the first president of the American Section of the International Association for Testing Materials recounted the rapid evolution of standardizing materials from an activity of individual scientists and separate local laboratories to the point at which "the movement took on an international character."[30] As figure 3.2, a graph of the use frequency of "standardization movement" in published books, suggests, this term rose to a peak in the early 1930s, falling off with the Great Depression and World War II.[31]

From the beginning, leaders of the standardization movement claimed it had an ethical imperative. Looking back late in life, Comfort Adams, le Maistre's contemporary and the AESC's first chair, summarized its purpose this way: "Progress of civilization depends to a large degree on successful standardization in many fields and on the cooperation necessary to develop standards."[32] That is why the standards movement worked to perfect cooperative standard setting. He and his colleagues had said much the same thing from the beginning. At a March 1919 symposium entitled "The Engineer as Citizen," Spencer Miller, vice president of the American Society of Mechanical Engineers, proclaimed, rather grandiosely:

> The engineer is assuming an ever-larger position in public life, and, in spite of himself, he is at the very center of life. The more we realize this great truth, the more seriously do we contemplate our responsibilities. . . . Eliminate the engineer from the world and civilization would soon pass through other "Dark Ages" comparable with savagery and barbarism. It is clear, therefore, that we, individually and collectively, should make every possible effort to mold [the] public in the right direction, especially at present to counteract the propaganda of those stirring up class hatred.[33]

Comfort Adams reinforced Miller, asking that engineers, "whose normal work is so much concerned with organization in industry, accept as a part of our responsibility as citizens the broader problems relating to the organization of society." He went on, in colorful metaphoric terms highlighting the dangers of class warfare, to ask "that we face the facts fairly and prepare to take an intelligent step forward, rather than wait until the great Tank of

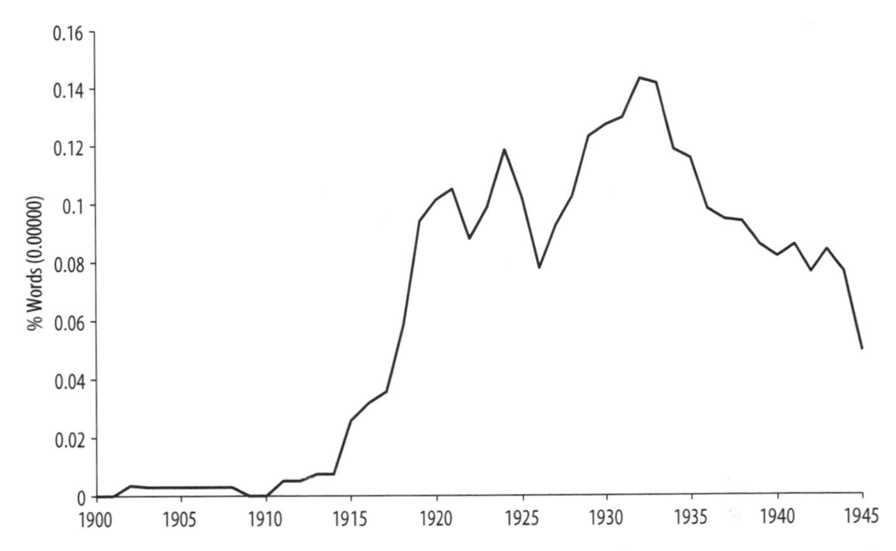

Figure 3.2. A Google Ngram of the frequency of use of "standardization movement" or "Standardization Movement" in Google Books, 1900–1945, shows that this movement language rose sharply during and after World War I, peaking in the early 1930s, then declined through the Great Depression and World War II.

Discontent, tired of waiting for us to move, has gained momentum enough to crush us and all that we represent."[34] Engineers needed to "so improve methods and machinery that the productivity of labor would increase until it is possible to pay labor a real living wage and still have a fair return for capital."[35] If that did not happen, Adams worried about the continuation of the civilization that the engineers' own work had created.

They saw promoting *standardization* as a primary way to do this. When Miller and Adams spoke about the standardization movement, they presented it as part of a larger mission that included optimizing mass production through *scientific management* at the level of the firm and the *rationalization* of the economy as a whole. Each was the focus of related early twentieth-century evangelizing movements in which engineers joined progressive industrialists in arguing that mass-production principles could solve contemporary social problems.[36] As we will see, such self-described "movements" involving engineers continued to proliferate in the 1920s, with the standardization movement remaining in the lead.

The Standardization Movement and Other Progressive International Movements

In their enthusiasm for such movements, the engineers were not alone. This was an era when members of the respectable middle classes sometimes formed transnational movements aimed at fundamental social change. The international nineteenth-century antislavery movement, the movement for women's suffrage and property rights, the early human rights movement (e.g., the campaign against foot binding in China), and the humanitarian Red Cross movement were largely middle-class campaigns often connected to mainstream churches and to Christian revivalism.[37] Middle-class transnational movements whose goals overlapped with those of the evangelizing engineers came a little later, including the late nineteenth-century peace movement with its focus on international law, the related early twentieth-century League of Nations movement and businessmen's peace movement, and the international movement for labor legislation.[38]

The early standardization movement was connected to the pre–World War I free trade movement, which, in Britain at least, centered on final consumers and united working-class and middle-class women, industrialists, and anti-colonialists.[39] The lack of shared standards was just as much a barrier to trade as any tariff or quota. Incompatible standards impeded trade at all levels—local, regional, and national as well as international. Using strongly utilitarian language, le Maistre made this argument in a 1922 speech to English engineers and industrialists, "Standardization: Its Fundamental Importance to the Prosperity of Our Trade." Common standards were needed not for "gaining individual dividends" but for "bringing about the greatest good for the greatest number."[40] Herbert Hoover made similar appeals when he was commerce secretary from 1921 through 1928. In 1922, he told a receptive AESC board, "The American manufacturer is no doubt attempting to correct his difficulties in foreign competition with tariffs," but it made more sense to focus on "our own productivity"; the real need was "the elimination of waste generally in the whole industrial system." That is where standards came in.[41]

Standardization, its advocates believed, could help increase productivity, seen by some as the key to the labor problem. In 1919, just four months after the intergovernmental International Labor Organization (ILO) was set up to improve working conditions around the world, Comfort Adams explained to members of the AESC that they, too, had a central role to play in solving the problems of the working class: "We must either face the possibility of a Bolshevik movement in this country or devise some means for increasing the average productivity of labor. This can be done by cooperation and standardization, which go hand in hand." He extolled the value of standardization to industry and the nation, and urged engineers to convince industry and national engineering organizations to undertake "broad and comprehensive cooperation in the production of standards"; by doing so, he claimed, "I think we will have accomplished one of the biggest jobs which has ever been undertaken in this country. It would do more to solve the present problems of the United States than anything else we could do."[42]

The standardizers' faith in ever-increasing productivity could even overcome the aversion to Bolshevism that Comfort Adams expressed, as le Maistre's 1923 extended tour in Russia and its consequences demonstrated. Standardization, rationalization, and the resulting increases in productivity promised to tame the Bolsheviks. The British *Electrical Review* reported that "Mr. le Maistre finds that the technical world in Russia is seething with activity and full of quiet enthusiasm; moreover, it is in high favour with the Soviet Government, and is encouraged to put forth every effort to set the wheels of industry in motion." The *Review* also claimed that le Maistre "believes that it will play the leading part in restoring prosperity and progress in Russia—a view which, it may be remembered, has also been held by its rulers."[43] And the Soviet engineers who came to Berlin to translate DIN standards no doubt held similar views.

In fact, the standardizers argued that the capacity of engineers to see beyond such political divisions was a contribution to peace, an argument that le Maistre had made when he visited the United States at the end of the First World War: "If we can bring together the engineers of the English-speaking races, it will shortly be one of the greatest helps towards the peace of the world."[44] He said that he believed this not because such cooperation would contribute to an Anglo-Saxon hegemony but because it would act as a seed around which a global community could crystalize.

Some members of the broader global peace movement seemed to agree with this assessment, even if the engineers did not agree with the peace movement's views about the way a global community should form. In 1913, some Belgian internationalists hosted a "World Congress of International Associations" that brought together representatives of many transnational movement organizations, including the IEC. The congress recommended that the peace movement immediately pursue an international system for "industrial standardization in manufacture." Omer De Bast, the IEC representative to the congress, reported back that, "without in the least desiring to minimize the important role played by this World Congress in promoting peace," he believed that "its enormous scope as well as its constitution made one doubt whether this Congress was capable of dealing with problems of a purely technical nature as it was with social or moral questions." The IEC (through De Bast) supported a substitute motion for appointment of "an International Commission for the purpose of undertaking international technical and industrial standardization generally, the Commission limiting itself to the study of principle articles, the standardization of which would economically benefit the industry." De Bast concluded that he was "glad to be able to assure" IEC members that "should this new international organization survive the pangs of birth, it should not in any way clash with nor hamper the progress of the work of the IEC."[45]

This congress was not the only time that the standardization movement was in contact with representatives of other early twentieth-century progressive international movements. Charles le Maistre sat in close proximity with those movements every day. When le Maistre's French colleague, André Lange, eulogized the father of international standardization in 1955, Lange declaimed, "It was from London, at 28 Victoria Street in the City of Westminster . . . that the influence of Charles le Maistre radiated for more than 40 years throughout the world."[46] During the same years, this was also the address of the British Labour Party, the Cobden Club (the main free trade movement organization in the UK), the International Free Trade League (the peace movement with which H. G. Wells was connected), and many other social movement organizations.[47] A 1913 French chronicler of "L'Angelterre radical" described 28 Victoria Street as a well-known building adorned with "red facades pierced by multiple bay windows, with many floors, bare and cold stairs, where associations—industrial and trade, political and charitable—all set up their offices and hold their meetings."[48]

The radicalism of many of the building's occupants, at least in the eyes of some, led to a 1917 police raid to seize "enemy propaganda." In fact, the only politically radical leaflets seized were copies of *A Reasonable Man's Peace*, by H. G. Wells. When questioned in Parliament, the home secretary admitted, "The large quantity of documents seized on the occasion of these raids may have included some not connected with enemy propaganda."[49] Perhaps published technical standards were among these latter documents.

The Standardization Movement and Business

The wartime home secretary and the detectives at Scotland Yard were not the only ones who mistook the work done at 28 Victoria Street as a threat to the existing order. In 1904, the same year electrical engineers conceived the IEC, the American economist and social theorist Thorstein Veblen began developing a theory that suggested the potential radicalism of professional engineers, including those involved in standardization. In *The Theory of Business Enterprise*, Veblen argued that, for modern businessmen, "an unsophisticated productive efficiency" was never "the prime element of business success."[50] They wanted to dominate markets, something often achieved through technical inefficiency. According to Veblen, to dominate markets, they had to prevent "predation" by other firms. The productive work of "making possible and putting into evidence" these opportunities for predation was that of "the inventors, engineers, experts, or whatever name may be applied to the comprehensive class that does the intellectual work involved in the modern industry," and this work was something quite distasteful to the "man of pecuniary affairs."[51] Veblen stood on the side of the efficiency and inventiveness of the engineers against the established businessmen who preferred stagnation and inefficiency if it helped maintain their profits. In 1919, as the war ended and fears of a postwar recession began, he published a series of articles in the *Dial* magazine (later republished as a book, *The Engineers and the Price System*) making a case for the revolutionary potential of men like Comfort Adams and Spencer Miller who called on engineers to save civilization.[52]

In the October 1919 IEC plenary, the first since the war began, the IEC's president, Maurice Leblanc of France, gave his version of Veblen's argument. As the official English-language summary of his keynote address reported, "Competition rendered it necessary for men to spend the greater portion of their time and energy in disputing their prey (the client) instead of in production." Yet "increased production was the watchword of to-day," since the labor force was so depleted by the war. Thus, "from henceforth only productive work would be deemed honorable, and any trade which enriched him who plied it in making money pass from the pockets of others into his own without any resultant benefit to the community, would be despised, if not prohibited." He blamed "egotism in the individual, in society, and in the state" for "the greatest catastrophe of all time," and argued that only "altruism, or, in other words, the evangelical spirit" could now save the world. He then proclaimed his faith that "evolution in the right direction would take place rapidly." He ended by arguing "that standardization in all its domain was preeminently the democratic reform of production. It was particularly necessary to-day, for it was incumbent upon all to ensure the maximum production with the minimum labor."[53]

At the close of the conference, Arthur J. Balfour—who had long supported the IEC and had been the wartime foreign secretary embarrassed by the raid on 28 Victoria Street—repeated many of the same claims. He condemned "that waste which arises from quite unnecessary individualism," called the full application of science to the problem of increasing production "the great panacea for the material ills of life," and challenged the standardizers to "carry on in increasing volume" the program outlined by Leblanc the day before. Balfour concluded "that the work done by a Society like this, modest, unostentatious, not appealing to the multitude, not occupying the attention of great mass meetings or eloquent orators, is going to do more than much that occupies the time and patience of and distresses the tempers of national legislators."[54] He spoke on October 21, the day on which the Treaty of Versailles ending the war was registered by the new League of Nations.

Despite Balfour's statement about the "modesty" of the standardizers, he and other interested observers recognized that the postwar engineers were rebelling not only against the price system but also against many of the works of government, including the consequences of the Treaty of Versailles. In his pioneering 1929 book, *Industrial Standardization*, Robert Brady, who was a colleague and admirer of Veblen,[55] explained the immediate goals, and the success, of the postwar standardization movement:

> The reparations debts to be paid by Germany to the Allies could be paid in the long run only in goods . . . which Germany could only achieve by undercutting its rivals—Great Britain, France, Italy, and the United States. These countries, the future recipients of reparations . . . imposed tariff barriers against the flood of cheap German goods. . . . [I]nternational competition took the form of concerted national movements to keep present markets by . . . realizing the economies of mass production through rigid standardization and simplification. . . . To meet this need, national standardization committees have been organized . . . in practically every important industrial country.[56]

While the proliferation of national standardization bodies was more than a pyrrhic victory for the movement, it was a partial one, so that even after they were established, the engineers continued to push for an international institution for rationalizing the global industrial economy.

Meanwhile, in many countries, the establishment of the national standard-setting body marked the beginning of a rapprochement between business and the engineers. Brady's second major study, *The Rationalization Movement in German Industry*, published in 1933, explained how this happened in Weimar Germany. Drawing on Veblen, Brady argued that, since the 1870s, Germany's industrial "development had been enormously accelerated by the consistent habit of borrowing from abroad or improving at home and putting into practice the latest and most scientific methods, organizational forms, machinery, and equipment."[57] In the context of the unprecedented demands for efficiency that all German companies faced immediately after the war, the logic of science became even more embedded within firms, including in willingness to use technical experts and to adopt all new discoveries that will improve efficiency, "provided, of course, that such alterations promise to justify themselves by improving the profit-making prospects of the enterprise in question."[58] Thus, a significant part of German management came to identify with what Veblen and Brady considered to be the *engineers'* goals "of technical efficiency, long-run improvement rather than the owners' concern for quick gain."[59]

The Standardization Movement and Other Engineering Movements

The career of Waldemar Hellmich, the founder of Germany's DIN, illustrates the relationship among the movements founded by the evangelizing engineers. Born in 1880 (six years after Charles le Maistre), Hellmich studied engineering in Berlin. He became "particularly interested in the design of apparatuses for chemical processes,"[60] but, three years after completing his studies, he took a job on the administrative staff of VDI, the German engineering society. In 1915 he became VDI's deputy director, briefly held a wartime job as administrator of the Artillery Workshop in the Berlin borough of Spandau, and became the first director of the German standards body DIN in 1917.[61] In 1919 he became director of VDI but continued to play a central role in DIN.

In DIN's early years, Hellmich's attitude toward business was often critical. A decade later, he argued that DIN helped society move "from self-interest to public spirit." He proclaimed, "German standardisation has been the work of the engineer in his best sense, sustained by the spirit which creates, not because it is obliged to, but because it must." The engineer's compulsion, Hellmich claimed, was his "instinct" that "strives from untidiness to order, from capriciousness to commitment, from coincidence to rule."[62]

Hellmich believed that engineers should teach business how to be more rational and, thus, in his mind, more public spirited. For that reason, he promoted scientific management, rationalization, simplification, and all the other "engineering" movements that could engage owners and managers. When Hellmich became head of VDI in 1919, the engineers' association published a book promoting scientific management, *Was will Taylor?* (What does Taylor want?).[63] Throughout the 1920s, VDI continued to be a major source of research, translations, and original publications on scientific management and related subjects.[64] If le Maistre was the father of the standardization movement around the world, Hellmich was a patriarch at the center of all of the related movements involved in the rationalization of German industry, and that is how he is depicted in a friendly caricature (figure 3.3)—as an overburdened father surrounded by more than a dozen children (organizations) including Germany's organization for Taylorite time-motion studies. At the center are DIN and VDI, and their younger brothers, the organizations' publishing companies, Beuth Verlag, the publisher of standards, and VDI Verlag, the communications hub of all the German engineering movements.

Although he worked more on the national than the international stage, Hellmich, like le Maistre, was a standardization entrepreneur. As was the case with most social movements, almost all of the engineering movements had such entrepreneurs: Frederick Taylor for scientific management, various national leaders including Hellmich and Axel Enström in Sweden for the broader rationalization movement, Hoover and George K. Burgess (the man he chose to head the US National Bureau of Standards) for associationalism and "simplified practice," Burgess for the safety movement, and even the economist Robert Brady (along with the AESC's Paul Agnew and a few other engineers involved in standardization) for the modern consumer movement in the United States, the movement that established *Consumer Reports* in 1936.[65]

The growth of the interrelated engineering movements, like that of many other economic and social movements, responded to new political opportunities.[66] In the early twentieth century, the most important opportunities for standardizers and rationalizers were the demands for high productivity cre-

Figure 3.3. This cartoon, from an elaborate birthday card, portrays Waldemar Hellmich as the father of all the German engineering movement organizations, with VDI (the German engineering association) and DIN (the German standards body) at the center. Courtesy of DIN Archives.

ated by World War I and its aftermath. Recall that the war itself gave the final push to establish DIN. Even in the victorious countries, the war created an elite consensus that the future of international commerce, and perhaps of international relations in general, would be achieved by a form of rational management. At the war's end, the wartime private-public partnership, the Allied Maritime Transport Executive, controlled 90 percent of international shipping. A significant part of that shipping served the massive relief effort to postwar Europe, especially to Belgium and the Soviet Union. That charitable enterprise made an international celebrity of its leader, a wealthy American mining engineer with worldwide experience, Herbert Hoover.[67] In September 1918, advisors to Britain's wartime coalition government suggested that this managed economic system was "the germ of future international organization,"[68] a view that helps explain the Conservative Arthur Balfour's surprisingly radical view of the engineers' future role at the IEC meeting the following year.

The 1920s resurgence of business elites along with traditional laissez faire ideas would eventually assure that neither Great Britain nor the United States would fully embrace the vision of a rationalized, managed global economy.[69] Nevertheless, beginning in 1921, Hoover became an activist US secretary of commerce under the postwar Republican administrations. He used the Commerce Department to revive the country's waning "war corporatism" by forging cooperative voluntary partnerships between government and business based on a "new associationalism." For example, in 1924, Hoover directed one of his aides to tackle the "stupendous problem of standardization" in the lumber industry, aiming toward "a new era in organizational development, a time when regulation would become internal yet would be directed for the "combined interest of the manufacturer, distributor, and the public."[70]

Hoover's vastly expanded Commerce Department aided and helped empower American standardizers and their allies in movements for product safety and for simplification (reducing "unnecessary" variety in products), but there was little political opportunity in the United States for those who championed the corporatist economic systems that the rationalization movements helped establish in Germany and Sweden. The war debts faced by Weimar Germany and the labor-led politics of Sweden and other Nordic countries assured that their rationalization movements would become central to a tripartite (labor-business-government) economic policy that remains influential today.[71]

Pace Veblen, the engineering movements were never inherently *against* business or the price system—rather, they were *for* various aspects of engineering. Consequently, the movements could flourish only when business executives were receptive to the rationalization, efficiency, standardization, and other things that the engineers could offer. The European rationalizers, Hoover's associationalists, and standardizers throughout the world defined themselves more in terms of the *processes* they advocated, processes that they believed could overcome divisions within industrial society. The rationalizers and associationalists supported direct political cooperation among businesses, government, and labor, something that was often difficult to achieve. The standardizers advocated something a bit more modest: the process of setting voluntary standards through the consensus of technical experts who represented interested parties (producer and user companies) as well as the collective, nonbusiness interests.

A Movement about a Process: Private, Consensus-Based Standard Setting

While the rationalizers and associationalists devoted a great deal of their effort to influencing the policies of governments, the standardizers aimed to create new, nongovernmental institutions to carry out the services that they advocated. The engineers involved in the standardization movement believed that standards created through an inclusive and balanced consensus process were superior to standards created by legislatures because technical committee members had the necessary expertise and interest that legislators did not. They considered them also superior to market standards created by one company because the broad, balanced participation that the process demanded assured that the resulting standards would be widely viewed as legitimate and therefore likely to be adopted. Finally, working on such committees and taking part in the larger movement encouraged participants to see beyond personal interests to general interests. Thus, they believed that the seeds of a better world resided within this kind of standard setting itself.

Stakeholder Participation and Balance

Although the standardization movement limited participation to technical experts, the early standardizers nevertheless developed the norm of including experts representing what they saw as all interested parties, and a second norm of balancing the participation of parties with conflicting interests to better serve a common good. Chapters 1 and 2 showed parts of the process evolving in the first organizations, but now we pull together and extend these threads.

"Stakeholder" was not the word that English-speaking standardizers used—it became a term of art only in the 1980s—but the basic idea was already present in the debates within engineering standardizing associations a century before.[72] Recall from chapter 1 that in the late nineteenth century the IATM, unlike engineering professional societies, did not limit membership to technically qualified *individuals* but included engineering societies, government departments, and even firms and industry associations as members, since all had a stake in standardization problems. Most importantly, the IATM's founder, Johann Bauschinger, asserted the need for representatives of firms to participate, and for them to come from both the *manufacturers* and *consumers* of materials, to avoid one or the other set of interests dominating. The early conflict between the IATM and its American Section over the latter's addition of many more producer representatives to its subcommittee of the (balanced) IATM iron and steel committee showed how seriously the association took this principle of balance. Because of this and other controversies, the Americans broke away to form ASTM; nevertheless, the norm of balanced representation was quickly solidified in ASTM, too, when it elected the strong advocate of balance, Charles B. Dudley, its first president. Debate about representation of commercial interests also appeared at this time within standard-setting committees of traditional engineering societies such as the American Institute for Electrical Engineers. That body settled on the balance of engineers from producer and user firms as well as consulting or unaffiliated engineers representing the general interest. Balancing the three constituencies provided a safeguard against interested behavior.

Recall from chapter 2 that a similar balancing of producer and consumer interests occurred in Great Britain in the 1901 formation of the first national general standards-setting body, the ESC, though in this case government departments were initially the largest consumer organizations. The ESC created a system of committees that included representatives from other associations, industrial trade organizations, government departments, and firms in order to have balanced input from all interested groups. By the late 1910s balanced representation of all interested parties was sacrosanct in the ESC, with Sir John Wolfe-Barry listing it as the first of its principles. Secretary le Maistre reinforced the principle's importance just after the ESC became the independent BESA: "From its inception, certain definite principles have governed the work of the committee, amongst which may be placed in the forefront the community of interest of producer and consumer, which is, in fact, the corner stone of the organization." He also referred to BESA as providing "the neutral ground upon which the producer and the consumer . . . have met" to serve the common interest.[73]

Although evidence of the IATM's influence on the ESC around this participation and balance principle is indirect, direct influence is clearer in its subsequent spread. Le Maistre advised in the formation of many national standardizing bodies, encouraging adoption of the ESC model. His influence was further reinforced when he organized the 1921 Unofficial Conference of the Secretaries of the National Standardising Bodies, the minutes of which reported "that the greatest care was taken in all countries to ensure that every interest, including that of both manufacturer and consumer, was consulted, though possibly not in exactly the same manner, in the setting up of the Standards."[74] Reflecting on the same meeting, the AESC president remarked, "It is interesting that, notwithstanding great difference in the details of procedure, the same general method of work is followed in the very different countries." He summarized that method in terms of stakeholder representation and balance: "Technical decisions concerning any specific piece of work are in the hands of a working committee which is so constituted as to be broadly representative, from both the technical and managerial points of view."[75] Of course, the exchange of views at this and subsequent meetings of the national secretaries further spread the practices across the network of organizations.

Even in the IEC, an organization where representation was by country and determined by each country's electrotechnical society, the need for participation of all those most interested was articulated from the start. At the 1906 organizing meeting, delegations from national electrotechnical societies discussed in some detail who should appoint the national delegations and what their composition should be, emphasizing especially the importance of having manufacturing interests represented. Colonel Crompton and le Maistre both gave closing remarks in which they asserted that standardization could not be successful without representation of manufacturers, consumers, and professional engineers with their presumed general interest.

On the international level of standard setting, the representation of all interested countries was also important. At the end of World War I, the German committee had been officially excluded from the IEC (due to the sensitivities of some Belgian electrical engineers, supported by the French), although German engineers had been invited to sit on advisory committees and Germans attended the IEC's plenary meetings, ostensibly as guests of other delegations (especially the Swiss) or as observers invited by the IEC Executive Committee. In March 1924, the Germans officially told the IEC that this hypocritical arrangement had gone on too long, especially since, after the 1923 secretaries meeting, Gustave Gerard, the general secretary of the Belgian national standards body, had joined Hellmich and le Maistre to plan the next meeting.[76] Le Maistre wrote to IEC president Guido Semenza on this issue: "It seems to me that . . . you should decide to have a little informal lunch in Paris [with his French colleagues], which, bye the bye, the Central Office could well pay for, to discuss this very important matter, which must obviously come to a head now."[77] This plan, if Semenza followed it, was not successful, as in April the IEC's newly created Executive Committee (Semenza, the heads of the British, French, and Swiss committees, Colonel Crompton, and le Maistre) decided that, "until such time as Germany is admitted to the League of Nations, the German Committee shall be advised of the meetings of IEC and of its Advisory Committees to which it will be entitled to send observers."[78] At the next Executive Committee meeting, le Maistre reported that he "found the German Committee very ready to assist the Commission in every way possible" but it was not willing to send mere "observers" any longer.[79]

In a January 1925 personal letter to the head of the German IEC delegation, le Maistre wrote that he was doing all he "possibly can to mend the breach but it is uphill work." He ended with affectionate greetings and a remark about how kindly he had been received by Hellmich on a recent visit to Berlin.[80] In March, le Maistre received a letter from a German colleague, M. Kloss, about a German meeting where "exceptionally strong feeling was shown in regard to the fact that the I.E.C. treated us, so to speak, as second class members . . . [because of] the morbid obstinacy of the French." The group was debating a resolution to stop all cooperation with the IEC, but Kloss convinced enough participants to vote against the resolution, "chiefly by pointing out that you, dear Mr. le Maistre, were making earnest efforts to remove the present difficulties." Ultimately, "it was agreed that under the present painful circumstances we would not approach the I.E.C. officially but would once more wait the result of the efforts of our friends."[81] A few weeks earlier, le Maistre had written to the IEC's members and asked for their votes on German readmission. Under the IEC's still-informal six-months rule, which allowed six months for votes of the IEC membership, the votes were tallied at the end of July. All but two members had either agreed or had not responded, attaining the required majority. (Belgium and Poland insisted on waiting until Germany was part of the League of Nations.) The Germans were officially welcomed back at the next plenary—the historic April 1926 meeting in New York—four months before Germany was admitted to the League of Nations.

This incident, and the somewhat extraordinary lengths to which le Maistre went to assure that German colleagues could participate in as many meetings as possible, illustrates what became a peacetime norm of the standardization movement at the international level where the key stakeholders included national societies, especially those from nations that were most involved in the industrial sectors in question. Their representatives needed to be included in international standard setting to represent and balance the interests of all countries. No country's delegation, even one invoking intergovernmental norms (as the Belgians had done), should be able to veto the full participation of others. Le Maistre seems to have followed a very similar policy in the early unofficial and then informal meetings of secretaries of national standard-setting associations, when the minutes of the 1921 meeting did not mention a German delegate but the American secretary's report to the AESC mentioned Hellmich's unofficial presence. By the 1923 meeting, he was listed in the minutes like all the other secretaries. The standardization community and the movement were larger than any international alliance, and good standardization demanded widespread international representation.

The Seeds of a Better World: A Process Encouraging Cooperation

Although the rules and norms of participation and balance in voluntary industrial standard setting were intended to prevent a single interest group from dominating the process, standardization entrepreneurs such as le Maistre believed that the process that evolved in the early twentieth century encouraged participants to transcend self-interest in favor of the public interest. Nalle Sturén, wife of the longest-serving general secretary of the post–World War II ISO, once commented, "Either only nice people work in standardization, or standardization makes you nice."[82] Of course, she was not in the room for technical committee deliberations, only meeting the standardizers on social occasions. Nevertheless, the early standardizers believed that they had created a process that actually helped make stakeholders nice.

In describing BESA's methods in 1919, le Maistre observed that the process required "the sinking of much personal opinion, but if its goal, through wideness of outlook and unity of thought and action, is the benefit of the community as a whole, standardization as a coordinated endeavor is bound increasingly to benefit humanity at large."[83] Ian Stewart, who began a long career with the Standards Association of Australia in the early 1930s, later described the "dialogues associated with standardization" as more than just a means to achieve optimal standards, but rather as "a liberal education for all who participate in them." They brought together people who would not otherwise encounter each other, with differing experience and expertise, as well as potentially conflicting interests. Nevertheless, he claimed, "All who emerge from such dialogues are wiser because mutual understanding has been strengthened." Consequently, "participation in the preparation of standards is not a chore to be endured but an opportunity to be used, to benefit from the process of mutual education and to influence the content of standards that will determine future practice, both national and international."[84] Stewart, like most of those committed to the standardization movement, was convinced that the process followed by the national and international standard-setting bodies of this pre–World War II era succeeded so well because it both allowed participants to pursue their interests (or those of the groups that they represented) but also allowed them to see beyond those narrow interests and even beyond their disciplinary blinders.[85]

Ivar Herlitz, an eminent Swedish electrical engineer who began his signature work on controlling gigantic power grids in the early 1920s, later summarized the early standardizers' view of their international process in a lecture honoring le Maistre:

> If we strike too high a tone of idealism, it would probably be easy for a critic to find cases where narrow-minded national points of view have been dominating and so forth. No, let us accept the fact that most of us are here to take care of our own interests. But let us hope that everyone in so doing is broad-minded enough to realize that, in the long run, his own interests are best served by a spirit of give and take, by a desire to find what is common rather than that which is conflicting. Then our work will be constructive in tearing down barriers between nations.

Herlitz went on to contrast engineers to politicians, who "may sometimes like to burn bridges and build barriers; engineers should not lend themselves to such tendencies but devote their efforts to the more constructive work of building bridges and opening doors."[86]

At its best, the process served as an apprenticeship in deliberative democratic (or at least *technocratic*) decision making (the elitist engineers certainly were not pursuing mass democracy). "Deliberative democracy" is a term used by social scientists to characterize the process followed in many social movements.[87] In meetings that approach this ideal type, all participants (1) are given the opportunity to speak and are not constrained by the behavior of other participants, (2) address each other with respect and listen with care, (3) provide detailed justifications—rooted in shared principles or taking into account the common good—for choosing one course of action over another, and (4) determine collective decisions by the force of the better argument.[88] Each of these elements characterized the technical committees of the IATM, ESC/BESA, IEC, and the other national standard-setting bodies formed before the Great Depression. The norms of giving all committee members respectful consideration and a full opportunity to express their views grew from the sense of community and common purpose discussed earlier. The norms of providing detailed arguments rooted in shared scientific and engineering principles and of reaching decisions based on the force of the better technical argument reflected the epistemic foundation of the community, the profession of engineering, reinforced by the universal practice of seeking wide consensus among participants. The attention to shared interests and the common good was central to the standardizers' self-identity as a "movement" that had ethical import.

These factors reinforced one another in ways that encouraged the standardizers to be, in Nalle Sturén's terms, nice. The benefits of participation gave the standardization movement something that other engineering movements may have lacked: regular, positive reinforcement of the standardizers' commitment to their common goals and a mechanism to avoid the dissatisfaction that can arise from conflicts within the group.[89] The movement's concern with the common good encouraged members to listen for common ground even when others were arguing primarily on the basis of self-interest. Respect for their shared profession—and for the rational debate that was central to scientific progress—reinforced the respect that committees' decision-making rules demanded. The norms of science demanded that minority opinions be duly considered and decisions, including published standards, be recognized as fallible and subject to reconsideration when new information and new arguments arose. Moreover, the skepticism that was central to the scientific approach encouraged participants to recognize that their own preferences might not be fixed or might simply be unknown as a result of the complexity of the technology involved and the uncertainty about how technology would develop in the future.[90]

A final contributor to the standard setters' willingness to collaborate came from the learning and connections that the process fostered. Even when standards were not achieved—for example, when some companies chose to profit by not standardizing and allowing what le Maistre (sounding like Veblen) considered the "wasteful" proliferation of different products—the learning and connections were useful later when many of the same engineers sat down together to solve a new problem.[91] In addition, the new connections could be valuable to the engineers or to their employers and even provided the basis for corporate alliances that allowed specific groups of companies to cooperate among themselves to set less than universal, but still valuable, standards.

Such a corporate alliance emerged in 1925 when the IEC's committee on electric lamp sockets failed to make much progress. Two main standards were in use, American and German. Because of very slight physical differences, US standard light bulbs would not go into German standard sockets, but German standard bulbs would fit loosely into the US standard socket. The US wanted to create an international socket standard, but the IEC committee was unable to achieve consensus.[92] Partially in response to a suggestion from the IEC, engineers from bulb manufacturers formed "a technical committee of makers' representatives" (what today would be called a "consortium," see chapter 7) to come up with a standard that those companies could adopt. At an April 1925 meeting of the IEC committee on electric light sockets whose work had been superseded, the IEC president and the chair of the IEC committee, along with many delegates who worked on both committees, welcomed the new group and suggested that it work on additional topics, including "the question of screw threads for fuses" and safety issues created by nonstandard connections (e.g., putting a standard German bulb into an American socket), which often resulted in electric shock.[93] This makers' committee worked on these problems for five years. At the 1930 annual meeting, the IEC endorsed a Europe-wide set of standards that the makers' committee had developed, and in a separate meeting in 1931 the relevant IEC committee endorsed a set of related standards for the Americas.[94]

The IEC's difficulty with setting these standards had to do with who was sitting at the table. The interests of Americans and Germans, and those of lamp manufacturers and bulb manufacturers, differed, albeit in complex ways. This is a common problem faced by advocates of deliberative democratic decision making, some of whom argue that it should be solved by demanding that participants focus *only* on the collective good and common interests. The practice of the standards movement was to adopt the more realistic approach of accepting that differing interests exist, creating the rules for balance discussed above and informally supporting alternatives such as the makers' committee only when necessary.[95]

Voluntary industrial standard setting differed from the ideal of deliberative democracy in a second way. Most ultimate users—the people who would screw light bulbs into fixtures in their homes—were not, themselves, sitting at the table. Instead, *experts* representing their interests were there, including engineers whose job was to represent the "general" or "common" interest. This was an elite movement, a movement that aimed to create what could be termed a deliberative technocracy.[96] The standardizers' commitment to the collective good came from their desire to reinforce the professional identity of a community that was inherently an elitist group of experts.

Setting National Standards in the 1920s

In the 1920s, the initial goal of the standards movement was to create organizations that would be permanent bases for the technical committees that would set and maintain standards. While international standard setting was the ultimate goal for most movement leaders, their immediate focus was on national bodies, organizations operating at what was still the scale of most industries. Le Maistre, Hellmich, and their colleagues successfully promoted the establishment of such bodies in all the industrialized countries within the twelve years after World War I; most were established by 1926. The next goal, for those bodies, was to establish technical committees and set national standards. Some, especially the Germans, were more successful than others. In some countries, especially the United States, many standards continued to be set by engineering societies, other standard-setting bodies, or, very occasionally, even the government, without ever being certified as national standards. In other countries, especially those just beginning to industrialize, businesses and government agencies relied on translating or adopting standards created elsewhere.

In most cases, the national bodies served as the peak in a hierarchical national network whose members included other standard-setting committees in trade associations, professional societies, and even government agencies. The coherence of these networks varied from country to country, with the Americans and the Germans at the opposite extremes. The AESC's founders had been more concerned with setting national standards than with trying to wrangle all the many standard-setting bodies in the United States into a fixed, hierarchical system.[97] In addition to competition from the older, though domain-limited, ASTM, the AESC also faced the reality that standards in some fields were dominated by standards committees in engineering societies or industry-specific groups. In contrast, in Germany, DIN's "prestige" and "its support by the German industries" convinced a senior official in Hoover's National Bureau of Standards that it was "inevitable that all such standardization work will eventually be placed under its general auspices."[98]

This difference helps explain why, in the 1920s, the German national standards body produced many more voluntary standards than any other national body. From 1919 to 1930 DIN averaged more than 300 standards per year; at the same time, the British and American standards bodies were averaging only 14 each (table 3.1). The explanation for this difference did not lie in variation in the way the consensus process operated.[99] Rather, in part, it reflected the nature of the national networks. The German system was much more orderly and precise, a nest of organizations with more specific, nonoverlapping responsibilities, like a set of elegantly fitting Japanese bento boxes. At the other extreme, the AESC was more like an overused garbage can at a picnic area filled to overflowing with smaller containers—cups, plates, and lunch bags facing every which way and spilling into each other—with some 80 percent of the smaller containers outside the main can, and an extra-large bag at its side that seemed to fill a similar function (ASTM). In the early years, the very decentralized structure of the US system meant that ASTM had produced many more standards than the AESC had, and, in marked contrast to Germany, businesses and the government in the United States treated ASTM standards with the same respect as they treated AESC standards.[100]

The difference in structure and output of the German system also reflected the dire economic situation created by that country's enormous war debts that remained a burden into the 1930s, a problem that neither the United States nor Britain faced. The resulting demands for extreme efficiency that all German firms faced made the cost of achieving rapid standardization seem low. The major cost was supporting the many engineers who volunteered to work on technical committees, but, in Germany, that cost could be borne because, in the 1920s, German management had come to identify with the *engineers'* goals of technical efficiency as the only viable long-term strategy.[101] By 1930, more than 1,000 firms in Germany had "systematically organized standards departments," and a virtuous circle connected the production of more standards and DIN's ability to take on many more projects than other national bodies: it had a workable business plan that generated ever-increasing revenue from the sale of standards. Before the Depression, DIN was already selling more than 100,000 standards (docu-

TABLE 3.1.
Standards Published by National Bodies since Founding and Average Yearly Rate from Founding through 1930

	Completed	Rate per year	In process
Under 10 per year			
Belgium	47	1.5	17
Hungary	14	1.6	78
Canada	32	2.9	25
Czechoslovakia	57	4.8	No data
10–25 per year			
Japan	109	10.9	45
Great Britain	395	13.6	No data
United States	168	14.4	173
Italy	132	14.7	116
Denmark	66	16.5	91
Switzerland	300	16.7	50
Australia	135	16.9	30
Netherlands	300	21.4	150
20–50 per year			
Norway	227	32.4	79
Sweden	288	36.0	184
Finland	217	36.2	500
France	190	38.0	155
Austria	434	43.4	309
Poland	272	45.3	496
Over 50 per year			
USSR	1,446	289.2	No data
Germany	4,412	339.4	1,100

Source: Data from Samuel B. Detwiler, "American and Foreign Standardization on a National Basis," *Commercial Standards Monthly* 7, no. 9 (March 1931): 297.

ments) each month.[102] DIN even turned its catalogue of standards into a major source of revenue by allowing manufacturers of products conforming to a standard to put their information directly under the standard's listing, at a cost of about "$38 per line" in 1928 dollars.[103]

The factors that explain Germany's unique early commitment to standardization explain much of the variation across national cases. Consider Sweden, where early twentieth-century fears about economic backwardness—being left behind by the most industrialized countries—led both major industrialists and national trade unions to embrace Taylorism, simplification, and standardization even before the Great Depression. One of the most popular early publications of the Swedish Standards Institute (SIS) was a 1928 pamphlet, *Wastefulness*, a celebration of simplification that proved popular among both industrialists and union leaders.[104] Throughout the interwar years, even before the Depression, the widespread public support of standardization in Sweden made it natural for government to subsidize part of SIS's work, allowing it to produce more standards.[105]

Even if we take into account cases like Sweden where the factors supporting national standardization were similar to those in Germany, the German and Soviet cases still stood out from all the others. In the Soviet Union, Lenin's planned economy looked to Taylorism and "the German wartime economy" as economic models, which is why the country maintained engineers in Berlin to monitor DIN standard setting and propose many of its standards for adoption in the USSR.[106] DIN's extensive standardization made the Soviet case possible.

At the same time, the Soviet case also helps us understand the German one. When Robert Brady wrote about Germany in the early 1930s, he considered it a planned economy, albeit one whose methods were very different than those of the Soviet Gosplan—one connected with capitalism and representative government, but a kind of planned economy, nonetheless. Elites in both countries considered economic rationalization to be a matter of existential significance. The social democratic Weimar government gave German industrialists unusual leeway to engage in cooperative economic rationalization. In 1921, the government authorized the Reich Board for Economic Efficiency (RKW, Reichskuratorium für Wirtschaftlichkeit), a group of industrialists under Carl Friedrich von Siemens (a longtime supporter of national and international standardization like his father Werner and his British uncle, Carl Wilhelm) to use public funds to implement a program of simplification and rationalization that was much broader than anything Herbert Hoover imagined in the United States. The RKW disbursed funds to organizations that determined the most efficient way to use machine tools and office machines; it set national standards for bookkeeping, technical education, and raw materials; and it gave "substantial subsidies" to DIN "for working out standards not serving individual but general requirements of the German industry and offering special advantages to consumers."[107] These subsidies focused on areas such as household devices, hospital equipment, and laboratory supplies. DIN's Waldemar Hellmich led the German engineers on the board and, while it is true that the engineers and the industrialists on the board quarreled over which group should dominate the rationalization process, the elite consensus on the economic project supported a national-level experiment in "organized capitalism" that fostered, and depended upon, wide agreement on the importance of national standards.[108]

DIN's efficiency at producing national standards set a goal for other national standardization bodies, and it may have given Germany (and perhaps the Soviet Union) some advantage over other interwar industrial economies. "In a highly industrialized society," asserted the leading mid-twentieth-century expert, Olle Sturén, "the total need for national *and* international standards [to assure necessary compatibilities and reduce costs without reducing innovation] is on the order of 15,000, or a maximum 20,000." Even as late as the early 1970s, "the vast majority (probably 90 percent) of those had to be national ones simply because the international standards organizations had not yet produced them."[109] By 1930, only Germany and the Soviet Union were well on their way to having their own national standards body provide that number. In the United States, a similar number of standards were probably being set within the country, but by multiple and competing agencies not organized under a completely unified national system. Britain resembled the United States more than Germany but produced fewer total standards than either. In the rest of the world, firms or government agencies that wanted to use many such standards might have to look for foreign standards to adopt—something that would not have seemed inherently problematic to many people involved in the standardization movement since they believed that, ultimately, it would be best to have one international set of standards, standards that could originate in any country.

Creating an International General Standards Body in the Late 1920s

Despite the internationalism of the standardization movement and the cooperation that the standardizers believed was encouraged by the process itself, the IEC was the only international standard-setting body at the end of World War I, and it was limited to a single technical domain. No international general standard-setting organization yet existed, and le Maistre's meetings of the secretaries of national standards associations (which began in 1921) faced many challenges in getting such an organization established.

One challenge to taking the next step was that, despite the movement's commitment to internationalism, many members of national bodies were still affected by the international commercial rivalries of the day. In 1922, the newspaper of the Chamber of Commerce of the United States in the Argentine Republic highlighted a study on German standardization by American engineer Oscar E. Wikander, who worried that "the day may not be far distant when the American manufacturers will receive inquiries from overseas countries to furnish goods according to German national standards." He argued that "one of the main reasons why forward-looking Germans force their standardization work is because they want to introduce German standards in the great importing countries, and possibly in the whole world."[110] The Germans had the same worries about the Americans. Hellmich believed that Herbert Hoover's sponsorship of an Iberian and American standardization conference over the 1924–1925 winter holiday was a declaration of a new "wirtschaftlichen [economic] Monroe-Doktrin."[111]

Le Maistre's BESA office had similar complaints about *both* Hoover and Hellmich. Around 1925, a cash-strapped BESA appealed to the Treasury for support to translate and disseminate British specifications and standards abroad, since "British Industry, due to the lack of British Engineering Standards, has in many cases suffered through the adoption of American and German standards, by purchasers abroad." Specifically, he complained about South American countries adopting US standards and about Americans setting up standards offices in Argentina and Brazil, "with the sole object of disseminating American standards and influencing purchasers in those countries to base their enquiries on American Standards against British Standards and giving advantage to American Industry."[112] Le Maistre also worried about the United States providing technical assistance to help the Latin Americans set up standardizing bodies of their own that would likely reject British standards. "The policy of the American government," he complained, "is one of backing [its own] business, rather than doing [international standards] business."[113]

Nevertheless, in its early years, the AESC had an even less sustainable business model than BESA's due to lackadaisical support from its business members and no financial support from the government except dues from member departments. Member dues and the sale of standards did not provide the funds needed to carry out ambitious programs of standard setting, let alone promote national standards abroad. The American situation was eased by reaching out to supporters of the standardization movement among executives of leading corporations including New York Edison, General Electric, AT&T, US Steel, and Bethlehem Steel.[114] In 1928, leadership of the AESC tried to make itself more attractive to such supporters by removing the words "Engineering" and "Committee" from the organization's name and renaming itself the American Standards Association (ASA). As members of the Executive Committee pointed out, the term "Engineering" proved "a serious handicap" in gaining commitment from business leaders (presumably from those who were not engineers); moreover, "if support from other sources is required, such funds can be more readily obtained for an 'Association.'"[115] The leading corporate executives provided such support by creating an "Underwriters' Fund" that covered 25 percent of the expenses of ASA in 1928 and 93 percent in 1929.[116]

Meanwhile, despite the commercial rivalries and financial problems, at the 1926 international meetings in New York (see figure 3.1), leaders of the national standards bodies provisionally agreed to form the International Federation of National Standardizing Associations, which would be referred to by the letters ISA. A committee that included le Maistre (as secretary) and Hellmich (in his role as head of DIN) then worked to overcome the substantive and administrative conflicts that stood in the way of a final agreement. The other members of the committee were representatives of the national bodies of Belgium, Czechoslovakia, the Netherlands, Sweden (Axel Enström), Switzerland (C. Hoenig), Great Britain (Sir Archibald Denny, a shipping magnate and celebrated naval engineer who was then serving as president of BESA), and the United States (Charles E. Skinner, the leading electrical engineer at Westinghouse, who had served as chair of the AESC). The French were informally represented by the incoming president of AFNOR, just formed at le Maistre's urging.[117]

The main administrative issues concerned the ISA's staff. The interim agreement reached in New York declared that Britain's Denny would chair the ISA, and le Maistre would run it from Denny's London office, around the corner from 28 Victoria Street.[118] However, after returning from New York, a number of leading German standardizers strongly recommended that a German-Swiss engineer working at the headquarters of the Swiss standards body, Alfred Huber-Ruf, join le Maistre to help him with all his international work, including the newly formed ISA. Karl Strecker, the head of the German IEC committee, argued that Huber-Ruf was essential because "the present working fashion being circumstantial and wearisome, an organization working more quickly is urgently needed."[119] Strecker believed that it really was a bit much to expect le Maistre to handle the additional job of secretary of the ISA on top of serving as secretary of BESA and general secretary of IEC.

Unfortunately, Huber-Ruf was not the best man to work alongside le Maistre, who found the younger man pushy and impolitic. Early in his career, Huber-Ruf had enjoyed living in England, working in Manchester on the construction of large electrical machinery. He wanted to move back to Britain, in part because of conflicts he had with his boss at the Swiss national standard-setting body, C. Hoenig, someone with whom le Maistre had a close working relationship and who had represented Switzerland at the 1926 meeting.[120] IEC president Semenza informed le Maistre that Huber-Ruf had been contacting the heads of all of the national bodies to lobby for the job, concluding, "This man has a lot of qualities including a remarkable obstinance and constance so I think he will fight till the extreme for obtaining this London position."[121] Le Maistre replied:

> He has many excellent qualities which would make him most useful to us in all the international work. His persistence, however, has been so out of the ordinary that it has not endeared him to those people who have had the difficult job of getting this international work going. He is too persistent, and Sir Archibald Denny, who feels his responsible position very much and wished to do nothing which could alienate the American section, finds him most difficult. . . . If only Mr. Huber-Ruf could be more quiet and show a little more diplomacy, he would be fitting himself much more, I think for the demands of the international situation.[122]

The fundamental problems that concerned the Americans were in part internal matters and in part substantive disagreements with the vision of the ISA that Huber-Ruf championed. The first substantive disagreement was that the United States wanted the ISA only to *coordinate* national standard setting (by sharing national standards and best practices) and not to start *setting* international standards, a goal the United States knew le Maistre supported just as Huber-Ruf did. Nevertheless, le Maistre's more nuanced view of what such standards might involve was more acceptable to the AESC leadership that had to satisfy important members, including ASTM, who thought that even national standardization bodies should be considered only the "first among equals" and not the primary site of national standard setting. A second substantive problem, to which le Maistre was more sympathetic, was the American interest in establishing international standards based on the Anglo-American measurement system alongside any international standards based on the metric system, a proposition that seemed wasteful and redundant to most engineers on the European continent.

In the short term, however, the AESC's internal problems, including the financial ones, proved more important than these substantive concerns. While Huber-Ruf was importuning the leaders of the prospective ISA members to give him a permanent job, the Americans were suffering financial problems and conducting a related debate about the role of a national association that stood above all other American standard setters. The AESC was not yet such an organization, at least not in the way that BESA had been from the beginning. For example, in 1926 the AIEE, not the AESC, still appointed the US delegation to the IEC. Moreover, ASTM treated the AESC as an equal, or perhaps even a younger (but now unexpectedly stronger) brother.[123] Neither the US delegation to the IEC, ASTM, nor many other member organizations within the AESC wanted a global organization in which the AESC would be treated as above other US standard setters, so "only a loosely organized international association would be acceptable" to them.[124] Because the United States was unwilling to join the ISA until these issues were resolved, participants in the 1926 meeting that was supposed to launch the work of the ISA agreed to delay the launch for a year.[125] Due in part to American reluctance to join, BESA and other "inch country" organizations delayed their membership in the ISA, as well.

The new organization, therefore, began on shaky financial grounds and with an uncertain future, especially around its administrative staff. Even with all his dogged persistence, Huber-Ruf was able to achieve only part of what he wanted. He eventually became general secretary of the international organization, but he did not leave Switzerland. At the September 1926 meeting, the United States had proposed, as a cost-cutting measure, that the Swiss national body serve as the emerging ISA's "informal secretariat as they had been doing prior to the New York conference." The idea received no support, notably not even from the Swiss, but all agreed that Huber-Ruf should "give sufficient time to act as his [Denny's] technical secretary." Whether Huber-Ruf's Swiss employers should compensate him for this or whether he should volunteer his time was left ambiguous, but no ISA salary was voted for him. The delegates agreed that, in addition, "for economy, much of the technical work would be farmed out to the different national bodies which would thus serve as informal secretariats or 'work centers.'"[126] In October 1928 the AESC (soon to be renamed ASA) had still not completed the necessary changes in its own organization, and the ISA was started without the Americans and the British. Huber-Ruf became general secretary, though his relationship with Hoenig apparently deteriorated so much that he was no longer working at the Swiss standards body. The new federation began operating from an address in Ennetbaden, Switzerland, a residential suburb of Baden about a 15-minute walk from what was then Huber-Ruf's main employer, Brown-Boveri. It is very likely that the ISA's headquarters was always Huber-Ruf's home, first in Ennetbaden, then, after World War II began, in Basel.[127]

In October 1929, the US body, now ASA, finally joined the ISA, but only after notifying the organization, on the suggestion of ASTM, that American standardizers held a position that was "both unanimous and emphatic that the present limitation of the ISA constitution providing for international cooperation in standardization work rather than attempting to set up formal international standards was sound, and was most desirable from the American point of view."[128]

In the ASA discussion about this reservation, the American objection to the way the ISA might operate became linked to the second substantive concern: a shift to unitary international standards might require the United States to convert to the metric system. The same concern delayed British entry until 1937.[129]

Le Maistre's vision for the new international organization had been for the simultaneous publication in English and French of both metric- and inch-system versions of every standard, but this vision was not shared by all the national bodies. Many supporters of the metric system argued such standards would be difficult to set, especially when the British and Americans did not even agree on precise measurements of the fundamental units of their system; the British inch and the American inch were still different lengths. The French applauded le Maistre's dual approach (and his efforts to make sure that Francophone opinion was represented in discussion of all major topics), but, to the Germans, the idea of an international organization that would help preserve the outdated Anglo-American customary system seemed absurd and an attempt to keep continental firms out of those international markets that were dominated by Britain and the United States.[130] The conflict became so bitter that in the fall of 1927, le Maistre withdrew as prospective co-secretary of the ISA (leaving Huber-Ruf on his own), writing to his American friend Clayton Sharp, "I felt it not possible to go on, particularly after the remarks and observations of one member of the Committee" created to establish the ISA.[131] That member may have been Hellmich, who believed le Maistre had a personal commitment to the Anglo-American system of measurement that amounted to a conflict of interest.[132]

A final concern of both le Maistre and the Americans was the status of the IEC after the formation of the ISA. This, too, created a conflict of interest for le Maistre. In 1926, his friend Sharp had waxed eloquent to the AESC's Executive Committee about how the Americans had saved the IEC from the ISA, which had been "like a heavy train with [a] powerful locomotive under full steam going at a high rate of speed." The AESC's refusal to join the ISA "had stopped it, which in his [Sharp's] view was a good piece of work."[133] The IEC's Semenza and others certainly did not want their organization to be subsumed under the ISA any more than the American committee of the IEC wanted to be subsumed under the AESC. In 1927, the AESC and the US IEC committee even briefly pursued an alternative proposal that the new global organization be built by extending the IEC into other fields.[134] Ultimately, ISA members simply formally acknowledged the IEC's precedence and unique competence in the electrical field and limited the ISA to working in other areas, while the US National Committee of the IEC gradually accommodated itself to working under the rules of the reconstituted ASA.[135]

Thus, by the end of the 1920s an international general standard-setting body, the ISA, had been established, but it had lost the leadership of the key movement leader of the time, le Maistre, and was dependent on the support of Huber-Ruf, who lacked le Maistre's diplomatic skills.

Hybrid Public-Private International Telecommunications Standardization Bodies

At the same time that national standard-setting bodies were struggling to establish the ISA, a different, hybrid public-private model of international technical standard setting also arose. As discussed in chapter 1, in 1865 representatives of national governments and national telegraph administrations had founded the International Telegraphic Union, a treaty organization of signatory governments, to negotiate right of transit and tariffs for the telegraph network as it crossed national boundaries. At the 1925 International Telegraph Conference in Paris, the ITU established two hybrid public-private standard-setting committees, the International Long-Distance Telephone Consultative Committee (CCIF) and the International Telegraph Consultative Committee (CCIT).[136] Two years later, in 1927, the Radio Telegraph Union (soon to be folded into the ITU) established a third such body for radio. We will focus our discussion on the origin of the third body, as its work is relevant to subsequent chapters.[137]

The problem of radio wave interference initially became salient in ship-to-ship and ship-to-shore radio communication at the turn of the twentieth century.[138] In 1906, 29 nations met to sign the resulting International Radiotelegraph Convention, creating the RTU.[139] At its 1927 conference in Washington, DC, RTU revised its existing International Radiotelegraph Convention, its General Radio Regulations, and its Additional Radio Regulations and called for signatory governments to consider merging the ITU and RTU.

The 1927 radio regulations addressed choice and calibration of apparatus, classification and control of radio emissions from electrical equipment, and the distribution and use of frequencies. The first two of these areas overlapped with the work of existing private standard-setting bodies, including the IEC, and they were matters that required the attention of experts in radio engineering. The national governments decided that the administrative allocation of broadcast frequencies to avoid international interference (a particular problem for European nations) would remain an issue for them to solve, but they created a committee that would become the RTU's equivalent of a standardizing body to deal with the other, more technical issues. Article 17 of the 1927 convention (signed by 74 countries) created the International Technical Consultative Committee for Radioelectric Communications (the Comité consultative international technique des communications radioélectriques), soon known as the International Radiocommunication Consultative Committee, or CCIR, "for the purpose of studying technical and related questions having reference to these communications."[140]

Although CCIR was attached to a public treaty organization rather than a private voluntary standard-setting organization, it became an important body involved in standardization relative to radio interference. It generally focused on interference between broadcasting services of different nations rather than on interference of industrial equipment with broadcasting, the subject of much national and international private standardization (as discussed in chapter 6). The 1927 radio regulations stated that the CCIR's "function is limited to giving opinions on the questions which it has studied," which the International Bureau communicated to members of the RTU treaty organization.[141] CCIR created expert study groups that were effectively technical committees similar to those of a private standards organization such as the IEC. Although these study groups were designed only to give recommendations, the national delegations to the RTU generally followed the CCIR's lead, so its recommendations often became part of the radio regulations or the convention itself and thus were often incorporated into national laws of signatory governments.[142]

The CCIR's hybrid, public-private nature was signaled by its composition and voting rules, as well as its process. The committee was made up of technical experts from the governments signing the convention and from "authorized private enterprises working radioelectric stations, which desire to participate in its work and which undertake to contribute, in equal shares, to the general expenses of the meeting in question."[143] Each member administration (the government entity in charge of radio communications in a country) was given one vote on the committee, and if a country was represented by experts from private firms, those experts together were given one vote (a provision that handled the US case, in which private firms rather than a government department provided radio transmissions). Essentially, every national delegation was given one vote on a proposed international standard, as in the IEC, the ISA, and the RTU itself; the CCIR aimed to achieve consensus among all the national delegations. In this case, however, the delegates, though technical experts, were typically from and representing government departments, not engineering societies and firms, giving it a more bureaucratic orientation. In terms of process, the CCIR adopted the methods developed two years earlier by the CCIF, which included setting questions, gathering and writing reports with partial answers to the questions, and submitting draft recommendations for approval first to the standardizing committee and then to the treaty organization.[144] The documentation for this process was quite extensive, as chapter 6 will demonstrate.

The 1927 meeting that established the CCIR had also considered merging the ITU and RTU. In 1932, the two bodies merged into the International Telecommunication Union, and the new ITU maintained the three separate committees for radio, telegraph, and telephone.[145] Despite its differences from the private and voluntary international standardization bodies like the IEC and ISA, the CCIR was similar in its search for consensus among national delegations made up of technical experts, and it played a role in standardizing within the television and radio industries after World War II, as chapters 5 and 6 will discuss. No direct influence of the private standardization movement on CCIR was evident, though as electrical engineers, the technical experts were likely aware of the existence and possibly the process of the IEC.

The Standardization Movement at the Forefront of Movements of Engineers

The high point of the first wave of standardization in the early twentieth century, and of all the transnational engineering movements, came in 1929, after the formation of the hybrid telecommunications committees and the ISA. From October 28 through November 7, 1929, engineers met in an unprecedented and never-repeated "World Engineering Congress" in Tokyo. A total of 4,494 men from 43 countries attended, along with "Mrs. L. M. Gilbreth (U.S.A.)," the "cheaper by the dozen" scientific management movement leader and "the only woman member of the Congress."[146] The engineers presented

813 papers (371 by non-Japanese attendees); enjoyed (along with accompanying wives and children) 61 "entertainments" including "dinner parties, luncheons, soirees, tea parties, garden parties, theatre parties, etc."; and had the opportunity to take part in 50 excursions throughout Japan and its empire, including trips to Manchuria, Korea, and Taiwan. The congress published 40 volumes of proceedings, more than 16,000 pages, extolling the past and future contribution of engineers to global prosperity, culture, and peace.[147]

The proceedings reorganized the papers given on different days to tell a story about the contribution of engineers. The first volume of papers began with a history of the congress's main purpose, to create a World Engineers' Federation. Dr. Stanislav Špaček recalled that, since 1921, Czech, American, and other engineers had been discussing the idea. As his first piece of evidence, he read a supportive 1922 letter from the AESC's Comfort Adams to his Czech colleague.[148] Špaček explained the purpose of the proposed federation:

> Although individual engineers may hate, hatred is not an engineering trait, but fraternity is. Engineers roam our country over, visit one another's works, exchange information and venture into foreign land. It is the common experience that everywhere they greet one another as brothers. But the uniformity with which students of present world conditions point to hatred among nations, races and classes as the greatest present barrier to peace and prosperity is impressive. It is a mental barrier. Scientists, inventors and engineers by cooperation in various ways have made possible the wonderful material advance beginning at the close of the Napoleonic wars. Now they must help the world get the intellectual and spiritual benefit of this physical progress.[149]

The proceedings' second paper, "The Engineer as a Factor in International Relations," by Viscount Tadashiro Inouye, president of the Japanese "Association for the Promotion of [Solutions to] Industrial Problems"[150] continued the same theme: "There exists," he argued, "a sincere desire for the concrete improvement of all peoples' living conditions, which in turn depends on permanent peace between nations." It was sad and ironic, he noted, that engineers had made the greatest contribution to permanent peace by making war no longer "chivalrous and romantic" but, rather, something "thoroughly repugnant. . . . [W]e engineers with formidable and powerful weapons of physical and chemical war on the one hand and with constructive and progressive ideals for peace on the other, are going to lead this world into the golden age of real peace and happiness." Engineers would also provide the basis for universal prosperity, because "engineering is essentially international in character. The progress made in any district or country will soon be spread to the whole world, and contribute to the advancement of human welfare."[151]

The next four papers provided the most important practical examples of how this had already been done—through standardization. Moreover, despite all the conflicts involved in creating the global organization, the ISA was presented as engineers' greatest achievement. In the first of the four papers, Alfred Huber-Ruf explained the work of the ISA, a possible model for the proposed federation of engineers, noting that its conferences so far illustrated that "co-operation is facilitated in proportion with the increase of the personal contact possible between the representative of the different standardizing bodies." He claimed that "in the majority of cases where personal contact can be realized, ways and means of reaching an agreement on the subject of basic principles are found possible."[152] Huber-Ruf's paper included a strong statement about the desirability of creating standards that did not discriminate against either the metric or the inch systems,[153] perhaps in deference to engineers from the United States, which had finally been able to join the ISA two weeks before the conference opened.

Appropriately, ASA's Charles E. Skinner provided the next paper, titled simply, "Standardization," to which he attributed most of the benefits that engineers had already achieved: "Much of the remarkable progress that has been made in the whole industrialized world during the last few decades has been due to the degree of standardization which has permitted . . . the production of thousands or hundreds of thousands of units with all essential features alike."[154] The next two papers detailed the standard-setting work of the German and Japanese national bodies, the most exemplary national organization followed by that of the hosts.[155]

Four more papers covered other examples of the theory and practice of international cooperation among engineers. Their general overly enthusiastic tenor was reflected in the abstract of the presentation by John Hays Hammond, a wealthy American mining engineer who had made his first fortune 30 years earlier as manager of Cecil Rhodes's gold and diamond mines across southern Africa. He claimed, "The profession is now universally regarded as the benefactor of the world at large," and "the engineer will become to a greater degree an 'Ambassador of Good Will.'" He went on to declare that an engineer devoted to the common good "is pre-eminently an 'Apostle of Peace.'

In the material development of civilization the engineer has played a most important role and in no period have his activities been more important than during the past few decades."[156]

Finally, four more sober papers on the status of the engineering profession in particular countries rounded out the introduction to the work of the congress, repeating similar arguments highlighting the example of standardization. Axel Enström, the rationalization movement leader and founder of the Swedish national standards body, noted that, because of the contributions that engineers had made to material welfare, their social standing had increased, but neither social standing nor universal prosperity were ends in themselves; spiritual development (for engineers and for society) was more important. He offered a "personal creed" as a guide for the work of all the world's engineers:

> I am proud and glad to be an engineer, for this occupation is in quite a special degree a service rendered to my fellow man. Technics are fitted to create a happier future, in the first place by rendering possible better material conditions of life, and in the second place, on the foundation of a developed material culture, by contributing to the rise of a true spiritual culture characterized by truth and good will between men and nations.[157]

His words were an appropriate summary of the congress's overall message and of how the engineers attending it saw their place in the world.

The "World Engineering Congress" and its message were widely praised by its participants, but the goal of creating a World Engineers' Federation was never realized.[158] As the steamships carrying the world's engineers approached Tokyo, the New York stock market had begun to fall. The congress opened on Black Tuesday. The European and American engineers returned home to the Great Depression, which had disastrous consequences for all the engineering movements, including the standardization movement and the standard-setting bodies in every industrialized country.

Standardizers throughout the world continued to meet with colleagues in other organizations and report back to their own members, even after the stock market crash, and they continued to translate and publish information about efforts taking place elsewhere. Thus, in a single 1931 issue of *Commercial Standards Monthly: A Review of Progress in Commercial Standardization and Simplification*, an American engineer could learn about the recent work of ASA, ASTM, and the US Commerce Department's National Bureau of

Standards, and could also read long articles on standards work in Australia, Canada, Italy, and Sweden and a separate comprehensive discussion of national bodies in 21 countries.[159]

For a few years, the standardization bodies of the countries that had industrialized earliest also continued to provide technical assistance to engineers interested in improving standardization in newly industrializing countries. For example, in 1932, ASA offered "short courses in standardization for Soviet Russia"—maintaining the sort of cooperation that le Maistre had pioneered with his early visit to the Soviet Union and that Hellmich developed with DIN's support of the Soviet translators resident in Berlin.[160] ASA also briefly continued its assistance to Latin America, where new national bodies were created in Argentina (1935), Uruguay (1939), Brazil (1940), and Mexico (1943), but the official history of the Brazilian organization emphasizes that, while the Latin American standardizers had been part of the global movement and they learned from European and US models, the new organizations were created as part of *national* responses to opportunities for newly industrializing countries created by the Depression that was affecting those countries that had industrialized first.[161]

Despite all the promise of the Tokyo Congress, the international standardization movement sank from its first great peak into its deepest trough.

part II

THE SECOND WAVE

The Great Depression debilitated the standardization movement as the standardizers struggled with financial problems. But World War II was a turning point, both improving the financial situation of national standard setters and initiating a second wave of innovation, this time focused on international standardization. The interwar ISA died out during the war, and in its wake ISO arose. With the loosely affiliated IEC, ISO stood at the peak of a new international network of organizations, with national organizations feeding into it.

Chapter 4 traces the standardization crises of the 1930s, ending in the collapse of the second wave, then the war itself. Immediately after the war Charles le Maistre and engineers from the Allied and some neutral countries launched ISO, giving rise to the third wave. Chapter 5 traces how Swedish standardization entrepreneur Olle Sturén, who became ISO secretary-general in 1968, led ISO to work with emerging nations to establish national standardization bodies and pushed ISO to develop international standards in critical areas. The chapter discusses the successful international standard for shipping containers that internationalized trade and the unsuccessful attempt to standardize color television. Finally, chapter 6 provides an in-depth case study of standardization in radio frequency interference and electromagnetic compatibility from before the war to the end of the 1980s, introducing Ralph M. Showers, an American standardization entrepreneur who brought the United States into the international standardization community for this technical arena.

4

Decline and Revival of the Movement, the 1930s to the 1950s

The international standardization movement faced severe obstacles in the 1930s, but many national bodies remained active and surprisingly committed to international goals until the outbreak of World War II. Engineers set a small number of international industrial standards that are still used today and thousands of national standards. Nevertheless, the worldwide depression proved a major challenge, rocking the shaky economic foundations of some of the major national standards bodies and increasing economic nationalism in all the major industrialized countries. By the outbreak of World War II, the international standards movement was in disarray, and even prospects on the national level looked bleak in many countries.

The war helped revive the standards movement, and the second wave of organizational innovation began in its wake. Wartime demands for rapid standardization solved the national bodies' financial problems and demonstrated how quickly technical committees could work when government financial support allowed them to meet continuously. The exigencies of the war also convinced the standardization bodies of the wartime United Nations—the anti-fascist alliance, not the postwar intergovernmental body—to create a new international organization, the wartime United Nations Standards Coordinating Committee (UNSCC). After the war, the UNSCC sponsored negotiations that led to a new global body, the International Organization for Standardization, which replaced the International Federation of the National Standardizing Associations. Despite ISO's origin in the victors' alliance, it quickly reengaged the standards communities in the defeated countries and resisted the newly emerging Cold War tensions. It became an organization that worked to keep the great political conflicts of the postwar world at bay while engineers concentrated on supposedly purely technical issues. ISO, in cooperation with the postwar intergovernmental United Nations development system, extended the standards movement globally. Throughout the 1950s, the two organizations promoted standardization in newly independent and less-industrialized countries, helped create national standardization bodies there, and encouraged businesses and governments to welcome and use international standards.

Some Prewar Achievements in International Standard Setting

Most of the engineers who attended the ill-timed Tokyo conference expected that the ISA, like most of the national standard-setting bodies before it, would eventually become the peak of a nested system of lower-level standard-setting organizations, superseding the national bodies as the central organization of the standardization movement. It never did. Nevertheless, although the American Standards Association had officially insisted that the ISA establish no international standards (as discussed in chapter 3), during the organization's decade of actual operations, from 1929 through 1940, it established 32 sets of international standards, many of which were championed by ASA in its official publications.[1] Despite the Depression and the increasing international tension after the Nazis' rise to power in 1933, the members of the ISA's technical committees continued to correspond and meet while Alfred Huber-Ruf took care of "the drafting, translation, and reproduction of documents with the help of his family from his home in Basle," recalled a still-astonished Willy Kuert more than a half century later. Kuert had worked for the Swiss national standards body, which had been given "stewardship of the ISA" after the war broke out in 1939.[2]

Ironically, given le Maistre's original concern with establishing both inch and metric standards, probably the ISA's most important accomplishment was to reconcile the differing British and American measurements of the fundamental units in their nonmetric system. Not surprisingly, the agreed-upon common inch was defined in terms of the millimeter.[3] The ISA also affirmed as an international standard today's familiar simplified system of international paper sizes (A2, A4, etc.) adopted everywhere except the United States and Canada and originally developed by the Germans.[4] Howard Coonley, the American industrialist who became the first president of the postwar ISO, liked to point out that the ISA established an international standard for the placement of sound on motion picture film, something of importance to one of the most successful, internationally oriented US industries of the 1930s, and most of us are familiar with the ISA standard measurement of sound, the decibel, a term that originated at the Bell Telephone Laboratories in the 1920s.[5] The ISA is also still with us when we worry about the number of *giga*bytes in our computers' memory, the safety of the *nano*particles in our sunscreen, or anything else that is pico-, micro-, kilo-, mega-, or the like; our standard way of referring to very large and very small units was one of the ISA's last standards.[6]

The pre–World War II work of the other international standards body, the International Electrotechnical Commission, also proved somewhat significant. Immediately before the outbreak of the war, the IEC had 28 technical committees (still called advisory committees) in operation, and it cooperated with intergovernmental and other nongovernmental organizations involved with broadcasting, railways, and electrical transmission in the Comité international spécial des perturbations radioélectriques (CISPR, discussed in chapter 6), a special committee of IEC.[7] Taken together, the IEC and CISPR committees had worked on hundreds of international electrical standards, most of which continue to be used and updated.

Crises of the 1930s

During the Second World War, national standardization organizations of the other highly industrialized countries began to catch up in productivity with the national bodies in Germany and the Soviet Union, but only after overcoming a series of crises in the 1930s, including the ineffectual response of most standardizers to the core problem of the Depression, funding crises, and the threats to all internationalist movements that came from trade competition and the rise of the Nazis.

The Limited Promise of Economic Growth through the Application of Standards

You might think that the entire standards movement would embrace the challenge posed by the Depression. At least as far back as John Wolfe-Barry's 1908 speech extolling the way standardization could lead to the prosperity of both industry and labor, standardizers had believed that their work was a key to economic growth. They embraced an ideology that said it directly contributed to productivity, reduced waste, and was the leading edge of the larger movement of engineers to rationalize the economy and resolve the conflict between capital and labor. Standardizers around the world still made these arguments after 1929, but they increasingly rang hollow in most countries where the more successful solution to the problem of the Depression proved to be a radical rethinking of macroeconomic policy, not further economic rationalization along the lines long championed by the likes of Herbert Hoover or Britain's Conservative leader, Arthur Balfour.

Le Maistre, a true believer, continued to make the old argument. In a 1931 presentation to the Royal Society of Arts, he gave an example of how agreements on standards could quickly increase jobs. The work in the granite quarries that produced the curbs of all of Britain's great roads, le Maistre noted, used to be intermittent, but, now that there was a national standard for such curbs, it would "enable the Quarry Masters to keep their men in constant employment making these kerbs, certain that the stocks will be used when the orders come in again; that is to say, such a Standard is stabilizing employment."[8] The argument sounded reasonable, but the employment benefit of standard curbs depended on a continuous demand from local governments throughout the country, something that was not forthcoming from the cash-strapped local authorities of 1931.

Often the views of the Depression-era standardizers sounded more like Puritanical diatribes against "waste" than reasoned economic arguments. In the same talk, le Maistre complained about the "wasteful" production of "too great a variety" of goods in order to entice every possible consumer, by using "mass suggestion through chain newspapers. . . . 'Have you a wireless? No, then buy one. . . .' 'Have you a two-year old wireless? Then buy a new one, it is better for you.'"[9] Standardization, he argued, would solve the current problems of overproduction and overcapacity (i.e., low demand) by reducing "the unnecessary variety of articles or commodities for one and the same purpose."[10] It was a bit unclear how reducing the variety of goods or restricting advertisement would lead to the greater employment needed to give consumers the money to buy all the goods that the empty factories could produce. In Britain and the United States, it was John Maynard Keynes's and Franklin Delano Roosevelt's policies of priming the pump—increasing demand through government spending and mass employment—that ended the Depression. It was not Hoover's, or le Maistre's, calls for simplification.

However, there were countries where simplification, rationalization, and standardization worked alongside government policies for massive spending and full employment. For example, in Sweden, the Depression only increased the popular support of standardization, as unions, industrialists, and the government embraced an even broader program of rationalization as part of a coordinated effort to reverse the economic collapse and stem emigration of skilled Swedish workers to the United States.[11] The Swedish Standards Institute began to work even more rapidly than it had in the 1920s, more than doubling its annual production of national standards from 1930 through 1937, and tripling it by 1939.[12] In the end, at the outbreak of World War II, it was producing national standards more rapidly than ASA was for the much larger US economy.[13]

Nonetheless, it is unclear what effect Sweden's (or Germany's or Italy's) deeper embrace of standardization had *independent* of their aggressive maintenance of full employment through government spending. In 1940, a reflective Paul Agnew argued that the somewhat grand arguments that standardizers made about the impact of their practice on economic growth was a long-term process, not something that happened overnight:

> It is commonplace that standardization has been an essential factor in the great increase in the real incomes of the populations of industrial countries since the latter part of the 19th century. This is because standards underlie all mass production methods, and because they facilitate the integrating processes necessary to large-scale production and distribution.[14]

In the twenty-first century, economic historians looking back at the period would agree. The great improvements in the material life of average people throughout the industrialized world came with the improvements in housing (especially the connections to networks of electricity, water, sewage, and communications), transportation, and the availability and preservation of food that came in the decades immediately before World War I and that continued to be extended more broadly in the 1920s and afterward. Private standards, including the earliest standards created by the private voluntary process, were very much part of this story.[15] But standardization did not provide the key to ending the economic collapse of the 1930s.

Paying the Bills

Even if standardization had been the key, most of the national standardization bodies would have been unable to provide it. During the Depression, some major national standards bodies had difficulty maintaining a sustainable business model. Recall that big businesses had saved ASA in the late 1920s by creating an underwriters' fund covering a quarter of ASA's expenditures in 1928 and almost all of its expenditures in 1929. That support dropped to 30 percent in 1932. Then it stopped.[16] ASA then faced two years when its regular sources of income (dues of members and sales of standards) left large deficits, followed by six years (1935–1940) of barely balanced books and significant belt-tightening.[17]

US government support for the standardization movement shrank at the same time. For example, in 1933 ASA had to assume responsibility for the nation's most comprehensive journal of the national movement's activities, *Commercial Standards Monthly*, which Herbert Hoover's Department of Commerce had begun publishing in the early 1920s. It merged with ASA's *Bulletin* to become *Industrial Standardization and Commercial Standards Monthly*, published by ASA "with the cooperation of the National Bureau of Standards," albeit the staff, office, and most of the finances were provided by the cash-strapped ASA.[18]

France's AFNOR suffered a similar financial crisis in the 1930s. In 1934, as part of a deal made to create a government of national unity, France's political parties on the far right demanded the end of the government subventions to AFNOR that had provided stability after the financial crash. The French body had to draw heavily on its reserves until 1938, when the new Radical-Socialist government restored and increased the government subsidy, created an interministerial commission to encourage rationalization and standardization, and authorized a national mark to show conformity with AFNOR standards.[19]

The British national body was also stressed by the new economic conditions. A 1931 reorganization gave it a new name, the British Standards Institution (BSI), and a new structure in which contributing trade associations, rather than the voluntary consulting engineers, increasingly came to dominate its agenda setting and the work of its technical committees. The newly prominent trade associations, along with a number of railway companies, were able to pay large dues to the BSI, which helped maintain the organization throughout the economic crisis, but this came at a cost.[20] Within standard-setting committees, the engineers' concern with the "best possible technical specification achievable" was increasingly traded off as participants attempted to reach a consensus that served their separate commercial interests. Moreover, while the "conservativeness of representatives of a trade association" on a committee was "counter-acted by the presence of government staff" when the Department of Scientific and Industrial Research was present, this hardly assured that the general interest would be served.[21] In a 1943 book on the era, Robert Brady reminded readers that, in

Britain, "trade association" and "cartel" referred to the same thing; in the 1930s the country had a cozy "feudalistic system of cartel controls" in which government served big business.[22]

Trade Competition and the Rise of the Axis Powers

The BSI's reorganization and the growing power of the cartels led to what the historian of British standardization, Robert McWilliam, called "the imperial illusion" that Britain would have a sufficient market for its industrial goods in its "white dominions" (Australia, New Zealand, and Canada) and perhaps in an economically dependent Argentina, and that a unified imperial standards community could protect that market from any competition from the Germans or the Americans.[23] While the resulting international tension over using standards to achieve international commercial advantage predated Wall Street's crash in 1929 (recall Hellmich's anger in 1925 over a "wirtschaftlichen Monroe-Doktrin"), it became more significant after the US Congress passed the 1930 Smoot-Hawley Tariff Act, the retaliation that followed, and the downward spiral of world trade.[24]

When le Maistre visited the rest of the empire in the 1930s, he sounded less and less like the champion of international standardization that he was when he first aided the Canadian and Australian standardization movements in the early 1920s. A typical title of one of his public speeches was "Empire Trade Requires Uniform Standards." As he explained to a Toronto audience in April 1932, "British export trade is [British businessmen's] great preoccupation at the moment," and that to maintain their markets they realize they must meet the needs of oversees customers. Consequently, "there is not the slightest desire on the part of Great Britain, as represented through our British Standards Institution, in the least degree to force British standards on to the Dominion." In fact, he claimed, "we desire to cooperate with the Dominions' standardizing bodies in order that by drawing up joint national purchasing specifications we may open new markets for reciprocal trade."[25] It is unclear how convinced the audience was, especially given the proximity of Canada to the United States. In 1936, le Maistre traveled to Argentina, intent on setting up a BSI version of the American office there, and, as late as March 1939, le Maistre's BSI deputy Percy Good visited Australia, New Zealand, and Canada to reinforce imperial cooperation.[26] By this time most British industrialists had come to doubt the promise of strong industrial growth within a closed empire; "Australia and New Zealand were not large enough markets," and Canada was adopting US standards to gain access to the enormous market there.[27] Moreover, all the talk of empire aside, the engineers had some difficulty discarding the standards movement's traditional internationalism. The historian of Standards Australia, Winton Higgins, provides a humorous example in the menu offered le Maistre during his 1932 visit: "Sandwiches (to Standard Specifications)" and "*Salade de la Société Nations*" followed by "'Simplified Practice—Billy Tea' and '*Gâteau à Boomerang*.'" Rationalization, the League of Nations, and Australia were on the menu, but not the empire.[28]

Nevertheless, by the late 1930s, the economic rise of Germany and Italy made many non-Axis standardizers perceive a threat that was more than the potential loss of traditional, and shrinking, markets. At the same time, German and Italian engineers, along with the Axis governments, appeared to become the great champions of international standardization, albeit with certain conditions. For example, in June 1938, three months after the Anschluss that incorporated Austria into Germany, German-speaking electrical engineers convinced their IEC colleagues to allow members of technical committees to speak in German. The German national committee even agreed to pay for translation of such remarks into either English or French, the languages in which IEC standards were published.[29]

Germany's growing importance in the ISA was even clearer. The ISA published German-language versions of all ISA standards from the beginning, and Italian versions of those standards produced from June through December of 1940 (the ISA's last full year of operation). ISA secretary Alfred Huber-Ruf was warmly received by Italy's Mussolini in March 1939, and he received special commendation from the then-Nazi-dominated DIN in 1940, after the German government had conquered a half dozen of the ISA's most prominent member countries. The ISA secretary was very proud of these commendations and included them among items he sent to wartime Swiss leaders asking their support for his work for the ISA and for plans to reenergize the international standardization movement after the war.[30]

DIN's Waldemar Hellmich took a very different path. In 1933, he stepped down from his official roles "after falling out with the masters of the Third Reich" and took a position in the German office of the Swiss pharmaceutical giant Hoffmann–La Roche.[31] This job required moving far away from Berlin to Grenzach, Germany, an inner suburb of the Swiss city of Basel, a short walk from the international border. Grenzach housed Hoffmann–La Roche's German division, while the firm's international headquarters was across the border in Basel. Roche's official history records that the company "helped numerous scientists from Europe, most of them Jews, to flee to the United States," no doubt aided by the proximity of the German headquarters with the main headquarters in neutral Switzerland. It does not say that Hellmich helped with that work, but it does credit him with quashing a 1940 attempt by "the branches within the Third Reich to split from the parent company and become independent." Had they become independent, the company's capacity to save Jewish scientists might have been greatly reduced.[32]

The Standardization Movement at the Beginning of the War

By September 1940 the international standardization movement was in a shambles. ISA members Austria, Belgium, China, Czechoslovakia, Denmark, Finland, France, Greece, Latvia, the Netherlands, Norway, Poland, and Romania were all either occupied or active battlegrounds. Great Britain and its dominions (including ISA members Australia, Canada, and New Zealand) stood against Germany, Japan, Italy, and Hungary. Britain would soon be joined by the USSR and then the United States, Brazil, Mexico, and Uruguay. The remaining ISA members—Argentina, Spain, Sweden, and Switzerland—were neutral, but throughout the war and immediately afterward they would often be treated with suspicion by the governments of the "United Nations," the anti-fascist Allies.

National standardization continued and even retained some degree of independence in many of the occupied countries. AFNOR, whose officers fled Paris during the German invasion, returned after Marshall Pétain signed the armistice. Wartime statutes and decrees that were reaffirmed by postwar governments strengthened the French national body and established a regular system of government subsidies.[33] Norway's government under Vidkun Quisling allowed the board of the national standards body, Norges Standardiseringsforbund, to remain in place, and the government ignored its refusal to support the Norwegian fascist party.[34]

In contrast, the original international bodies were simply mothballed for the duration of the war. In March 1941, the ASA's Paul Agnew reported to his board of directors that ISA files and records would remain in neutral Switzerland in the hands of Huber-Ruf, who should, in Agnew's view, be paid a retainer for his work.[35] Two weeks later, Huber-Ruf had cabled that "all efforts to hold ISA elections had been discontinued," and Agnew reported that the IEC had similarly gone into "hibernation."[36]

Wartime Standardization

While the international standards movement went into abeyance, the war allowed the productivity of the BSI and ASA to catch up to that of DIN. The key was standardization for the war effort. The British and US governments paid the national bodies to employ engineers to work almost full time on technical committees (rather than to volunteer part time). The committees continued to follow their long-standing rules of stakeholder balance and consensus, but the full-time participation allowed them to reach consensus among the various stakeholders more rapidly.

In the United States, the change began before Pearl Harbor with a set of American Defense Emergency Standards for quality control in factories producing weaponry for Britain under the lend-lease program that began in March 1941.[37] By September, ASA was also working on defense standards for the Office of Price Administration to limit the number of sizes of household appliances and standardize "denim, broadcloth, and percale sheets."[38] Many groups within ASA, including the National Electrical Manufacturers Association, worried that working for the government on such standards amounted to an illegal restraint on trade, but ASA's officers were able to deflect the debate by obtaining a legal opinion from the US attorney general and the various defense agencies before accepting the next round of requests. The government offices, of course, concurred that working for the army, the navy, and the lend-lease program was legal.[39] Much of the October 1941 issue of *Industrial Standardization* was devoted to publication of that correspondence, preceded by a didactic essay by Paul Agnew, "Legal Aspects of Standardization and Simplification: A Discussion from the Point of View of the Lay Worker." Agnew's (often repeated) main point was that "it is legal to make a standard—and it is legal to use the standard for many purposes. But to use a standard for *some* purposes may be illegal. The danger lies in parties at interest entering into agreements as to the uses of the standard that may be illegal."[40] That is, ASA could legally accept almost any government contract to set standards using its consensus process, but if companies *used* some of those standards to collude against their competition, that might be illegal; nevertheless, that would be the companies' problem, not ASA's.

From 1931 to 1939, ASA had produced, on average, 46 standards a year. In 1941, ASA almost doubled its previous yearly record and the next year produced 119, a "high but not record breaking" total. Of these, more than half were connected to the war: 41 for "war production," 21 "for use of the armed forces," and 5 for "essential civilian services."[41] In 1942, ASA sold four times as many standards as in any prior year, most of them war related. The same year, government reimbursements for ASA standardizing work raised about one-sixth of the association's income; in 1943 government contracts were providing about half. The amount ASA disbursed for salaries grew significantly, as well, reflecting compensation the association provided to engineers working on standards requested by the government.[42]

The British government and the BSI followed much the same procedure, completing about 400 wartime standards through 1942. In early 1943, le Maistre told an audience of engineers and shipbuilders that, in this work, the BSI had become "an instrument of the State, at the same time maintaining a very large measure of independence. It has retained the confidence of industry and this has been the keystone of whatever measure of success has been achieved." He also complimented "the splendid spirit of cooperation shown by the thousands of members who voluntarily took part in this national work for the benefit of the community as a whole," which "augurs well for the future."[43] Of course, the cooperation of industry may have been easier to gain when the government was paying the compensation of the volunteer engineers, rather than their companies.

Le Maistre described other ways in which the BSI's process had changed because of the war. For example, he noted that "a certain curtailment in regard to its peace-time procedure, particularly its previous very wide circulation of proposed draft specifications," had both saved paper and "much expedited the work without in any way sacrificing underlying principles." In particular, he claimed that it did not sacrifice stakeholder balance: "Moreover, as decisions reached have to be confirmed by post by the full committee responsible, there is no danger of rule by minority interests." He did, nonetheless, lament, "Our well known grey booklets [the typical form of BSI standards at the time] have, for the time being, had to give way to very nondescript yellow ones in order to meet the exigencies of the paper shortage and the pressing attentions of Paper Control."[44]

In ASA, the methods of accelerating decisions were largely the same. Every committee still needed to include representatives of "all groups primarily concerned with the undertaking—producers, consumers, distributors, and specialists in the field" selected by ASA's "War Committee" through "consultation with the groups concerned." Nonetheless, "to save time and unnecessary motions they are not formally designated as representatives of the cooperating groups." Draft standards were circulated for comment, but not as widely as under the regular procedure, and, as was the case in Britain, a time limit was enforced. Committees reviewed the comments and criticisms, amended the draft in light of them, and took a vote, which usually proved unanimous. The standard was then published, on "buff-colored paper rather than the usual white."[45]

The American and British "emergency" procedures in many ways corresponded to the procedures that the Germans had used from the beginning, with greater attention to efficiency and more careful identification of the specific stakeholders that needed to be involved, but with the representatives of the "consumer" interest changing. As the former president of the American Society of Mechanical Engineers recalled in 1951:

> While these War Standards were compulsory, the consensus principle, nevertheless, was maintained to marked degree in writing them. The area of agreement was narrowed to those directly participating. At the consultation table, private industry retained its position of the responsible producers and the government, with the war effort absorbing the position at once of buyer and user. When properly carried out, the procedure developed War standards that did not unnecessarily depart from current industrial practice and made good use of the technical and creative abilities of American industry.[46]

As the war wound down, ASA considered making many of the changes permanent: removing some of the "borderline" stakeholder representatives on committees and limiting the number of representatives from each of the main groups; using subcommittees to draft standards; placing tighter time limits on comments, which "had been found very effective in much of the war work"; and requiring committees to lay out their programs of work with schedules and progress reports. Some members of the ASA Standards Council worried about these revolutionary ideas. For example, "Mr. Harte . . . cit[ed] the American Railroad Engineering Association . . . who thought that a committee of less than 35 or 40 aroused immediate criticism"; after all, "if a committee had a good chairman, the size of the committee did not matter."[47] Mr. Harte's conclusion was emphatically *not* supported by the experience of more rapid and efficient standard setting throughout the war.

Most of the products and processes that the BSI and ASA standardized under wartime rules were indistinguishable from those that came before them in times of peace. Of course, there were the occasional "Specifications for Lighting and Perimeters of Internment and Prisoners of War Camps" (Percy Good wrote on the top of the first page, "Secret, not for publication").

But a 1943 American list of six new British War Emergency Standards included "Flushing Cisterns for Water Closets," "Timber Ladders," and "Test Pieces for Use in the Training of Welders."[48] A contemporary ASA list of wartime jobs included "Building Code Requirements for Iron and Steel," "Code for Electricity Meters," "Sizes of Children's Garments and Patterns," and "Approval and Installation Requirements for Gas Ranges and Gas Water Heaters," all standards that continued to affect the lives of most Americans who were born long after World War II.[49]

Unambiguously military demands for standardization also had a significant impact on the future of the international standardization movement. Because the alliance shared some military equipment and, especially, because so many Americans and their critically important equipment were based in Great Britain and its outposts in the Mediterranean, Asia, and the Pacific, cooperative wartime standardization would seem to have been essential. Unfortunately, as a US Eighth Air Force mechanic told Senator Ralph Flanders, "We can't borrow plane parts from the British. We can't even steal them. *They don't fit.*"[50] Sen. Flanders listed the war's greatest standards disasters: In 1940, "Belgian troops might have fought better and longer if British ammunition had fitted their empty rifles." In 1942, when the British first faced Rommel's forces outside Alexandria, the lack of interchangeable parts among Allied radios was "a contributing cause of the British defeat." And when Packard took over the production of Rolls-Royce aircraft engines, "it first had to spend ten months redrawing 2000 blueprints American fashion and translating the Whitworth thread forms into the American standard on 3200 nuts, bolts, and studs."[51] Near the end of the war the *Economist* reported, "The difference in British and American standards for screw threads has added a sum of at least £25 million to the costs of the war."[52] Of course, the nonmonetary costs were much greater when the lack of standardization meant that an American plane could not be repaired at a British base. The problem of standard parts even made it into American wartime posters, as shown in figure 4.1.

ASA, the BSI, and the other national standards bodies in the Allied countries became acutely aware of the problem as the war wore on. In December 1943, the BSI's Percy Good visited North America to work with US and Canadian colleagues on creating an organization "to 'spark plug' cooperation between allied belligerent countries in standardization matters."[53] Agreement on what by May of 1944 was being called the United Nations Standards Coordinating Committee was far from simple.[54] ASA members who objected to

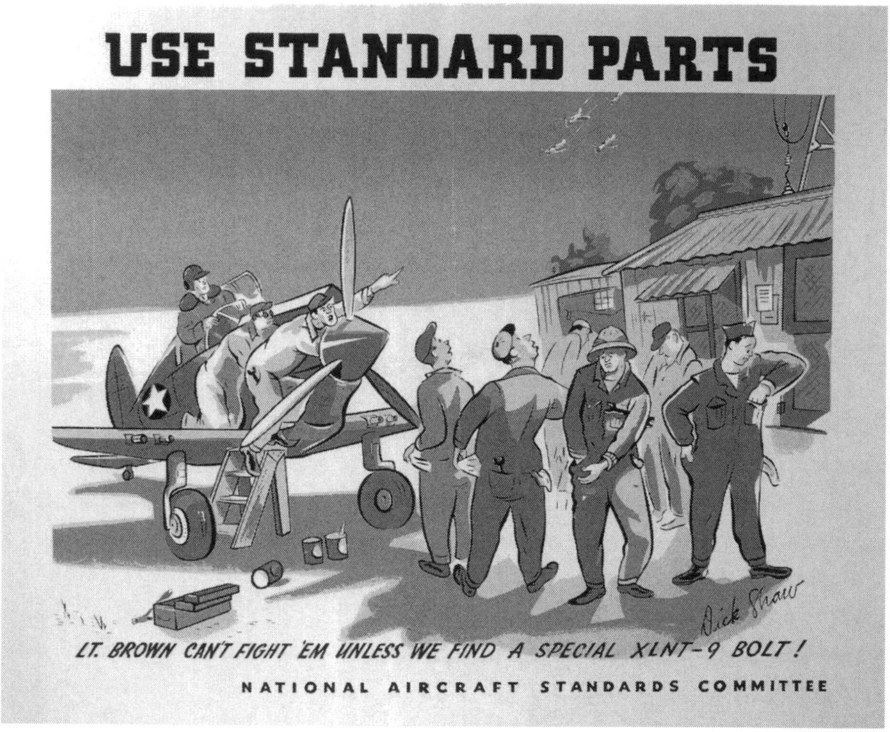

USE STANDARD PARTS

LT. BROWN CAN'T FIGHT 'EM UNLESS WE FIND A SPECIAL XLNT-9 BOLT!

NATIONAL AIRCRAFT STANDARDS COMMITTEE

Figure 4.1. This wartime poster illustrates the importance of using standard parts (specifically standard screw threads and bolts) in World War II combat overseas. Fly Now: The National Air and Space Museum Poster Collection, Smithsonian Institution.

streamlined wartime standardization methods as a "violation of the ASA Constitution" wanted the UNSCC to follow the more lengthy peacetime consensus rules and wanted those "acceptable to ASA" to be "published in accordance with ASA procedures for other American Standards," rather than according to the war standards procedure.[55]

While the UNSCC was eventually established with its own procedures that were closer to the wartime norm, there were other hurdles to overcome.[56] The core US-British-Canadian group had to expand, first to include the standardization bodies in other British Commonwealth countries (Australia, New Zealand, and South Africa), then in the Soviet Union and Latin American members of the alliance (Brazil, Chile, and Mexico), and finally among occupied members of the United Nations who joined "as they were liberated" (China, Czechoslovakia, Denmark, France, the Netherlands, Norway, and

Poland).[57] The first meeting took place in July 1944, before most countries were liberated and before Latin Americans agreed.[58] Le Maistre, naturally, was made UNSCC secretary and put in charge of a London office with an American co-secretary on the other side of the Atlantic.[59] Unlike in 1926, no one objected that le Maistre would be overburdened by his new international work. He had retired from the BSI in April 1944 but remained secretary of the "hibernating" IEC.[60]

The UNSCC arrived too late to make much difference to the war effort, and critics, including the *Economist*, worried that the real purpose of the organization was to try to gain a postwar trade advantage for Great Britain and the United States by imposing supposed United Nations standards that were actually British and American standards on all the goods that would flow to the newly liberated Allied countries and the soon-to-be occupied enemy countries.[61] As we will see in the next section, that concern proved unfounded. The UNSCC constitution limited its operation to only two years, and, while individual national bodies did undertake some standardizing efforts for its committee, most UNSCC meetings were devoted to planning a permanent international organization that would, at least initially, exclude the Axis powers but that, nevertheless, included no provision to impose the conflicting standards of the victors—Britain, the United States, and the Soviet Union—on the rest of the world.[62]

Negotiating ISO

The negotiations that would result in the formation of a permanent organization (ISO) took place through a series of international meetings from September 1945 through October 1946, although the idea that the UNSCC's major work might be to create a successor to the ISA was raised as early as December 1944 at the quarterly meeting of the ASA Standards Council. There, Paul Agnew rose to introduce H. J. Wollner, a chemist who had recently been in charge of the US Treasury's laboratories and who had just been made co-secretary of the UNSCC and head of its New York office.[63] Agnew asked Wollner to report on UNSCC's work, but he responded that "they were just preparing for future projects" and that, "actually, the committee was nothing more than a channel through which several national standardizing bodies could present and correlate their ideas." In response, a visitor from the BSI, John Ryan, explained that he was in the United States because he had been a British delegate to a conference in Rye, New York, at which "international standardization has been strongly recommended."

Indeed, he thought the UNSCC had an opportunity to play a role in such postwar work, suggesting that "Mr. Wollner might discuss the situation with the [London-based] Secretary General of the International Chamber of Commerce who was now in the United States."[64]

The meeting to which Ryan referred was a gathering of 400 businessmen from 52 countries. This International Business Conference had been called by the American Section of the International Chamber of Commerce, the National Association of Manufacturers, and the National Foreign Trade Council.[65] It was meant to be the private-sector equivalent of the conference on postwar planning held at Dumbarton Oaks in Washington, DC, from August through October 1944, and it followed the precedent of a similar meeting at the end of the First World War.[66] In 1919, a large group of international businessmen who would go on to establish the International Chamber of Commerce met in Atlantic City and reached agreements that had, for better or worse, influenced the interwar international economic system.[67] The main theme of the 1944 International Business Conference was developing strategies to break down economic barriers between countries and regions as rapidly as possible. The goal was a truly global economy, and, for that reason, the desirability of establishing international standards was raised repeatedly.[68] Ryan's intervention may have been influential with Wollner, especially since the BSI representative was vice chairman of a major wartime supplier and would become chair of the BSI in 1952.[69]

In the winter of 1944 or spring of 1945, le Maistre sent out invitations to nonmember national standards bodies to join the UNSCC and attend a fall 1945 meeting that would create a new international standardization body. As le Maistre explained:

> In the present circumstances the Member Countries feel the UNSCC is the proper Body through which the United Nations should consider the problem of post-war international standardization as a whole. Their view is that post-war conditions will make it necessary to evolve a new international organisation which will cover the international standardization in all fields and that the status of the I.S.A., I.E.C., and I.C.I. should therefore be reviewed in light of the new requirements.[70]

The meeting took place in New York on October 8–11, 1945, and was attended by about 30 men and one woman (Miss Harrison of the BSI), representing the national standards bodies of Australia, Austria, Brazil, Canada, China, Denmark, Mexico, New Zealand, Norway, South Africa, Switzerland, the United

Kingdom, and the United States, with a Soviet diplomatic observer attending on the final day. Le Maistre and ASA's Paul Agnew were the only attendees who had been with the standardization movement from the beginning.[71] Immediately before, from September 23 through October 6, the Americans, British, and Canadians held an agenda-setting session in Ottawa that continued in New York on October 7.[72]

The main meeting agreed to propose to UNSCC members the creation of an International Standards Coordinating Association that would be open to all countries and with which the IEC would be invited to affiliate. While the association's working languages and location were to remain open questions, the attendees followed the precedent of the recently agreed United Nations Charter and proposed an 11-nation governing body that would include representatives of the "Big Five," China, France, the Soviet Union, United Kingdom, and the United States, "to lend all possible strength to the permanent organization," but imposed this requirement only for the first five years. The attendees agreed to meet next in London in late spring of 1946, perhaps not coincidentally immediately after the first session of the UN General Assembly (also to be held in London), but they rejected the name "United Nations" Standards Association. In the middle of their meeting, the delegates enjoyed a reception at Radio City's Rainbow Room, and they concluded their deliberations with "a small sightseeing trip by boat around Manhattan Island."[73]

Three long-standing divisive issues, and a fourth, new one, absorbed most of the four days before the engineers boarded the Circle Line cruise. The new one was the role of the developing countries. The old ones were whether the organization should set international standards (or only coordinate national ones), how to accommodate the countries that still did not use the metric system, and what to do about Huber-Ruf and the ISA.

ASA's Harold S. Osborne, who chaired the first day, immediately raised the new issue by pointing out that only 10 percent of the world's countries *had* national standardizing bodies and asking about what was going to be done about the other 90 percent. "Limiting membership to those 10%," Osborne argued, "starts the new organization on the wrong foot."[74] Other delegates voted down Osborne's motion to remedy the problem by making it possible for a country's engineers to form groups solely for the purpose of being represented internationally. They argued (incorrectly) that his figure of 10 percent was surely too low. The number was "closer to 50%" and might even be 80 percent within the next year.[75] The conflict had to do with how one defined "countries." There were around 20 national standards bodies operating in

1945, and the new United Nations organization had begun with 51 members, but those members did not include the defeated powers or even those that were neutral at the end of World War II, and they included only two colonies, India and the Philippines, both of which were expected to be liberated by 1947.[76] Osborne's calculation took account of all the nearly 200 countries that would eventually enter the United Nations, most of which, in 1945, were less industrialized colonies.

Next, Osborne had focused the meeting on whether the body would "coordinate" standards internationally or would, as ISA had, actually set international standards. Percy Good, who had replaced le Maistre as secretary of the BSI, adamantly declared that the organization should only "coordinate," and he wanted that in the organization's name. No decision was taken, but the topic returned at the end of the day when Good railed against the suggestion that the "recommendations" of the new organization be called that; he preferred the term "reports." No doubt Good was thinking about the fact that the IEC's standards were still called "recommendations," but everyone treated them like international standards. ASA's Paul Agnew said that calling them recommendations would really cause no problem—it just had to be made clear that each of the organization's recommendations would be that of a particular technical committee that would also report on the status of various relevant national standards.[77] The chair, Osborne, concurred and reminded everyone that nothing was mandatory and the published recommendations would make that status clear. Good objected, "This body can't reach decisions and these decisions [on the use of "recommendations"] must be those of the national standards bodies and promulgated as their decisions when they have taken them."[78]

Osborne adjourned the meeting for lunch, and, when the issue came up again two days later, Agnew, who was then in the chair, led conferees to accept that no recommendations or standards would be issued over the objection of a member body and, some hours later, to accept this wording for the stated purpose of the organization: "the maximum coordination and unification of national standards."[79] Presumably hallway conversations over the intervening days had helped pave the way for this result. Thus, while Good may not have acknowledged it, the new organization would have the same power to issue international consensus standards that the ISA enjoyed.

Good was equally annoyed about a second perennial issue: accommodating countries that had not adopted the metric system. The Chinese delegation raised the suggestion that the new organization should simply adopt the practice of using a single system whenever possible and developing two standards as nearly consistent as possible when not. Good objected, saying that Great Britain "would not participate in any discussion of this." This was a major reason that the BSI had argued that the proposed organization should be considered only a "coordinating" body. Agnew, from the other major nonmetric country, disagreed and said that the Chinese proposal was a good way to overcome the inconvenience of two systems. The group, including Good, agreed to circulate the Chinese proposal to UNSCC members and make a decision at a later meeting.[80]

The French and Swiss raised the third divisive question of what to do about the ISA. The French delegate reported that many Europeans thought the United States, UK, and France were cutting out the old organization. The Swiss proposed that the ISA meet for "five minutes" right before the spring UNSCC meeting, just to dissolve itself, but some delegates were uncertain. Would it be possible for the ISA to meet without Italy, Japan, and Germany? Would it even be legal or ethical for engineers from the victorious powers to meet with them? What should be done with ISA records, presumably preserved by Huber-Ruf?[81] Ultimately, the meeting instructed the UNSCC's "Executive Committee to discuss the liquidation of the old ISA with the Swiss Standards Association" in its role as the official "custodian of the affairs of the ISA."[82]

The plan to hold a final meeting the next spring to establish the new organization proved optimistic. As Paul Agnew reported to ASA colleagues in April 1946, "Developments had come about as the result of a more thorough study of the situation in regard to the old International Standards Association which had made it practically essential that this meeting be postponed."[83] Since 1941, Huber-Ruf had been working full time as the lowest-paid of three consulting engineers in a small Swiss government office concerned with the needs of industry during the war. He was 62, in poor health, and unable to stop working because he served as a private engineer on temporary contracts and had no right to a Swiss pension. Agnew reported that the Swiss Standards Association had made an "informal" proposal that the ISA's remaining funds be added to by contributions "from each of the ISA members . . . so that his long and faithful, and, on the whole, very effective work for ISA could be fairly if not adequately compensated."[84] "There was a flurry of correspondence between ISA members, and they decided that the 1939 ISA Council was still capable of acting." The IEC had planned to hold its first postwar meeting in Paris in July, so le Maistre opportunistically called a meeting of the UNSCC at the same time and place, and convinced the ISA council members to do the same. This was six months after the new United Nations met for the first time in London in January 1946. In Paris, the UNSCC and ISA meetings directed le Maistre to meet with Huber-Ruf, who was reported as having been too ill to attend the meeting, and to seek out his views about the Swiss proposals.[85]

The Paris UNSCC meeting, the first in which the Soviet Union participated fully, primarily served to prepare for an October meeting in London at which ISO was established. The Soviet delegates raised questions about the working languages of the new organization, insisting that Russian, along with English and French, be an official language of the organization. The proposal to make it an official language was voted down. The French and American delegates met with the Soviet delegates immediately after the Paris meeting and reached a compromise that Russian would be named as an official language but that the USSR body would do all translating to and publishing in Russian itself.[86] At the IEC meeting, this issue was resolved more smoothly; a Soviet proposal that Russian be made a language of the IEC with the USSR taking on the cost of translation was immediately approved and praised by other members.[87] Still, in both cases, standardizers found a diplomatic way to keep the USSR in the standardization community, despite the rising tensions of what would soon become the Cold War. At the same Paris meeting, the Swiss delegate raised another political issue that would become a troubling topic over the next decade: Would all the formerly neutral countries that were members of the IEC be allowed to become members of the new international organization, and if they were not, would they then have to withdraw from the IEC? The Swiss were especially concerned about Portugal and Argentina. Percy Good assured them that everyone would be welcome in both organizations.[88]

The London meeting in October included the neutrals Sweden and Switzerland. Argentina had been invited but chose not to attend. The national bodies of Austria, Finland, and Italy, whose former governments had supported the Nazis, were also there. There were also two "observers," the national standards bodies of Yugoslavia and of the colony of Palestine. At the meeting of the conference's steering committee before the main meeting began, le Maistre reported on his unsatisfactory meeting with Huber-Ruf, who insisted that the terms of the 1939 ISA council members had expired and therefore that they had no right to take any action; consequently, he claimed that he was still general secretary of the ISA and should be appointed director of the new organization. The excessive persistence Huber-Ruf had shown in getting his job at the ISA was again in evidence in his attempt to keep it, but this time his Nazi associations, as well as his obstinacy, seemingly prevented his success. The committee concluded that it should neither officially answer Huber-Ruf nor invite him to the main meeting, as he had requested. Nevertheless, they invited le Maistre to "communicate with him in his personal capacity" and then decided "to recommend to the Conference the adoption of a resolution to the effect that ISA Member Bodies be asked to state that they agree to break their association with the ISA and to affiliate with the new Organization."[89] At a side meeting, the 14 attending member bodies of the ISA unanimously voted that the ISA "be considered as dissolved from April 1942."[90]

The main meeting, which went on for almost two weeks, was surprisingly amiable. The delegates took time for "all-day excursions" to Hampton Court and Windsor Castle. They visited the National Physical Laboratory and the new television studios of the BBC. The Indian delegation gave a reception at India House, and the BSI hosted a closing concert "at the Vinter's Hall in the City of London," with "a delightful selection of old madrigals, duets, and folk songs."[91] Paul Agnew reported back to the ASA that the major controversies had been quickly resolved and that "the name finally chosen for the new organization was 'International Organization for Standardization,' for short 'ISO.'" He explained, "The English speaking countries would have preferred to have the title a little different but it had been a question of complications in the translation into French, which was one illustration of the language difficulty."[92] Russian was officially accepted as a third language along with English and French, with the agreement, as in the IEC, that the Soviet Union would bear the cost. The delegates had to participate in a series of votes on the location of the organization's headquarters, but in the end they chose Geneva over Montreal by a vote of 12 to 11. The question of the ISA had been complicated. Agnew's later account to ASA differed from the official report of the meeting in saying that "Huber-Ruf was in very ill health and had been unable to attend" and in noting that the ISA members "unanimously agreed to use the moderate funds available as pension for the former secretary." In addition, several members of the ISA, including ASA itself, "had agreed to supplement this grant by an increased pension for Mr. Huber-Ruf in appreciation of his many years of loyalty and energetic service."[93] Le Maistre may have met with Huber-Ruf at his home in December 1946, but there is no evidence that the Swiss engineer ever received a pension for his ISA work.[94]

Meanwhile, the new ISO would lead in the postwar revival of the standardization movement, initiating a second wave primarily emphasizing standardization on the international level.

Taking Standardization to the Larger World

In 1947, le Maistre and the IEC moved from London to Geneva, where the venerable IEC shared a small private house with the new ISO. Le Maistre kept his IEC title but the IEC's Committee of Action "instructed [him] to support ISO as needed" as his "main work."[95] The first general secretary of ISO, Henry St. Leger, "an American with close French connections and a perfect knowledge of both English and French," who was also serving on the war crimes tribunal in Nuremberg, failed to appreciate le Maistre's help.[96] St. Leger concentrated on building relations with the many UN organizations that had their headquarters in Geneva, which had been the site of the League of Nations and which shared with New York the status of being a headquarters city of the new United Nations. St. Leger's ISO produced few standards (only about 100 from 1946 through 1960); it followed the extremely cautious approach of having each of its "recommendations" simply affirm an existing national standard and suggest that it be implemented in other countries, as well.[97] While le Maistre's BSI protégée Percy Good had advocated this approach, it was not the way le Maistre ran the IEC. Moreover, in working with the UN and in dealing with the sticky questions of reintegrating standard setters from the defeated powers and expanding the standards movement to former colonies and other less industrialized countries, St. Leger used his own well-developed set of diplomatic and administrative skills and thought he did not need le Maistre's help.

The international standardization bodies, the IEC as well as ISO, resumed normal relations with neutrals relatively quickly, typically a few years before they were admitted to the UN. The same was true with the standards bodies of the former Axis countries. In contrast, most of the developing countries that joined ISO from 1947 through 1960 did so some years after they became members of the UN, since they were required to have a national standardization body to do so, something that most countries gained with the support of UN technical assistance (table 4.1).

The status of the suspect neutrals—countries that the Soviet Union, Britain, or France considered to have been too close to the Nazis—became complex in part because of the close connections that quickly developed among the IEC, ISO, and the United Nations. In 1947, under St. Leger's leadership,

TABLE 4.1.
Year of Admission of Neutral, Defeated, and Less Developed Countries

	To ISO	To the UN
Argentina	1960	1945
Brazil	1947	1945
Bulgaria	1955	1955
Burma	1957	1948
Chile	1947	1945
Colombia	1960	1945
Egypt	1957	1945
Germany, DR	1955	1973
Germany, FR	1951	1973
Hungary	1947	1955
Indonesia	1954	1950
Iran	1959	1945
Ireland	1951	1955
Israel	1947*	1949
Italy	1947	1955
Japan	1952	1956
Pakistan	1951	1947
Portugal	1949	1955
Romania	1950	1955
Spain	1951	1955
Sweden	1947	1945
Turkey	1956	1945
Uruguay	1950	1945
Venezuela	1959	1945

Sources: Data from Caroline Le Serre, "Historical Record of ISO Membership Since Its Creation (1947)," accessed July 12, 2017, http://www.iso.org/iso/historical_record_of_iso_membership_1947_to_today.pdf; United Nations, "United Nations Member States," press release, July 3, 2006, accessed July 12, 2017, http://www.un.org/press/en/2006/org1469.doc.htm.

*In some ISO records, Israel is considered a founding member, perhaps due to the presence of colonial representatives of Palestine at the founding meeting.

ISO became the fourth of about 200 international private nongovernmental organizations granted "general consultative status" with the UN system since 1946; only the International Chamber of Commerce, the International Cooperative Alliance, and the World Federation of Trade Unions preceded it.[98] At the same time, during ISO's first year, the IEC agreed to affiliate with it. A meeting on the issue of neutral nations in ISO called by Hilding Törnebohm, who had led the Swedish Standards Institute delegation to the 1946 meeting that created ISO (SIS and the Swiss Standards Association were the only standards bodies from neutral countries to be invited), devised a strategy that suspect neutrals should continue "de facto collaboration" with the international standard-setting bodies until the "political pressure" to exclude them from the international organizations diminished.[99] The pressure diminished within three or four years, allowing (re)admission to the standards bodies; it took another five years after that, however, until all the neutrals and most of the former Axis countries were admitted to the UN, and almost two decades more until the two parts of Germany became UN members in 1973.

In keeping with the strategy for suspect neutrals, the president of the Spanish committee wrote to the IEC president, Emile Uytborck of Belgium, in October 1947, offering "to withdraw from the IEC if this step will assist in however small a way in promoting future good relations of former friendly Committees in the arduous task of setting in motion the new institution [ISO]." He expressed appreciation, "in this distrustful moment," for "the goodwill towards me of the Delegates of the IEC and of the Secretariat of the ISO and you yourself."[100] Uytborck accepted the resignation for the Committee of Action "with deep regret that this action should be necessary since the IEC being a technical and scientific organization is no way concerned with politics." He then reasserted the standardization movement's belief that "the work of the IEC to bring the fullest benefit to the world at large, needs the continuous and friendly cooperation of electrical engineers and scientists of all nations."[101]

While Törnebohm was working with the other neutrals, Percy Good and le Maistre worked on bringing back the Germans. In March 1947, Good wrote to a colleague in the British Element of the Control Commission for Germany (the occupation government) to inform him of a 1946 visit Good had made to DIN headquarters in Berlin and a planned follow-up visit by another senior BSI colleague, J. O. Cooke. Simultaneously, Good wrote to his German colleague to apologize "that I have not replied earlier to your letter of 16th September last, but it has been necessary for us to obtain formal approval by the Government to the resumption of our formal relations."[102] Meanwhile, Good's colleague on the Control Commission, E. G. Lewin, wrote to Cooke that, while the Russians professed to want DIN to begin working effectively, "the whole essence of their move was control, designed apparently to hamper and restrict the work in every way." He believed that Russia was seconded by the French in this and that "throughout our discussion up to date we have been fighting off one limitation after another on the DIN's activities."[103] Nonetheless, the British standardizers had won the day (perhaps by convincing the Americans, the only one of the four occupying powers that Lewin did not mention), and the German standards body was, as of March 1947, able to "re-establish itself on [a] practicable basis and to operate in a substantially normal fashion," except when it came to dealing with cooperation with standards bodies internationally.[104]

In May 1947 Waldemar Hellmich, not a conspicuous supporter of the Nazis during the war, successfully initiated a "searching reexamination" of the engineering profession throughout Germany that continued for many years. It resulted in VDI adopting a "credo of the engineer" in 1950, which

> stressed "reverence for values beyond knowledge" and "humility before the all-powerful" creator; "dedication to the service of mankind" in honor, justice, and impartiality; "respect for the dignity of human life," irrespective of origin, class, or ideology; rejection of technical abuse in "loyal work for human ethics and culture"; collegial cooperation for "the sensible development of technology"; and finally, placing "professional honor above economic advantage."[105]

In his widely discussed 1947 speech and subsequent article, Hellmich interpreted "the behavior of the German engineers during the Nazi period as Faustian"; German engineers had sold their souls to the devil.[106] One scholar described the effort and resulting credo as follows: "Chastened by complicity with the Third Reich, thoughtful engineers struggled for a broader self-conception, combining technical expertise with social responsibility."[107] After the adoption of the new credo and suppression of Nazi statutes and affiliates, the German engineering associations, Hellmich's children, reappeared, at least in western zones of occupation that became the Federal Republic of Germany.[108]

In 1951, West Germany successfully applied to join ISO, and to rejoin the IEC. The IEC's documents include records of the vote; the only no votes were Czechoslovakia, Hungary, Israel, and the Soviet Union.[109] Le Maistre, who had been working on Germany's readmission since the end of the war, asked the newly accepted delegates "as soon as possible to propose the name of a German town where the annual meeting of the I.E.C. could take place," which resulted in the 1956 IEC meeting in Munich, three years after le Maistre's death.[110]

Reintegrating Germany into the two international standards bodies was one of his last services to the movement he had worked for since 1901. Le Maistre died quietly in July 1953 at his home in Surrey, Lea Gate House, the place where he had hoped to retire. The building was a relatively small modernist masterpiece; hidden from the road and built in 1939 with standard metalwork window frames and balconies, it somehow still managed to look like a solid brick country house of the Victorian era, a place very appropriate to

Figure 4.2. The IEC's official portrait of Charles le Maistre shows him late in life when he focused on reuniting the pre–World War II international standardization movement and preventing the emergence of new Cold War–era rifts. IEC Photo File, courtesy of IEC.

the man.[111] At the Stockholm IEC meeting in 1958, Germany's Richard Vieweg lauded le Maistre's "far-sighted plans for international cooperation," saying, "To have the earnest desire, with the resulting forces of patience and conviction to reach the goal, was the lasting example of Charles le Maistre,"[112] whose last IEC portrait appears in figure 4.2.

In one case, the IEC was more than a decade ahead of both ISO and the UN; the electrotechnologists admitted the People's Republic of China in 1957, while the UN took until 1971 and ISO until 1977. A member of the British IEC delegation explained its positive vote to le Maistre's IEC successor, Louis Ruppert. He had asked the opinion of "Mr. Duncan in the Foreign Office" before voting but found that "he was somewhat embarrassed by being asked to express an opinion. He pointed out that as the IEC was a non-governmental organization it was not for the Foreign Office to express any views." With that disclaimer, he said "that he felt sure that we would bear in mind the fact that, in the political field, as soon as China came in, America walked out." But the British IEC representative responded that "whereas the U.S. National Committee had voted against admission of China, it was doubtful whether they would go to the length of 'walking out' if China were admitted by a majority vote."[113] And, indeed, the United States did not walk out, and only Belgium, Canada, France, and Turkey joined the United States in voting against China's admission.[114]

The admission of most developing countries to the IEC and ISO was less controversial, but it came about as part of a complicated relationship that the standard-setting bodies developed with the United Nations. Recall that one reason ISO and the IEC ended up in Geneva was to be in one of the UN's headquarter cities—arguably, initially, the more important one. In the early days, New York was just the talking shop, the site of the General Assembly and the Security Council, but Geneva was where most agencies of the UN system relevant to industrial economies were located: the General Agreement on Tariffs and Trade (GATT), UN Economic Commission for Europe (a central mechanism for postwar reconstruction), World Health Organization, International Labor Organization, and others. And Geneva was near Paris with the UN Educational, Scientific, and Cultural Organization (UNESCO) and Rome with the Food and Agriculture Organization.

Between 1947 and 1951, both ISO and the UN agencies attempted to assert their leadership over the entire field of international standardization. St. Leger proposed a plan whereby ISO would be made coordinator of all the standardization work *within* the UN because

> the Member Bodies of ISO are the recognized national standardizing organizations. . . . On the national level these organizations are, in some cases, departments or offices of national Government; in some other cases they are private bodies, but with Government participation in standardizing activities. Many of the organizations receive financial support from their Governments; others are wholly supported by Government. The Organization, therefore, occupies a quasi-official position in the sphere of its activities.[115]

Nothing ultimately came from St. Leger's elaborate plans for ISO nor his attempt to portray ISO as less private and more intergovernmental.

From the UN agency side, in 1951 UNESCO asked that ISO, the IEC, the International Bureau of Weights and Measures, and two other associations be brought together under UNESCO in a proposed new organization, the

"Union of International Technical Associations." ISO, the IEC, and the International Bureau refused UNESCO's invitation.[116] UN agencies continued to do all kinds of "technical" standardization that ISO's St. Leger believed that ISO should coordinate, including "statistical standardization," labor and health "safety codes," and "standardization of metrological services," and the Economic Commission for Europe began (and continues) a great deal of industrial standardization work relative to roads, automobiles, and other transportation infrastructure needed for postwar reconstruction.[117]

Despite this tension over ownership of international standardization, cooperation among ISO, its members, and the UN agencies took off in the 1950s when the UN began providing technical assistance to developing countries. This involved sending experts from industrialized countries to help governments in the developing world design new policies, create new organizations, and identify large, financially sustainable industrial projects such as electrical dams. In the 1950s, a little over half of the UN's extensive funding for technical assistance came from the United States, with the rest from the Soviet Union and the governments of other industrialized countries. All the UN agencies had a bureaucratic interest in providing experts because for every seven people they could put in the field, they were given funding for one position at headquarters. By the early 1960s, the vast majority of the staff and the budget of the entire UN system—even of "technical" agencies like the World Meteorological Organization and the International Civil Aviation Organization—was devoted to the developing world. In the 1950s, UN technical assistance focused primarily on supporting the institutions needed to create and maintain an industrial economy: electric power systems, engineering schools (including the Indian Institutes of Technology and Turkey's Middle East Technical University) and national standard-setting bodies.[118]

Olle Sturén, the Swedish engineer who would follow St. Leger as ISO general secretary, first became involved in international standard setting as the UN expert seconded from Sweden's SIS. He advised Turkey on how to set up its own national standards body for "three happy years" in the early 1950s. Turkey was where, he later wrote, "I got my interest in internationalism." He planned to stay in the UN, advising other countries on standardization, but SIS invited him back to Stockholm to become its director. Sweden had just been elected to the ISO governing council, and the job was one in which he could "combine standardization and internationalism in a very nice way," which he did, by promoting standardization worldwide on the ISO Council and later as its general secretary.[119]

In his keynote lecture at the 1961 IEC annual meeting, India's Mohammed Hayath explained why it was critically important for every developing country to be able to set its own national standards today, and why it would be increasingly important to set global standards in the future. Developing countries were inherently confronted with the problem of reconciling the different national standards that came with the equipment that they needed from already industrialized countries. Not only did developing countries need to cope with the different standards used by various countries that wished to provide them with aid, but the problem increasingly existed whenever they purchased equipment because "countries in Asia and Africa are starting to obtain equipment from many countries on the basis of competitive global bids, rather than from restricted trade areas as was the case in the past."[120] He noted, for example, that developing countries typically sourced turbines from Japan and generators from the United Kingdom, so national standards that, for example, required compatibility across suppliers using the metric and the English system of measurements were needed. Moreover, since much of the developing world was tropical, the tolerances for different equipment could not be the same as those for the temperate world.[121]

In the next chapter we show that the developing world's concern for gaining control of its own standard setting and moving toward international standards, the internationalism of engineers like Sturén, and the pressure of some industries for access to global markets combined to focus ISO and many of the national standardizing bodies on the problem of creating much larger market areas. The second wave of innovation in standardization consisted primarily of expanding the work of ISO and establishing new national organizations in developing countries, though regional standardization bodies would also arise.

5

Standards for a Global Market, the 1960s to the 1980s

The formation of the International Organization for Standardization launched the second wave of institutional innovation in voluntary standard setting. From the 1960s through the 1980s the community of standardizers and network of organizations established by the standardization movement shifted focus from setting national standards for producers and users of industrial products to setting international standards that facilitated commerce more broadly, including through incorporation into official regulations. In this chapter we look at two significant cases of international standard setting: one, shipping containers, in which effective global standardization took place; and another, color television, where it failed. While the first case shows how private international standard setting worked at its best, the second, a complex mix of private and intergovernmental standardization, shows some of the potential failure modes for international standardization. Before and after those cases we discuss institution building during this period.

The chapter begins with the work of a Swedish engineer, Olle Sturén, a standardization entrepreneur who played at least as significant a role in the second wave as Charles le Maistre or Waldemar Hellmich did in the first, particularly in shifting standardization activity toward the international. The chapter's final section discusses innovations that significantly changed the relationship between the private standard-setting bodies and national governments, including the General Agreements on Tariffs and Trade Standards Code and the establishment of formal links between public regulation and voluntary standard setting in Europe, the United States, and much of the rest of the world.

Olle Sturén: Standardization Entrepreneur

The man who served the longest term as ISO's secretary-general thus far was born in Sweden in 1919 and followed a leisurely and self-described "Bohemian" path to a degree in civil engineering in 1945. An avid and accomplished tennis player, he took a government position involved with the postwar housing boom at the same time he worked as secretary of the Swedish Lawn Tennis Association. In 1947, Sturén was looking for an engineering job that would allow him time to complete the work of turning a local tennis tournament into the Swedish Open. He found what he wanted in the construction sector standardization organization that was linked to the Swedish Standards Institute, as he explained in a talk to the ISO Council late in his career: "That is how I became a standardizer. I did not know anything about it. I could hardly spell it in Swedish and I joined it because it was the only employer who accepted my conditions for how I wanted to do my work." He quickly learned about standardization: "I started in the building division and I standardized building blocks. Modular coordination was something very new."[1] He moved on to standardize kitchen equipment and kitchen design, an interest he kept throughout his life—occasionally to the dismay of his wife, Nalle (whom we met in chapter 3, speculating on why standardizers were so nice), since whenever they moved to a new house or apartment, Olle would take it upon himself to redesign the kitchen.[2]

Sturén rose quickly in SIS, continuing his habit of holding down more than one job and becoming increasingly interested in international standard setting. In 1950, he visited the American Standards Association on a Marshall Plan fellowship.[3] From 1953 through 1956, he went back and forth to Turkey providing technical assistance in the establishment of the Turkish national standards body and even temporarily acting as head of the UN office in Ankara.[4] In 1955, Sturén served as local organizer and general secretary of ISO's Fourth General Assembly, held in Stockholm. It was much larger than previous assemblies, with 529 delegates from 37 national standards bodies, plus 120 accompanying family members. Sturén aimed to make the meeting public, memorable, and fun. He published a daily bulletin of conference gossip, with two extra numbers detailing the conferees' excursions around the city, and he convinced the local press to provide lighthearted (if sometimes suggestive and, to today's sensibilities, sexist) coverage of the whole event. For example, *Aftonbladet* published a photo of one of SIS's elder statesmen making a careful chalkboard drawing of screw threads for an attentive young woman.

The caption read, "Hilding Törnebohm shows the Russian interpreter what his favorite screw looks like."[5]

Partially based on the success of the ISO conference, Sturén became director of SIS in 1957. The next year he radically reorganized the national standards body and hosted another, much larger international conference in Stockholm. The general meeting of the venerable International Electrotechnical Commission attracted 1,232 participants. Again, there was the *Bulletin*, the pleasurable events (wild strawberries were served before the Charles le Maistre lecture because they had been le Maistre's favorite), and the sexist jokes, this time at the expense of a British colleague who explained standardization with reference to bra sizes.[6]

Three of Sturén's more serious activities, all in 1960–1961, give a clearer idea of what he did as head of SIS. They also anticipated the role he would play as ISO secretary-general in using standards to create a broad international market. He led the Swedish delegation at the first meeting of the ISO technical committee on container standards; he visited the newly opened Nigerian national standards body, created shortly after the independence of Africa's most-populous country; and, in his role as a member of the ISO Council, he initiated an ISO round table on computer standards in Geneva.[7] Shortly before his death in 2004, Bob Bemer, a software pioneer who helped define the ASCII character set (the connection between computers and written alphabets), argued that this ISO Round Table was one of 18 events that made the internet and World Wide Web possible, "a without THIS, no THAT." Bemer wrote that "Sturén's goal was to ensure that any difficulties the entire world might have with computer standards would be resolved before being adopted irrevocably." He claimed that the Round Table Sturén held "led to both the formation of ISO Technical Committee 97 on Computers and Information Processing Standards, and his own position as Secretary General of ISO for many years."[8] Ironically, Sturén was personally not particularly comfortable with information technology. After he retired from ISO, he compiled an extensive chronological file of materials to be used when he wrote his own memoir; he never wrote it, according to his son, because "he couldn't quite come to terms with using a computer."[9]

In fact, Sturén saved few documents about individual technologies; the focus of his memoir would have been his role in building international standards organizations. From 1958 to 1961 he concentrated on negotiating the formation of CEN, the European Committee for Standardization, a project he undertook with Roy Binney, director-general of the British Standards Institution, and Arthur Zinzen, managing director of the German national standards body (DIN). CEN was deliberately created to unify the standards activities of the original European Economic Community (EEC) members, the "inner six" (France, Germany, Italy, Belgium, the Netherlands, and Luxembourg), and the "outer seven" who formed the European Free Trade Association (Austria, Denmark, Norway, Portugal, Sweden, Switzerland, and the United Kingdom). Sturén explained the purpose of CEN in a May 1958 SIS report: "If a free European market means anything at all, it means more competition. It means opening national markets to the competition of all Europe, allowing the European consumer to buy from the most efficient producer in his continent." He saw national standards as a barrier to free trade within Europe: "But free competition might be eliminated and domestic producers be favored, if delivery is required to be made according to a national standard differing considerably from corresponding standards in other countries." That is why Europe needed "international standards of regional character." Moreover, he argued that Western Europe, unlike the more self-sufficient regions of Eastern Europe or the United States, had an interest in assuring that those standards were truly global, "an interest in promoting world standardization as a framework for its own standards."[10] A short time later the same report appeared under the names of Sturén, Zinzen, Binney, and Benedetto Cusimano (of the Italian national standards body) at a consumer-oriented conference of the International Committee for Scientific Management, presumably as part of the effort to promote the creation of CEN.[11]

The national standards bodies of members of the EEC and European Free Trade Association created CEN in 1961 (and Sturén served as its president in 1967–1968), but this was only a partial solution to the problem Sturén and his coauthors had outlined. It was in Western Europe's interest to promote world standards, but that was ISO's job. Yet ISO was not doing that under the leadership of the American diplomat Henry St. Leger. Unfortunately, as one of the first people to join ISO's secretariat put it, "What made the development of the organization impossible was the personality of the Secretary-General."[12] In the 1950s, St. Leger was happy to have ISO work slowly, at the rate of about 10 new standards each year, but the actual extent of the job was, as Sturén frequently argued, to establish and maintain about 10,000 standards, the necessarily international half of the 20,000 or so standards he believed were needed for an industrialized country to function well.[13]

Matters came to a head in 1964 when the BSI's Roy Binney began a five-year term as ISO's vice president—the officer who headed the committee responsible for the organization's agenda. At the November 1964 General Assembly, the Netherlands' national body presented a one-page statement demanding that the secretariat find ways to (1) stop the duplication of work by different national standards bodies; (2) speed up ISO's own work; (3) liaise more effectively with intergovernmental organizations, consumer groups, and industrial associations in order to avoid duplication, ensure that ISO was focusing on the most critical issues, and more effectively provide technical assistance and training to standardizers in the developing world; (4) work more effectively with the IEC; and (5) "spread knowledge and proper use of standardization in non-technical circles."[14]

Three months later, in early 1965, St. Leger began slowly implementing an ISO General Assembly decision to respond to this statement. He wrote to the heads of the national bodies for France, Israel, the Netherlands, Poland, Sweden, and the USSR, asking them to appoint one representative each to "the 'Committee for the Study of the Netherlands' Statement concerning ISO Liaisons and Activities' (NEDCO)."[15] Sturén immediately wrote back from Stockholm that the Swedish representative "will be Mr. Olle Sturén."[16] Over the next two months Sturén had himself selected as NEDCO's chair, held three preliminary meetings (in the Hague, London, and at Amsterdam's Schiphol airport), convinced the Dutch to write an extended "Elaboration of Statement of the Netherlands Delegation" with proposed solutions to many of the problems previously identified, and invited the members of the committee to attend the first official meeting of NEDCO in Stockholm, with "dinner at our home," on June 2 and 3, 1965.[17] An early June meeting was essential, Sturén wrote, if they were to have their report completed for "this year's ISO Council meeting which as you know takes place in Geneva 13–16 July."

NEDCO met twice more in Geneva prior to the July council session where it presented a report based on the "Elaboration of Statement of the Netherlands Delegation." ISO, NEDCO argued, needed at least four new divisions focusing on specific technologies that would follow the model of the IEC's focus on electrical engineering. ISO should also follow the IEC's more efficient and effective methods of standard setting, and the secretariat should produce more publications, edit all documents more thoroughly and consistently, conduct public relations, and increase technical assistance to the developing countries.[18]

After the preliminary report was accepted, the next question was what additional resources would be needed to make its recommendations possible.

Figuring that out led to some surprises. Sturén consulted with the small staff in Geneva and wrote a "private and confidential letter" to Roy Binney to report that secretariat employees "gave me some information" about ISO operations "which they stated had never before been given to any Council Member, etc. I would like to pass this information to you as vice president of ISO," and "am therefore prepared to visit you in London at a time when you are available." He attached a detailed report of all of the work of the small staff, from St. Leger down to "Mrs. S. Droz, Cleaner and occasionally help-photographer."[19] Without documenting the information that Sturén conveyed from the staff to Binney, the two men then exchanged two months of cryptic letters about ISO's secretary-general.[20] Then, in the words of Roger Maréchal, one of ISO's two assistant secretaries general, "Henry St. Leger gave up at the end of 1965, and for eight months there was no Secretary-General."[21] Maréchal, who was French, ran ISO with his Swiss-French colleague, W. Rambal. They confessed to ISO's president at that time, Sir Jehangir Ghandy, of India, that they had, in fact, been accustomed to do so "for many years now."[22] The ISO Council selected Charles Sharpston, a bilingual British statistician who had worked in the government of Tanganyika, to take over St. Leger's job in September 1966.[23] During the eight-month interregnum, the council adopted the final NEDCO report, the implementation of which would require a major expansion of the Geneva secretariat and many fundamental changes in the way ISO carried out all of its work.

Sharpston did not end up leading that transformation. Despite his facility with the French language, he lacked Sturén's natural diplomacy. In early 1968, M. Rambal communicated a variety of grievances to Sturén and to AFNOR's Vincent Clermont, for which Binney graciously took the blame because he was the new secretary's countryman and informal tutor in the ways of the unfamiliar world of standardization.[24] AFNOR's official history reports that the "Englishman" was soon "unanimously" asked to leave.[25] In ISO's official history Roger Maréchal puts it this way: "After two years, Sharpston gave up. Then Olle Sturén came in [December] 1968. The real development was when Olle Sturén came. There was another spirit!"[26]

In his 1969 speech to the ISO Council, Sturén laid out a radical agenda that combined the NEDCO initiatives and his 1958 analysis of why most new standards needed to be international. He declared, "We should now start to publish International Standards" because both governments and consumers expected and demanded such standards and in field after field business was becoming internationalized. The organization needed to set its own agenda that reflected the needs of the global economy; "We must cease to think of ISO as the fire brigade that responds to national requests." Priorities included "office equipment" and "freight containers," and "ISO should take on work on water and air pollution, and noise levels, without delay." He argued that ISO could easily speed up its work: "We must recognize that, de facto, much of the work produced by ISO Working Groups," its term for technical committees, "[ends] almost immediately [when] the document is drafted," but current rules meant that it was often "only several years later that we are able to publish an ISO Recommendation."[27]

In 1972 Sturén described the new ISO to an American audience. "More than 50,000 experts are engaged in ISO work." Two thousand new international standards were in the pipeline, and that rate would double in four to five years. And while most of the work could be done by correspondence, 18,000 delegates had attended face-to-face meetings of technical committees in the previous year.[28] Coordination of work with national and regional standards bodies had increased dramatically. Many national standards bodies planned simply to endorse ISO standards, and "standards to be issued by CEN . . . will also be merely endorsements of ISO Standards whenever they are available. Then the [European] Commission in Brussels . . . intends to refer to [them] . . . in technical regulations it issues on behalf of member governments."[29] The same thing was happening, he noted, in Eastern Europe through the Council for Mutual Economic Assistance.

Sturén mildly chastised the Americans for lagging behind their European counterparts in their commitment to international standardization. The US national standards body, unlike those in Europe, rarely volunteered to host meetings, provide the secretariat of an ISO technical committee, or sometimes even to take part unless the subject was of "immediate commercial advantage" to American firms.[30] Similarly, the Unites States had yet to adopt the practice of writing government regulations with "reference to standards," the new European practice (described above) where official regulators did "not produce their own specifications, or develop their own test methods, but rather rely on the [private] standards organizations for this work."[31] Sturén believed that the American practice risked duplication of effort and could lead to official standards drafted by civil servants who were technically incompetent, beholden to a narrow set of business interests, or both.

This practice of encouraging governments and intergovernmental organizations to reference international standards in law was, in part, another significant legacy of Sturén's first ten years at ISO, perhaps the only one that was not anticipated in the NEDCO report or his early analysis of the ever-increasing need for international standards. Ironically, the European practice arose, in large part, in response to American concerns about nontariff barriers to trade—in particular, European standards for electronic components purchased by governments, standards that appeared to discriminate against US producers. The United States raised the issue at a 1970 conference on standards called by the UN Economic Commission for Europe and simultaneously at negotiations under the GATT.[32]

A logical solution to the problem, one that Sturén endorsed, was to level the playing field by requiring governments to make use of international, regional, and national standards in regulatory work—the more international the better. That way, all the companies that produced something referred to in legislation could be part of the process of deciding on the standard, or, at the very least (in the case of regulatory references to national standards) the rules would be transparent and created by a balanced group of experts rather than by nonexpert legislators beholden to special interests. The development of such a standards code became a topic for the Tokyo round of GATT negotiations that ran from 1973 to 1979, but, in the meantime, Sturén convinced ISO members to "put its machinery at the disposal of all parties interested in getting an international standard" and to support intergovernmental standard setting (such as that of the World Health Organization (WHO) in the pharmaceutical field), to help assure that standards referred to by regulators would not become barriers to trade.[33] In 1973–1974, ISO and the IEC developed a joint "Code of Principles on 'Reference to Standards'" that pledged the two organizations and their member bodies to "give special attention" to governmental and intergovernmental requests to set standards that could be referred to in specific laws or regulations. They pledged that the committees involved in setting requested standards would be "fully representative of the views of all interested parties: government, public authorities, producers, distributors, users, etc.," and they asked cooperating governments to refrain from issuing new regulations until a committee's deliberations were completed. Moreover, "wherever international standards exist, national authorities and intergovernmental organizations should refer thereto in their regulatory texts either directly or through harmonized national standards."[34] With the publication of the code, Sturén led the two international standard-setting bodies and their members to propose that they take on tasks far beyond what had been proposed by NEDCO. It also shifted private standard setting from its strongly voluntary roots toward creating standards that often became, in effect, mandatory by their incorporation into regulation.

Sturén's main method of pursuing his agenda as ISO secretary-general was face-to-face conversation, sometimes at home in Geneva over dinner cooked by his wife, Nalle, and served by his two sons, but more often with national standardizing officials in their own countries. He traveled almost every month of his first decade in office, visiting 60 countries, many of them multiple times. Fifteen were in Western Europe, five in Eastern Europe, nine in the Americas, three in sub-Saharan Africa, six in North Africa and the Middle East, and the rest throughout Asia and Oceania. He not only visited every major industrialized country and the most populous countries in the developing world, he also visited smaller places like Cuba, Mongolia, and Singapore (long before the latter city-state had become an economic powerhouse).[35] He was almost always accompanied by his wife, and often by his two sons, all of whom had diplomatic roles to play, something that Nalle loved. She recalled one party where the Russian delegates were standing off in a corner by themselves. She rounded up another woman at the party and the two of them began playing chess, although neither of them knew much about the game. Within minutes, the Russians were all gathered round and the evening saved.[36]

Perhaps Sturén's most important trips were to the People's Republic of China in 1976 and 1978, the second coinciding with Deng Xiaoping's rise to power and China's admission to ISO (figure 5.1). China's rapid engagement with international standardization was one of the first signs of the globally focused Chinese manufacturing economy that would begin under Deng.

At the end of Sturén's first decade in office, he reflected again on the situation in international standard setting. ISO now had a "de facto monopoly" on all international standard setting outside of two sectors—electrotechnical questions, handled by the IEC (and the standardizing committees of the International Telecommunication Union, which he did not mention) and pharmaceuticals, handled by WHO. Safety standards and other issues of direct interest to government regulatory bodies were outside their original scope, but ISO and the IEC were now putting their machinery at the disposal of governments trying to set international standards in these areas. In that way, ISO was becoming something more than a federation of its member bodies; it was an international organization that "stretches beyond the more traditional of its member bodies."[37] It had published more than 5,000 international standards, but given the many new fields that were opening up, including environmental protection, ergonomics, and energy conservation,

Figure 5.1. Visiting Beijing in 1978, Olle and Nalle Sturén pose with their host, Vice Prime Minister Kang Shi-En, who stands between them, and Chinese standardization leaders including Guo Lisheng (back row, far left), to whom we are grateful for identifying the date and context. Photograph courtesy of ISO.

Sturén thought 5,000 more were needed immediately.[38] Soon ISO would be producing 800-1,000 standards a year with the work of 100,000 experts on some 1,700 technical committees and their subcommittees, twice the numbers of committees and standardizers that Sturén had cited just five years earlier.[39]

One challenge that could prevent ISO from achieving its goal of an additional 5,000 standards within the next few years was the failure of some countries to harmonize their own views. Sturén gave as an example a UN "conference last year on the standardization of freight containers where some participants, particularly representatives of the developing countries, expressed views which do not harmonize with the attitude taken in ISO by their national standards bodies."[40] The industrialized countries presented a similar problem; a few standard-setting bodies such as the UN Economic Commission for Europe, whose mandates on topics such as automobile safety and building codes overlapped with those of ISO, created standards that could compete with those of ISO.[41]

The pressure for more international standards work would only continue. It was the consequence of the international economy that had been created by a revolution in transportation. "Today [in 1978], almost nothing is too cheap—or too expensive—to transport," because of ISO standard shipping containers, the subject of the next section.[42] But, of course, a global market cannot be created by new transportation technologies alone. Governments must establish common rules, and producers and consumers must find common standards—hence the increased demand for ISO standards. "Let me only say," Sturén concluded, "that we can claim, confidently, that no other set of international documents has such an influence, and such importance, for international trade."[43]

A Global Transportation Infrastructure: Standard Containers

Standardizers often proudly put forward ISO's standardized, intermodal shipping container as a great achievement of international standardization in the decades following ISO's founding.[44] Scholars of standardization have also frequently cited it as an example of a standard that has had global impact and that exemplifies various aspects of the standardization process.[45] In the 1970s, the Containerization Institute defined containerization as "the utilizing, grouping or consolidating of multiple units into a larger container for more efficient movement."[46] In 1973, a US container and transportation expert, Eric Rath, provided a slightly more elaborate definition: "Containerization consists of the simple application of temporary portable storage facilities loaded with cargo made mobile as a unit for intermodal unified transport."[47]

Rath argued that "containers did not revolutionize transportation because of a unique invention in technology." Indeed, quite the opposite: "First came the need for integrated transport coordination, and then the container was developed to fulfill this mission." Consequently the container is important not as a technology but as a "common denominator of a multitude of types of cargo," from raw materials to manufactured goods, and from rubber ducks to clothing to refrigerators.[48] To achieve this common denominator, all parties involved had to agree on the characteristics of containers and of the supporting infrastructure, a task suited to the voluntary, consensus-based standardization process. As Rath put it, "Container technology requires a community of interests. . . . It requires a system of mutuality . . . adjusted to international cooperative interests . . . to carry one product without re-handling over several systems of transportation mobility." Because containers were only one part of a larger system, he described containerization as "a systems approach to transport service." Waxing eloquent in his enthusiasm, he claimed, "It represents the dawn of an era, in which all modes of transportation will be subjected to integration into a single, worldwide system."[49] To make such a system, standards were necessary. Although even today, as in the 1970s when Rath wrote, a single monolithic system has not driven out all alternatives, ISO standard containers dominate intercontinental cargo transport worldwide.[50] The widespread adoption of standardized containers revolutionized world trade. This section outlines the process by which engineers worked, first in the American Standards Association and then in ISO, to achieve this important standard, and with what consequences.

The notion of combining cargo in standard-sized containers for more efficient intermodal transportation was not, of course, entirely new in the 1950s. Even before World War II, railroads had experimented with containers to reduce their costs, and in Europe the International Chamber of Commerce had formed the Bureau international des containers as early as 1933.[51] During the years immediately following the war, a few shippers experimented with ways to improve cargo-handling efficiency, including using roll-on/roll-off containers made from repurposed amphibious landing ships, but none of the experiments reduced costs significantly.[52] In postwar America, however, two groups had strong reasons to push toward containerization: the US military wanted to protect its cargoes from theft and damage, and US commercial steamship companies wanted to increase efficiency and reduce costs. In 1947, the US Army had developed small containers (8.5 feet by 6.25 feet by 6.83 feet) for transporting military supplies more rapidly, more safely, and with less pilferage. In the 1950s and 1960s, military CONEX (Container Express) boxes with similar overall dimensions (plus the option of two smaller modular units taking up the space of a single box) were used for moving the household goods of military families overseas.[53] The military found such containers highly efficient and effective, and it supported research and later standardization in containerization.[54]

Commercial interest in containerization among steamship companies came primarily from a desire to reduce bottlenecks and pilferage in harbors that made their shipping fleets less productive. In *The Box*, Marc Levinson vividly describes how "gridlock on the docks" slowed the loading and unloading of ships, and consequently world trade.[55] Inefficiency, graft, and strikes by union longshoremen loading and unloading a break-bulk or non-containerized cargo of more than 200,000 items for a typical ship resulted in port costs estimated to account for 37–49 percent of shipping costs, and shipping costs in turn accounted for up to 25 percent of the product cost. Ships spent up to half of every round trip sitting in dock being loaded and unloaded, thus decreasing efficiency of capital investments. This situation held for ships engaged in both domestic and foreign trade, but domestic competition from trucks and railroads created a serious competitive threat for domestic shipping first.

The breakthrough that led to the revolutionary modern containerization system came during the 1950s, with US shipping firms as pioneers. The two best known of these firms were Sea-Land Services (initially known as Pan-Atlantic Steamship Corporation) and Matson Navigation Company, though some smaller and less known firms also played important early roles.

One such firm, the relatively unknown Ocean Van Lines, has been credited with creating "the 'real' predecessor of the modern container."[56] In 1949, it had a contract to transport US Army supplies from Seattle to Camp Richardson, Alaska, and decided to establish regular service along the route using van-like containers on trucks and stacked two-high on steamships. Since no such containers existed at this time, the company turned to Brown Industries in Spokane, Washington, a maker of aluminum trailers, and ordered two hundred aluminum containers. Brown engineer Keith Tantlinger worked with its existing trailers, which were 8 feet by 8.5 feet by 30 feet, strengthening them for stacking and developing ways to attach the stacked containers. Ocean Van Lines finally launched the service in 1951, but it ran for only two years. The importance of this experiment lies primarily in the experience it gave Brown's Tantlinger, who developed containers and systems for one of the better-known firms, Pan-Atlantic Steamship (later Sea-Land Services), a few years later.[57]

In 1955 Malcolm McLean, president of McLean Trucking, bought out Pan-Atlantic Steamship Corporation with the idea of making the trucking business more efficient by shipping truck bodies up and down the East Coast more cheaply than they could be driven, then putting them back on trucks for the last part of their journeys.[58] Because the Interstate Commerce Commission (ICC) would not approve McLean Trucking's ownership of a shipping line, McLean sold his trucking firm and also bought Pan-Atlantic's parent company, the international carrier Waterman Steamship. McLean had initially imagined rolling entire truck bodies, wheels and all, onto ships but quickly became convinced that the wheels would take too much space. In casting about for a way to achieve McLean's vision of transportation by both land and sea using the same containers, one of his executives contacted Keith Tantlinger, still working for Brown Industries. Tantlinger met with McLean and proposed a way of doing it. In 1955 McLean received and approved an initial order of two 33-foot containers, a length chosen to allow good use of the leftover World War II T-2 tanker ships that the government was selling off at low prices. Then, as Levinson puts it, "McLean ordered two hundred boxes and demanded that the reluctant Tantlinger move to Mobile to be his chief engineer."[59]

Tantlinger became vice president of engineering and research at Pan-Atlantic later in 1955. By 1956 Pan-Atlantic was shipping trailer bodies filled with freight up and down the coast, though the ICC would not allow him to own the trucks to which they were then transferred for the land-based part of their journeys. By 1957, Pan-Atlantic had converted several C-2 steamships to pure container ships by developing its own 8-foot-by-8-foot-by-35-foot containers, cells for holding them, and shipboard cranes with corner lifting devices for loading them, and it began running its Sea-Land Service up and down the East Coast and along the Gulf Coast.[60] In 1960, Pan-Atlantic changed its name to Sea-Land Service, and a year later its cargo business based on its proprietary integrated 35-foot container system was profitable.

A second well-known pioneering firm, Matson Navigation Company, was based on the West Coast and shipped sugar, oil, and general freight between Hawaii and California. After successfully converting the sugar shipments to an automated bulk process with considerable cost savings, Matson became interested in reducing port and local transportation costs of its general freight shipping, too. It developed its own proprietary system based on 8-foot-by-8-foot-by-24-foot containers, designed by its internal engineering operation (headed by Leslie Harlander) to optimize space on its C-3 cargo ships, and built by Trailmobile, a manufacturer of truck bodies. It also designed and built dock-based (rather than shipboard) cranes, and its own corner fitting and lifting devices. Matson put this proprietary system into service in 1959.[61]

Meanwhile, other firms in the United States and elsewhere were also experimenting with containers. Alaska Steamship Company, which had carried Ocean Van Lines' 30-foot containers between Seattle and Valdez, Alaska, in 1951 and 1952, began to carry 24-foot containers between Seattle and Seward, Alaska, in 1956. Grace Lines developed 17-foot containers and an entirely different loading system to carry cargo between the United States and Latin

America on two refitted ships. When its first converted container ship arrived in Venezuela in 1960, however, the local longshoremen initially refused to unload it and made it clear that they would not allow container ships in the future. Another failed attempt, this time to use the ships on a route from the Great Lakes to the Caribbean, ended Grace Lines' efforts.[62] In 1964 the Australian firm Associated Steamships Ltd. launched a specially designed containership, *Kooringa*, with 17-foot containers, providing a service that included land transport at either end of the sea voyage between Melbourne and Perth. British Railways experimented with 27-foot containers that could travel by train or truck within the UK, and the British and Irish Steam Packet Company introduced container service initially between Ireland and the UK. By the late 1960s both would extend their services to continental Europe.[63]

By this time, many shipping lines saw containers as the future of intermodal cargo shipping, but it was obvious that the proprietary nature of the existing systems limited the growth of containerization and thus its cost savings. Standards were needed to address that problem, and the impetus again came from the United States. By the end of 1957 Vince Grey, a member of the American Standards Association, Herbert Hall, a retired Alcoa engineer, and Fred Muller, an engineer representing the Truck Trailer Manufacturing Association, began discussing the need for container standards.[64] The first formal move toward standards in the United States, however, was made not by ASA but by a powerful government agency, the US Maritime Administration (Marad), which administered subsidies and laws related to ships. Supported by the navy, in June 1958 it established committees to develop standards for container size and construction.[65] One month later ASA created its own committee to address intermodal container standards, Materials Handling Sectional Committee 5 (MH-5), with Herbert Hall as chair, Fred Muller as secretary, and 75 members.[66] This committee immediately asked Marad to dissolve its committees, as any standards created for containers would affect far more than just maritime interests, and representatives of these interests would need to be involved; moreover, standards would need to be international to be most useful, and ASA was the appropriate body to work with ISO toward that end. Marad refused to end its effort, but the ASA standardization process became the more important one going forward.[67]

ASA's MH-5 committee followed ASA rules and norms that required the participation of all affected stakeholders, and its 22-person Van Container Subcommittee included representatives of trailer manufacturers, trucking companies, railroads, government agencies, equipment manufacturers, and associations, in addition to two ocean shipping lines.[68] Maritime shipping was only one of several users of the container, and it ended up having relatively little influence on the dimensional standards debated by MH-5 and later by ISO, though demanding maritime conditions were important in determining strength requirements. The varying lengths of truck trailers allowed on roads in different parts of the United States as well as railroad regulations for car length and height shaped the subcommittee's size recommendations.[69]

In early 1959 at the first meeting of the MH-5 Task Force on size (part of MH-5's Van Container Subcommittee), six standard container lengths were initially suggested: 12 feet and 24 feet, 17 feet and 35 feet, and 20 feet and 40 feet. They represented two lengths then in use (24 feet and 35 feet), one new length based on the maximum length of closed railroad cars then allowed by US regulation (40 feet), and half lengths of all three.[70] In subsequent meetings of the task force and Van Container Subcommittee, the list was whittled down to 20 feet and 40 feet, and then 10 feet and 30 feet were added. Task force member and MH-5 chairman Herbert Hall argued that the ASA Standards Review Board would not accept the original diverse set of lengths. Moreover, Hall thought that 10-, 20-, 30-, and 40-foot containers would allow maximum flexibility and that the relationships among all four lengths, which could be combined in various ways, made them preferable to the other proposed lengths. As a retired employee of Alcoa, Hall was familiar with constructing but not loading and transporting containers, so he was insensitive to these costs, which were relatively larger for using more, smaller containers than for using fewer larger ones. This set of lengths, with an 8-foot-by-8-foot cross section, was approved by the ASA Standards Review Board in 1961.[71] ASA only published these dimensions as American Standards in 1965, along with standards for strength and lifting agreed to by other subcommittees of MH-5, but in the United States most players treated the dimensions as standards starting in 1961.[72]

Thus, the pioneering firms, Sea-Land and Matson, had invested heavily in what turned out to be nonstandard container sizes. They were only two firms within one of many industries that would purchase and use containers. The broad set of interests involved with intermodal containers, along with the requirement for stakeholder balance, prevented the two firms from determining the dimensions arrived at through the voluntary consensus process, which Tantlinger would later call a "dog fight."[73] Because the standards were voluntary, the two firms could continue to use their nonstandard containers within their closed systems; nevertheless, these containers could not become part of the global flow of standard containers. To avoid losing the ability to bid for US military contracts, the two firms persuaded Congress to amend the Merchant Marine Act to prevent it from giving preference to container services using ASA's new American Standard dimensions.[74] This concession was an important one, particularly for Sea-Land, which in 1966 won major contracts to transport supplies for US troops in the Vietnam War, first to Okinawa and then to Vietnam. It won these contracts, in great part, because McLean himself went to Vietnam and oversaw the establishment (at great cost to Sea-Land) of terminals, cranes, and trucks that would allow container use.[75] His risk paid off, as by 1973 Sea-Land was earning annual revenues of $450 million from the Defense Department, using its 35-foot containers.[76] In the 1970s Sea-Land gradually moved toward 40-foot containers, initially carrying a few 40-foot containers on the decks of its ships filled with 35-foot containers, to ease its ultimate transition to the new standard.[77]

At the same time that the MH-5 committee was debating dimensions in the United States, Vince Grey was urging ISO to establish an international committee on container systems. In 1961 ISO's newly created Technical Committee (TC) 104 on Freight Containers, with ASA as its secretariat, held its first meeting in New York.[78] Recall that even though Olle Sturén certainly had other things to do as head of SIS and a leading promoter of CEN (but not yet secretary-general of ISO), he took part in these initial meetings, a clear sign of how important Europeans considered this work. TC104 was interested primarily in performance and interoperability standards; thus, dimensions necessary for interoperability were its first order of business. US regulatory limitations on size (especially length) of containers to be carried over roads were more restrictive than those in Europe. Thus, the US delegation proposed the dimensions they had already agreed to in MH-5: lengths of 10/20/30/40 feet and a cross section of 8 feet by 8 feet.[79] Continental European railroads already used shorter containers that were slightly wider than 8 feet (and measured in metric units) and wanted them to be declared standard, too. An initial compromise added two shorter lengths (5 feet and 6.5 feet) to 10/20/30/40-foot lengths in the Series 1 standard, which was published as ISO/R 668 in 1968. Ultimately, however, the two shorter lengths were dropped.[80]

Two other series were approved by TC104, as well. In 1968, the smaller standards of the International Union of Railways were declared ISO Series 2 but published as a technical report, rather than a standard, since they "were 'intended essentially for internal continental systems.'"[81] In the late 1960s TC104 began work on Series 3, based on small containers used in the Soviet Union. Series 2 and Series 3 were never taken up outside their original regions, however, and were ultimately dropped from the standard. A 1997 interview with Vince Grey, the US standardizer who was initially secretary of ISO's TC104, then head of the US TC104 delegation, and finally chairman of TC104, as well as being involved in ASA's MH-5 committee at the very beginning, illuminated TC104's strategy around the three series of dimensions. He explained that "one of the first problems we ran into was various countries trying to have the international standard reflect their own national practices. We really didn't want to do that. We weren't just looking to affirm what existed, we were creating something new." He thought the leaders handled this problem well. When the European railroads and trucks proposed their existing container sizes, "instead of locking horns, and saying 'no way!', we accepted these container sizes and called them Series 2 containers." Similarly, when "the Russians wanted their Eastern European sizes put in . . . we called them Series 3." The initial ISO standard included all three series, but there was no market for Series 2 or 3 containers, so they were ultimately dropped. He argued, "That was a most tactful way to let something like that happen! It's better to get on with the work as long as you can achieve the basic goal and let the merits of each series be judged by the users."[82]

The battle over container lengths was not the only standards fight in TC104 and MH-5, and arguably not even the most interesting one. The second major standards issue concerned the corner fittings used to stack containers securely—whether on shipboard, train, truck, or in storage—as well as to lift and move them. Standard corners were essential to interoperability. As Levinson explains, "Every company had financial reasons to favor its own fitting. Adopting some other design would require it to install new fittings on every container, to buy new lifting and locking devices, and to pay a license fee to the patent holder."[83] Pan-Atlantic/Sea-Land had patented its corner fittings and system for locking its containers together and lifting them, while National Castings Company had patented its Speedloader system with a corner fitting resembling the one used by Matson.[84] Grace Lines adopted the National Castings system, as did, for a short period, the British Standards Institution.[85]

The issue of what to do about patented technology was a tricky one, as standardizers avoided incorporating patents in standards when possible, to guarantee broadest possible use.[86] According to Levinson, MH-5 chair Hall told the task force struggling with the corner fittings that patented technology could be included "so long as it was in widespread use and was available to all for a nominal royalty."[87] The MH-5 task force on corner fittings was chaired by Keith Tantlinger, the engineer who had developed Sea-Land's patented fittings when he worked as an executive for McLean several years previously and who had since moved to Fruehauf Trailer Company. After extensive argument, a majority of the task force members thought that the Sea-Land corner was technically superior and should become the standard if McLean would release the patent unconditionally so they could alter its dimensions.[88] Tantlinger then approached his old employer and asked him to release the patent so it could become part of the standard.

In 1963 McLean agreed to relinquish the patent, allowing ASA to change it as needed.[89] He had no reason to do any favors for ASA, given its 1961 rejection of his 35-foot containers in favor of the 10/20/30/40-foot standard. In addition, because the fittings would have to be changed dimensionally for the standard, Sea-Land's equipment would still need to be modified; moreover, Sea-Land continued to use its nonstandard containers through the 1970s. Consequently, in the short term McLean's release of the patent did not benefit him at all. Standards scholar Tineke Egyedi notes two other cases of parties giving up patent rights for container standards (a patent for a twist-lock lifting mechanism released by a container manufacturer, Strick Trailers, and one for a system for container identification made available on a nonexclusive, royalty-free basis); she also states, "Such company reactions are sparse in other standards processes."[90] Although it is impossible to know how often such actions took place in other standards negotiations, the surrender of the Sea-Land corner patent (and the other two patents) shows that, at its best, the consensus standardization process and the standards movement ideology encouraged players and even non–committee members like McLean to act for the common good. Having a standard—even when it is not the one that would most benefit a particular company—would allow even a short-term loser in a standardization process ultimately to benefit from the market created by the standard. And, indeed, Sea-Land became a large player in a flourishing industry in the long term.

Even with the release of the Sea-Land patent, the process was not over, either in ASA's MH-5 or in ISO's TC104. National Castings continued promoting its own corner fitting until it was bought out by another firm, and although a majority favored the Sea-Land corner, the MH-5 committee had not yet reached consensus as the next TC104 meeting approached.[91] In September 1965, to reach a recommendation in time for the upcoming meeting of TC104 that opened at The Hague a few days later, ASA's Standards Review Board acted without full MH-5 committee consensus (an unusual procedural move that required waiving rules) and approved the Sea-Land fitting with modifications.[92] Normally, the recommended design with technical drawings would have been distributed to TC104 members at least four months before the meeting, but TC104 agreed to waive that requirement at the meeting at The Hague. "Three high-ranking corporate executives—Tantlinger, Harlander, and Eugene Hinden of Strick Trailers—then retreated to a railcar factory in nearby Utrecht, where they worked with Dutch draftsmen for forty-eight hours nonstop to produce the requisite drawings."[93] Then the UK delegation withdrew its opposition (which was based on its recommendation of the National Castings design for its national standard), allowing TC104 to approve it.[94] Several engineers in a hotel room at the TC104 London meeting two years later performed yet another ad hoc redesign, this time to support the loads and stresses TC104 had since defined as performance requirements for intermodal containers. This subsequent redesign required containers built since 1965 to have new corner fittings welded into place.[95] Finally, however, in 1967 TC104 had agreed to dimensions and corner fittings. In June they were sent out to member countries for approval.[96]

Only in 1970 did ISO finally publish the first complete set of container standards—including dimensions, corner fittings, and other elements—as an international standard.[97] But adoption momentum preceded publication. In the late 1960s, the Maritime Subsidy Board of Marad introduced an efficiency measure into the criteria for awarding ship construction subsidies, encouraging American shipping firms to embrace containerization if they had not yet done so; consequently, by 1970, most American shipping lines (with the exception of Sea-Land and Matson) had begun to use the new standard containers.[98] By this time, international shipping firms on global trade routes were also adopting standard containers as a result of competition. In the late 1960s Sea-Land had entered the transatlantic shipping trade with its proprietary container ships, creating a serious competitive threat to international shippers from the United States and other countries, driving adoption of standard containers in global shipping.[99] By 1978, 500 container ships plied the waters worldwide (of which 104 were US ships), and 1.5 million containers circulated.[100] Even after converting most of its fleet to standard containers, Sea-Land continued to use its 35-foot containers in shipping to the Caribbean and South America until Brazil refused to handle the nonstandard containers, at which point it finally converted the last of its ships to standard 40-foot containers.[101]

The correlation between container adoption and the skyrocketing trade growth from the mid-1960s to the mid-1980s has led many to attribute much of the enormous growth in global trade since World War II to containerization. Between the 1950s and the 2000s, cargo-loading costs dropped by a startling 93 percent.[102] Reduction in loading costs was not, however, the only factor powering the revolution in global trade. Indeed, some economic studies have argued that oil price rises more than made up for cuts in cargo loading costs, resulting in overall rising ocean shipping rates, and that tariff reductions increased trade by three times as much as declines in transportation costs.[103] More recently, however, economists Daniel Bernhofen, Zouheir El-Sahli, and Richard Kneller estimated the actual impact of containerization on world trade, using detailed data on trade between pairs of nations and sophisticated statistical methods.[104] They found that adoption of ISO standard containers accounted for much of the increase in trade in the 1970s and 1980s, and much more of the increase in trade than GATT membership did. The developed countries experienced the biggest gains, and the intermodal use of containerization was particularly important. This result suggests that, by developing this standard, the engineers of ASA and ISO enabled the creation of a global cargo transportation infrastructure, leading to a great expansion of global trade, albeit one that benefitted some countries more than others.

Egyedi identifies standard containers as a *gateway technology* connecting multiple transportation subsystems into a major system.[105] She argues, however, that container standards embody modal and geographic bias. Consistent with the economists' findings that the greatest effect was on North-North, multimodal trade, Egyedi argues that the ISO container standards favored US and intercontinental transportation that included deep sea transport (reflected in ISO container strength standards, which led to heavier containers) over truck-train land transport.[106] In continental Europe, rail and truck transportation tended to forgo ISO containers for much smaller and lighter containers known as swap bodies, which Egyedi identifies as a competing gateway in continental Europe; however, newer swap bodies have adopted corner fittings that work with container handling equipment for ISO standard containers, indicating emerging compatibility between the two gateways. Moreover, she points out that developing countries in the UN Conference on Trade and Development's Ad Hoc Intergovernmental Group on Container Standards, founded in 1974, complained that TC104 was dominated by experts from developed countries and always met in those countries, factors that led the committee to ignore the needs of many developing countries where investment in the roads and ports needed to benefit from containerization was unlikely.[107] Thus, voluntary adoption of ISO containers was slower and less complete in developing countries, lessening the effect of containerization on North-South and South-South trade.

In recent decades, nonstandard containers for special purposes have proliferated. For example, the Stora Enso Cargo Unit for bulk goods and Geest Line containers are both 45 feet long and have nonstandard heights and widths, and United Brands uses 43-foot containers to transport fruit. With changes in US regulations, trucks in the United States and Canada regularly use 53-foot containers now, and one shipping line, OCEANEX, operates a shipping vessel designed for this size.[108] These longer containers make for less-expensive domestic shipping, but most of the oceangoing container fleet is locked into the less efficient (for North American domestic shipping) 40-foot containers.

How important is very widespread adoption of standard intermodal containers? Hans van Ham and Joan Rijsenbrij, transportation experts at Delft University, recently argued that use of these nonstandard containers and systems reduces global cost efficiency and interchangeability, and that the cost for handling them is covered by the economies of scale realized by standard containers. "Maybe, after some 50 years there is again the need for some powerful visionists who set standards for the next 50 years."[109] They also observe that the grand compromises that led to the ISO standards came from committees including top executives in charge of technical or business operations at major firms, and they speculate that the absence of such high-level individuals on these committees today may contribute to the loosening hold of international standards. They exhort, "It should be remembered: '*Standardization is nobody's best, but better for all.*'"[110] A standards coordination manager at British shipping company Overseas Containers Ltd. had a slightly different perspective in 1977, however, that reflected a more complex understanding of the trade-offs between standard and nonstandard containers:

> At their best, standards are imperfect compromises and there will always be cases for which a given standard is less than satisfactory. . . . There will always be "special cases" for which a non-standard treatment will be the best solution for all concerned and it is to be hoped that nothing will be done to make the non-standard treatment more difficult than it need be.[111]

Figure 5.2. This 2016 photo of a workaday container ship epitomizes the world economy enabled by ISO standards. Shared on Pexels.com by Israel Garcia, accessed August 31, 2018, https://www.pexels.com/photo/container-container-ship-ferry-gate-69540/.

Whatever the optimal balance between standard and nonstandard containers, Bernhofen and his colleagues have shown that the adoption of standard containers increased trade significantly. Although global trade has not always been seen as an unalloyed positive, it (and consequently the container) has certainly shaped our contemporary world, as the familiarity of figure 5.2 attests.

A Failure of International Standards: Color Television

If the intermodal shipping container demonstrates a successful example of international standardization creating a global market, the failure to achieve an international standard for color TV during the 1960s demonstrates what can go wrong in such an attempt. This case involves a larger set of standard-setting bodies, including governmental and hybrid public-private bodies, as well as private ones, illuminating some of the strengths of private voluntary standardization by contrast. To understand the failure to reach international color TV standards, we must briefly consider monochrome television standards, as the lack of an international monochrome standard contributed to the later failure to achieve color television standards. Both failures reflected problems that emerged when individual nations created standards and installed quasi-public network technology first, before international standardization was attempted. Issues of backward compatibility created further problems during the move from black and white to color. Finally, in a hybrid, public-private international standards body, the International Telecommunication Union's International Radiocommunication Consultative Committee, national and international politics became a factor, preventing the attainment of even a Europe-wide standard for color television.

Failure to Achieve Monochrome TV Standards

The key component of early television sets was a cathode ray picture tube with an electron gun at the back that systematically scanned a phosphorescent screen at the front, going back and forth across the screen from top to bottom creating hundreds of thin lines that formed the picture. One of the primary issues for standardizers was how many lines should be used to fill the screen. More lines gave you a sharper picture, but each line might make the tube cost a little more.

Monochrome television was initially developed and standardized within a few countries before and immediately after World War II. In 1936, Britain's noncommercial, government-chartered British Broadcasting Corporation (BBC) began broadcasting in black and white using a 405-line standard for scanning.[112] After several years of experimentation with and dissension on the optimal technical parameters in the United States, in 1940 the Federal Communications Commission, which had regulatory authority in this area, assembled the National Television System Committee (NTSC), an expert technical committee, to consider various proposals and arrive at a consensus recommendation—in effect, a hybrid of voluntary, committee-based standard setting and government regulation. In 1941 NTSC recommended a standard including 525-line scanning, and the FCC approved it, though World War II intervened, delaying implementation.[113] In 1946, a Soviet engineer developed a 625-line system that built on many aspects of the American 525-line system. France had tried several scanning standards before the war, and during the war the occupied country used the German 441-line system, but in 1948 French television pioneer Henri de France developed a system with 819 lines.[114] With this variety of national systems, standards were needed to allow transmission across national lines.

Attempts to achieve international standards for monochrome television systems occurred within the ITU's CCIR. After World War II, that hybrid public-private standard-setting body expanded its coverage to include television as well as radio broadcasting. When the CCIR held its Fifth Plenary in Stockholm in 1948, the 405-line, 525-line, 625-line, and 819-line systems were all presented to Study Group XI, newly formed to focus on television broadcasting, as potential international standards. The French delegation supported its distinctively French 819-line standard. Philips, the Dutch firm, along with its ally RCA (Radio Corporation of America), promoted the Soviet 625-line system because it was close enough to the American 525-line standard (differing only in the frequency of the current) to allow for a quasi-transatlantic standard that would allow those manufacturers to produce for a much broader market.[115] No agreement on the scanning system was reached at that meeting, nor at the Zurich Study Group XI meeting in 1949. At the Zurich meeting, the US delegation backed its own 525-line standard (despite RCA's lobbying), Great Britain supported its 405-line system, France supported its 819-line service, and most other nations supported the 625-line system.[116] Considerable research and exchange of information as well as a London meeting in 1950 did not resolve the impasse, with Study Group XI recognizing "the impossibility of arriving at a unanimous agreement on some of the questions which were on its agenda given the existence in many countries of public services of television using different standards and the great number of sets in service."[117] Even a last-ditch effort at the 1951 Geneva CCIR plenary failed, and the study group chair, Erik B. Esping from Sweden, later reported the disappointing outcome: "I made every effort to get the Study Group to decide on a recommendation on one single standard for black and white television. But my efforts were in vain."[118]

Color TV Standards: The Emergence of the US NTSC Standard

Although efforts to set international standards for color television started at an earlier stage of technical development than those for monochrome television—indeed, immediately after the failed attempt at monochrome standards just related—they were also unsuccessful in creating a single standard. Again, different standards emerged in different countries, competing in a story that reflects desire for backward compatibility with black and white television, corporate interests, and national and international politics. The first national color TV standard to emerge, the subject of this section, was the US NTSC standard, which itself was the product of considerable contestation. In the United States, the FCC again held regulatory authority over standards for this network technology. Historian Hugh Slotten has argued that the story of the NTSC standard reflected "the relationship between standardization and commercialization, the tension between advocacy and objectivity, and the role of technical experts and technical evaluation in policy making."[119]

In color TV, the major problems of standardization involved not only the number of lines that made up the picture but also how a complete picture would be built up from three separate images, one each in red, green, and blue light (e.g., by transmitting a rapid sequence of three full screens or "fields" of different colors, a sequence of different colored lines, or simultaneous neighboring dots of different colors). Finally, compatibility of color signals with monochrome televisions was also an important issue—that is, whether color broadcasts could be viewed in black and white on monochrome TVs.

In the United States, the Columbia Broadcasting System (CBS) proposed a field-sequence color television system as an alternative to the monochrome system (which was supported by RCA and other equipment manufacturers) and requested UHF (ultrahigh frequency) channels on which to broadcast it, but the FCC did not authorize its request in 1941, when the agency endorsed the 525-line monochrome standard.[120] After the war ended, CBS again requested approval of its system, and in 1946, after it demonstrated the field sequence system's capability to broadcast live, the FCC agreed to consider it. Meanwhile, in anticipation of CBS's renewed request, RCA had worked to create an alternative simultaneous-scanning color television system, this one compatible with the already accepted monochrome standard. It presented this system to the FCC, to evaluate at the early 1947 hearings along with the CBS system.[121] RCA used the demonstration of its system, which could not yet broadcast live and still needed much development, to raise doubts about and show that there were alternatives to the CBS system, successfully preventing its approval by the FCC. This rejection—along with an FCC freeze on expanding VHF (very high frequency) channels because of interference and the consequent opening of UHF channels to monochrome broadcasting—led CBS to pause development of its color system in order to fight its way into the monochrome market before it lost that market, too.[122]

Meanwhile Congress, particularly Senator Edwin Johnson, Democratic chairman of the Senate Committee on Interstate and Foreign Commerce, pressured the FCC to stop delaying color television, accusing it of giving in to RCA's influence and failing to prevent its monopoly control.[123] Under this pressure, the FCC called another set of hearings in 1949–1950 to determine which of three systems now proposed should be made standard for color television: CBS's improved field sequence system, a new RCA proposal for a dot sequence system, and a line-by-line system developed by Color Television Incorporated (CTI). During the same period, Sen. Johnson also convened an independent advisory group chaired by Edward U. Condon, director of the Bureau of Standards, to study the issue; this advisory group issued a comprehensive report to the FCC evaluating the three systems on nine different dimensions without making a final recommendation.[124] In 1950, disturbed by where the hearings were headed, industry experts decided that they should reestablish the NTSC committee to advise the commission on a standard. Now, however, the FCC, under congressional pressure, was not willing to give this industry-oriented body the authority that it had granted it in establishing monochrome standards; indeed, the FCC questioned its chairman, Walter Baker (who had also chaired NTSC during the monochrome standards effort), quite skeptically.[125]

The FCC considered the evidence from its hearings and the Condon report over the summer of 1950, and in September it released its own report favoring the CBS proposal as standard.[126] It rejected RCA's proposal for unsatisfactory image color and texture, exceedingly complicated television receivers, and greater susceptibility to interference; and it rejected the CTI's proposal for line crawl and texture problems, inadequate compatibility with monochrome, and unduly complex equipment.[127] Many of the industry engineers who testified opposed the CBS system because of inferior resolution (compared to monochrome) and incompatibility with monochrome television.[128] In response, the FCC simply noted that although modifying monochrome televisions to allow reception of color programs in black and white would be expensive, the longer they waited, the more monochrome televisions would need to be modified. In October 1950, the FCC declared the CBS system the new standard, a declaration that RCA and other industry manufacturing firms actively opposed.[129] Soon after, members of the Radio and Television Manufacturers Association met without CBS and agreed not to support the CBS standard, with RCA announcing that it was filing suit to overturn the FCC decision.[130] RCA unsuccessfully fought all the way to the Supreme Court; in May 1951 the court upheld the FCC's authority to set standards in this arena and rejected charges of bias and violating the public interest.

Barred by a lower court ruling from broadcasting under the new standard until the Supreme Court case was decided, in June 1951 CBS established color broadcasts and set up a manufacturing subsidiary to make color television receivers. It was alone, however, and lacked buy-in from the rest of the industry on this standard. Within months, Korean War materials restrictions prevented CBS from ramping up production of color televisions, and in October 1951 it discontinued color broadcasting.[131] The FCC's declared color television standard was now effectively dead, demonstrating why stakeholder consensus was so important for private standardization and even, in this case, for governmental regulation. RCA and other firms had been able to block the efforts of CBS to establish its color TV standard, simply by staying out of the market.

Meanwhile, although the FCC did not support reestablishing the NTSC expert body, industry members reorganized it themselves and began work on a new color television standard, this time with the support of RCA and other industry firms that opposed the FCC's standard.[132] They worked with and improved RCA's compatible (525-line) dot-sequential system, eventually impressing the FCC to the point that, in July 1953, it agreed to evaluate the newly proposed NTSC standard. After two major successful demonstrations, and after CBS agreed to follow the rest of the industry, the FCC approved the NTSC standard and set detailed rules and regulations for it in December 1953.[133] Color broadcasting using this standard began soon after. The industry-based NTSC, which was disbanded in February 1954, was not an ongoing voluntary standards body, although it had elements of such bodies, including membership of engineers representing manufacturers and users of the standard, and multiple technical committees focusing on specific technical issues. Although short lived, NTSC gave its name (or at least its acronym) to the American color television standard as the issue moved into the international arena, where consensus would be impossible to achieve.

Color TV Standards: The Search for a European Standard

Europe lagged the United States by about a decade in adopting color television. In 1955, at a Brussels interim meeting of CCIR Study Group XI (the technical committee on television, chaired by Sweden's Erik B. Esping), Charles J. Hirsch, one of the architects of the NTSC system, gave a presentation on it to interested members, jump-starting the conversation in Europe.[134] In 1956, study group members visited the United States to see demonstrations of the NTSC system, then visited several European countries to observe their experimentation with color television.[135] Subsequent developments would lead to a three-way competition within the standardization process that ultimately failed to achieve a single European standard, much less a global one.

As European media historian Andreas Fickers argues, "What started as a scientific endeavor to determine the best colour television system for Europe slowly but surely mutated into a fierce techno-political controversy between the major stakeholders."[136]

France entered the competition first within Europe. In the late 1950s, Henri de France, the inventor of the 819-line monochrome system, developed a variation of NTSC, called Séquentiel couleur à mémoire (sequential color with memory), or SECAM. By then, France had decided to switch from 819 to 625 lines in monochrome television to ease its problems of conversion within Europe, so SECAM was designed to work with 625 lines.[137] It transmitted sequentially by line, with the previous line staying in memory to be viewed with the new one. At this time Gaullist France, which sought to create a strong Cold War identity independent of the United States, backed SECAM as a national champion. The government wanted to invest in developing French capabilities in this technical area and to use standards as nontariff barriers to nurture France's color television industry. France would also gain political prestige and royalties from licensing its patents if a French system became an international standard. Compagnie française de télévision (CFT) was formed to develop the SECAM patents. The French state unified internal support for SECAM by cultivating the other French television equipment manufacturers to overcome their opposition and by ordering the French government broadcasting agency to support SECAM, despite its doubts.[138] Then France and CFT began promoting it elsewhere in Europe. France claimed SECAM had technical advantages over the NTSC standard (an argument disputed by RCA and other US supporters of NTSC) and was compatible with European 625-line monochrome broadcasting. In addition, it presented SECAM as the European alternative to the American NTSC standard, appealing to "the shared European fear of overdependence upon American technology."[139] Soon, however, this latter strategy was challenged when Germany developed another European alternative.

In 1962, Dr. Walter Bruch of A.E.G.-Telefunken developed yet another variant of the NTSC system for 625-line broadcasting, strongly influenced by SECAM and incorporating an additional German patent, called PAL (for phase alternating line).[140] This system threatened France's strategy of positioning SECAM as the only European alternative to NTSC, as well as threatening the royalties France hoped to get from SECAM if it were adopted as a European or world standard. Technically, the differences among NTSC, SECAM, and PAL were relatively small—95 percent of the components were the same.[141] Indeed, PAL drew on SECAM as well as NTSC patents, although the CFT and A.E.G.-Telefunken would only reach a settlement for relatively small royalties to the CFT several years later.[142] Yet the danger to France's SECAM champion strategy and its national pride was severe. This high-stakes competition unfortunately played out in the CCIR's standardization process.

In late 1962 the European Broadcasting Union, trade association for broadcasters in Western Europe, established an ad hoc commission to evaluate three systems—NTSC, SECAM, and PAL—for adoption as a standard by Europe. The commission reported to the CCIR on the strengths and weaknesses of all three technologies under various conditions but did not recommend one; it believed that having a single standard was much more important than which one, and that no one system was best in all situations.[143] In early 1964, a subgroup of CCIR Study Group XI, led by chair Erik Esping, met in London to try to select a single European standard from among the three standards, as a first step toward a broader global standard. The only decisions to come out of this meeting, however, were those by individual countries to continue research.[144]

The next opportunity to reach agreement was the CCIR interim subgroup meetings in Vienna in March–April 1965. France was worried that SECAM would have difficulty competing with PAL, given the strong reputation of the German electronics industry. Thus, in the lead-up to the Vienna meetings, France negotiated a (not-very-favorable) techno-scientific agreement with the Soviet Union to convince it (and its Eastern European satellite states) to support the SECAM system, strengthening its position in the CCIR.[145] The USSR saw the political advantage of defeating the American system. Moreover, because components of the American system's videotape recording equipment had potential defense applications, the US government had delayed and restricted RCA's demonstration of NTSC in the Soviet Union and had initially refused to allow the system to be sold there. Less than a week before the Vienna conference (and well after the Franco-Soviet agreement had been made) the US government informed the USSR that it would authorize export of the entire NTSC system to that country in exchange for support of the NTSC standard in the meeting—no doubt too little and too late from the Soviet point of view. A few days later, the governments of France and the USSR announced their cooperative agreement to adopt SECAM.[146]

In the opening sessions of the Vienna interim study group meetings, CCIR director ad interim Leslie William Hayes urged Study Group XI to reach a decision, pointing out that the group had adopted a question and two study programs around color TV a decade ago, in 1955, and had unanimously agreed a year before that a final review and final recommendation would be made "at the meeting of Study Group XI to be held in Vienna in the spring of 1965." Hayes concluded, "Well, here we are, Gentlemen."[147] After this exhortation to action, he also reminded them that, "as C.C.I.R. is a technical organization, its recommendations must necessarily be based primarily on technical considerations." Unfortunately, the relevant technical considerations depended on what criteria delegations weighed most heavily, and those factors turned out to be economic and political. The announcement of the Franco-Soviet agreement just before the meetings conveyed that these two nations had made their decisions before and regardless of the CCIR deliberations, generating considerable resentment. Moreover, the French delegation to the meetings was led by diplomats and government ministers, a highly unusual occurrence for CCIR study group meetings, injecting a more political tone into the negotiations, challenging the role of the technical experts, and creating further resentment, even among the French technical experts.[148] In an attempt to unify the opposition to SECAM, the delegations from the United States and the Federal Republic of Germany proposed to merge NTSC and PAL, by highlighting their common elements and leaving the decision to use the uncommon elements up to the individual company. The result, whether among two or three options, was a very small majority of votes for SECAM but certainly no consensus, and the meeting ended in a standoff.[149] The next opportunity for CCIR Study Group XI to overcome this disappointing but unsurprising impasse was the 1966 CCIR plenary meetings in Oslo.

Before the Oslo meeting, the British (despite opposition from the BBC) and Dutch delegations had both decided that they would support PAL, because the three-way vote at the Vienna meeting had shown that within Europe PAL was a stronger competitor than NTSC against SECAM, which the British, at least, viewed as the worst choice.[150] Meanwhile, the Soviet Union had proposed to the French yet another variant, referred to as SECAM IV, which was in some ways a compromise between SECAM and PAL. France objected to the Soviet Union supporting the new (and less developed) system, claiming it was a violation of their agreement. After much debate, the Franco-Soviet alliance agreed to support SECAM III (the current version of the French SECAM) at Oslo, unless it appeared that SECAM IV could get consensus.[151]

Three months before the Oslo CCIR Plenary, Study Group XI chairman Erik Esping addressed an IEEE meeting in the United States, discussing the status of the debate and its current impasse. In this Cold War era, he argued "that it is in the interest of all nations participating in this work to let the unity of the world, or at least of Europe, win the game in Oslo." Languages always created barriers, but, "why should not, at least, scientists and technicians do their utmost to create one single and uniform system for color television in the world"?[152] Esping focused on achieving at least a common *European* color TV standard, apparently doubting the possibility of achieving a worldwide standard. At this point the United States had 5.5 million color television sets (expected to double by 1967), and other established users of the NTSC standard such as Japan already had so much invested in that system that they were also unlikely to change if NTSC was not chosen. And although it was technically possible to use the NTSC system with a 625-line standard, around which Europe was finally converging, European countries wanted to show their technological capability and realize economic benefits to their own manufacturing industries.[153] Even that was not to be.

In Oslo, political representatives again mixed with technical experts, especially in the French contingent.[154] Esping set up a small subgroup of key (mostly European) states to try to work out a compromise. It included France, Italy, the Netherlands, West Germany, the UK, the United States, Switzerland, Czechoslovakia, the Soviet Union, and Yugoslavia.[155] The subgroup first concluded that no compromise could be reached for a single worldwide standard, since "the system already in public service in countries using the 525 line standard was not generally acceptable elsewhere," thus eliminating NTSC as the standard and assuring at least two systems worldwide. Then the subgroup's members sought a compromise for Europe, especially since several countries planned to introduce color television service in 1967. France declared that it was willing to support SECAM IV if all other European states agreed on it as the European standard and agreed to put all their efforts into developing it to the point of commercialization. Over several days and in communication with their governments, Germany and the UK held firm to PAL, in spite of pressure from Esping and accusations that they were blocking a unified European standard. They were unwilling to delay their planned 1967 introductions of color TV until SECAM IV was ready and unwilling to commit to developing it rather than PAL. Thus, France withdrew its support of SECAM IV, and the subgroup effort failed, sending the debate back to the broader study group.[156]

Ultimately, the European countries split between SECAM III (France, the Soviet Union, and other Eastern European countries) and PAL (all the Western European countries except France), while the United States, Japan, Canada, and a few others kept their allegiance to NTSC.[157] No consensus was reached. France had assured itself a market in the USSR and Eastern Europe, preventing a recurrence of its isolation in the 819-line monochrome system, but it had not gained a Western European market for its industry.[158] As Esping's report was summarized to the plenary, although it "was one of the problems which had been most thoroughly studied within the history of the C.C.I.R.," the study group had been unable "to make one Recommendation on not more than two systems for the whole world or for one single system in Europe." Esping himself "very much regretted that this had not been possible."[159]

Clearly politics, especially France's decision to openly politicize the process, proved a barrier to compromise. This politicization was made possible in part by the hybrid public-private nature of the CCIR standards body, with its delegations chosen by national governments rather than by technical associations. Within Europe, France accused West Germany and the UK of blocking European unity, and Germany and the UK accused France and the Soviet Union of turning a technical issue into a political one through their alliance.[160] Backward technical compatibility with black and white televisions played a small role, and thus so did the earlier failure to achieve a single monochrome standard.[161] The Oslo subgroup determined initially that it was impossible to get a single worldwide standard because the monochrome 525-line standard had shaped the NTSC color standard, making that option incompatible with existing European monochrome systems. Of course, European unity on a 625-line standard, as well as better compatibility with the United States, Japan, and other countries using NTSC standards, could have been achieved had issues of national politics and economics not dominated the two key conferences and the actions leading up to them. Achieving worldwide standards was challenging in the postwar climate, though chapter 6 will show that it was possible. This failure highlights how impressive the many successes, such as with shipping containers, actually were.

National Regulation and International Standard Setting

The similarities and contrasts between the container and color TV cases are striking. The similarities include the central importance of competing firms in American standardizing and the significance of European interests, whether conflicting or shared, national or regional. Also, in both cases, Swedish standard setters (Sturén and Esping) strongly supported the international standardization effort. Sweden was a highly industrialized country with a small national market that had a strong interest in frictionless trade with other industrialized countries.

The differences were significant, as well. In the container case, the competing American firms were the only ones that already had significant businesses based on different standards; if they could reach agreement, international standardization would be much easier. With color TV, in contrast, national interests came strongly into play. With containers, both national and international standards bodies were private and voluntary, with national delegations in ISO appointed by national technical bodies, not by governments. From the beginning, legislatures and national regulatory bodies were involved in setting standards for television, however, because in many countries governments supported the development of the technology and controlled broadcasting, and in all countries governments allocated the frequency spectrum that made broadcast television possible. Although the CCIR prided itself on operating as a technical body, the ITU was an intergovernmental organization, and thus delegations were ultimately under the control of governments, opening the way for the French to follow a nationalist agenda.

The color TV case demonstrates that governments and regulatory systems could impede international standard setting. Nonetheless, just because governments had the power to impede international standard setting, they did not necessarily do so. In the second half of the twentieth century, regulators often encouraged international standards. In a series of lectures delivered throughout the 1980s, Olle Sturén developed a perceptive analysis of what led to greater or lesser conflict between regulators and those promoting international standard setting in various countries. The global pattern resulted from different levels of national economic development, the experience of empire, and the size of the internal market of already-industrialized countries.

In the developing world, Sturén believed, governments embraced international standard setting to spare their nascent export industries the cost of building to different specifications for different markets and to allow their importers of high-technology goods to buy from suppliers anywhere in the world. Moreover, international standards facilitated the transfer of technologies to industrializing countries. ISO's consensus process provided many developing countries with technical assistance to improve the skills of their engineers who took part in standard setting, and ISO often covered the costs of developing country engineers who took part in technical committees.[162] Of course, *parts* of developing countries' governments sometimes did not agree with the development ministries' general interest in voluntary international standards. Sturén often complained that foreign ministries in the global South had preferred putting international standards for freight containers into a legally binding treaty that would have provided financial support for improving their own transportation systems. However, this kind of internal conflict did not undermine the general tendency of governments in the developing world to support private international standard setting.

At the opposite extreme was what Sturén called the "old colonial" attitude that could still be found in the UK and France in the 1980s. There, business and government often acted as if national standards were enough if they could be imposed on former colonies. This was changing in light of the developing countries' "wish to supplement political independence they have gained with technical freedom vis-à-vis the old colonial master"[163] and the growing British and French interest in Europe-wide markets. The same diminishing opposition to international standards was typical of national governments Sturén identified as having a "German-like" attitude. Germans were committed to their national standards, they had worked for generations to develop the largest and most integrated catalog of industrial standards, and their promotion of their own standards was "not linked to a particular business deal but aimed at preparing the future market" of its trading partners.[164] Like developing country exporters, however, the Germans were not interested in making products to conform to different national standards of different markets and were coming to accept international standards.

"Japan-like" countries, on the contrary, thrived through their ability to make different goods for different markets, according to Sturén's analysis. They preferred the cost saving that came with international standards, but they had no need to try to impose their own national standards on the world and a strong interest in assuring that no other industrialized country would gain an advantage by imposing *its* standards on others. Together with countries with the "Sweden-like" attitude—countries that can compete with the big industrialized exporters in a "limited range of industries"—"Japan-like" countries were part of the coalition that was most favorable to international standardization.[165]

Finally, there were "USA-like" countries—"the United States almost alone in this category"—that had massive national markets for most industrial goods. Their firms and government regulators had little interest in international standards in most fields and contributed "to international standardization" only in sectors in which markets are inherently global, "such as aircraft, computers, freight containers, and photography [including cinema and film]" where they push "strongly for international adoption of the American standards."[166]

In the 1980s, it was the politics among countries with these different attitudes that ultimately resulted in the embrace of international voluntary standards by most countries. Ultimately, the key contributions to that outcome were the regulatory politics within the United States and the European Union, because the developing countries had little power to affect the outcome, Japan was a natural ally of the Europeans, and the Europeans were increasingly united as Germany and the old colonial powers all moved toward the "Sweden-like" position. Moreover, recall from chapter 3 that, unlike the other early industrializers, the United States did not develop a fully hierarchical network of standard-setting organizations, nor, despite Herbert Hoover and the experience of national standard setting throughout World War II, did it have a tradition of close collaboration between government and private standard setters. Nonetheless, both the US government and many American firms consistently worried that countries with "old colonial" or "German-like" attitudes toward international standard setting might gain unfair trade advantages through their international promotion of their national standards.

From the perspective of many leaders of the ASA, the problem began with competition among standard setters. During and immediately after the war, at the same time that ISO was forming, ASA tried to become a body more like Germany's DIN, Britain's BSI, or France's AFNOR. ASA's leadership attempted to persuade its members (including professional associations and trade associations that set standards) to take their standards through the ASA process and make them "American standards." The leadership also asked the US government to recognize the superiority of such standards, arguing that they had revised ASA's constitution to take on this central public role. The constitution now assured "that all of those interested in a particular standard would have a voice in its development" and that there would be three members at large on the association's board "to assure a voice for consumer interests."[167]

The effort was only partially successful, as was a second attempt from 1966 through 1969. ASA briefly took the name United States of America Standards Institute and then settled on the American National Standards Institute

(ANSI). The US government's Federal Trade Commission opposed the first name because it suggested that the organization was somehow more "official" than any of the other 400 or so US voluntary standard setters. Effectively the government sided with the many standard-setting bodies that worried about losing their separate identities and their incomes; for example, sales of standards accounted for more than 80 percent of the revenue of ASTM but less than 30 percent of ANSI's.[168] Moreover, while almost all the organizations in the network of US voluntary standards organizations followed the principle of balanced representation of stakeholders in their various standard-setting committees, they did so in different ways. ASTM, for example, subsidized important stakeholders that would otherwise not be able to take part in committees,[169] the practice that Olle Sturén instituted in ISO to assure developing country participation in technical committees. Despite these attempts to become like the single dominant national standard-setting bodies in other industrialized countries, throughout the Cold War, ANSI remained able to orchestrate only about a quarter of the voluntary standards sector in the United States.[170]

Any attempt to create a truly centralized American voluntary standards system was further undermined by a May 1982 US Supreme Court decision that held the American Society for Mechanical Engineers liable for "damages cause by the anti-competitive activities of some of its members involved in the standard setting process," to wit, the chairman of ASME's committee that maintained its steam boiler standard. He had knowingly misinterpreted the standard to the disadvantage of the Hydrolevel Corporation, which had developed a new way to prevent boiler explosions that he had said was inconsistent with the standard. The Hydrolevel decision created great uncertainty among bodies engaged in American private standard setting, some of which left the field while others competed to satisfy the court's new, rather unclear, requirements for greater due process safeguards. As this was the beginning of the heyday of deregulation under President Ronald Reagan, Congress and the executive branch did not step in to clarify the matter.[171]

Even if both the US government and most of the voluntary standards sector opposed the creation of a US entity that was as powerful as the national standards bodies of countries with an "old colonial" or "German-like" attitude toward international standards, US trade negotiators and American companies agreed that British, French, or German standards should not govern international trade. As discussed earlier in this chapter, the United States pushed for the inclusion of the GATT Standards Code as part of the

1979 Tokyo round agreement. According to Olle Sturén, the code meant that where technical regulations and standards were needed and relevant international standards existed, GATT members would be required to use them. The code allowed exceptions, but it still set that rule.[172]

While the GATT Standards Code was being negotiated, the US executive branch's Office of Management and Budget (OMB) began working on a national standards policy, something that would be needed if the proposed code were to be accepted and implemented. OMB Circular A-119, "Federal Participation in the Development and Use of Voluntary Standards," was first proposed in 1976, and a final version was adopted in 1982. The circular encouraged and provided rules under which the US government would use European-style "reference to" private standards in legislation and government purchase requirements. It also specified conditions under which US government agencies should participate in the work of voluntary standard-setting committees, including that the standard setter must have wide stakeholder participation. However, the government failed to define the due process requirements (e.g., rules for dealing with opposition to a proposed standard) for a standard-setting body to be considered a "voluntary consensus standard setting" organization, omitting a draft definition from the final version because many of the hundreds of private US standard setters opposed it.[173]

A 1992 US Office of Technology Assessment report described the interagency body charged with implementing OMB A-119 as ineffective and designed to "bury problems rather than resolve them." Similarly, the report described the interagency groups created to pursue US interests under the GATT Standards Code as "responding only when need arises."[174] As a result, the United States was able neither to provide effective support for nor to block the internationalization of voluntary standard setting for industry.

The situation was very different in Europe, where, in the 1980s, almost all governments and firms began to accept the "Swedish-like" view. The ground for this change had been prepared over many years. Recall that Sturén had convinced his European colleagues to create a Europe-wide standards body in the late 1950s and early 1960s, long before Sweden or Great Britain were part of the European Union. CEN, formed in 1961, was joined by CENELEC (the European Committee for Electrotechnical Standardization) in 1973, and by ETSI (the European Telecommunications Standards Institute) in 1988—organizations that might have been useful during the debates over a color television standard in the 1960s.

As the membership of the European Union grew and as all traditional barriers to trade within it were removed in the 1960s and 1970s, the intergovernmental European Council became increasingly concerned with "non-tariff barriers to trade" such as incompatible national standards. Attempts to supersede national standards through directives from Brussels or through a cumbersome diplomatic procedure of negotiating "mutual recognition agreements" proved ineffective. The council simply had no way to enforce its directives when member governments ignored them, let alone to police the activities of every European firm.[175] In 1985, council members turned to CEN and its partners and announced what they called the "New Approach":

> Their scheme was simple—the regulators would define simple objectives—the so called "essential requirements"—and industry would develop the technical specifications that define the requirements to meet those objectives. The work would be done at the European level (CEN, CENLEC, and ETSI) by representatives of the national standards organizations of each member state and the [Europe-wide] regulators would formally adopt reference to those standards as a way (and in most cases the only practical way) to demonstrate that the product in question met the regulatory objective.[176]

The initial goal was to remove all the EU's incompatible standards by 1992. To get the work done that quickly, the council subsidized those involved in the process, something similar to what governments had done during the Second World War.

Representatives of European industry embraced the opportunity to "influence the details of the legislation" that governed their products.[177] The largest national standards bodies (DIN, the BSI, and AFNOR) shifted their focus to acting as secretariats for committees setting European (and, eventually, global) standards. A new culture of international standard setting was established throughout Europe.

In 2000, Unisys's director of standards management, Stephen Oksala, argued that, during the 1980s and 1990s, Europe's New Approach had also changed the standard-setting culture of those American firms that engaged in the European process through their European branches. Those firms began demanding that the European work be done at the *highest* international level through ISO and the IEC, in order to include US interests. Those demands, in turn, led to formal agreements in 1991 between CEN and ISO and CENLEC and the IEC to do just that.[178] Oksala lamented, "This process met its objectives, but the results have been mixed." That is, "for some industry sectors, European participation at the international level has overwhelmed American companies, leading to the belief that these agreements were done for the Europeans—although the intent was the opposite." It was, he explained, "another example of unintended consequences."[179] Oksala's overall argument was that, throughout the 1980s and 1990s, the internationalization of standard setting had occurred as an unintended consequence of government actions: the US push for the GATT Standards Code and the even stronger, more enforceable, code that was part of the 1995 agreement to form the World Trade Organization; and the 1986–1992 New Approach of the European Council.

Olle Sturén would hardly have been surprised by this conclusion, given that he had worked since the 1950s to create the institutional mechanisms that encouraged such an outcome. Nevertheless, in his last year as ISO's "official" head, 1986 (he remained secretary-general emeritus until his death in 2003), he often emphasized two additional factors. One was the (self-interested) leadership of standard setters from countries with "Swedish-like" attitudes toward international standard setting. The second was the internationalism of the engineers within the standardization movement.

He emphasized both factors in a speech to the Standards Institution of Israel (SII) in January 1986. Sturén explained that the countries that gained the most from international standardization, and that, therefore, could be most relied upon to champion the process, were "undoubtedly the small industrialized or fairly industrialized countries."[180] There was no need to mention that this group included Sweden and Israel. ISO's major work had always been to create the international markets and confront the global problems to which such countries had to be so attentive, so "when new methods of transportation were developed in the 1950s . . . ISO set up technical committees for preparation of International Standards for pallets, containers, packaging dimensions, etc.," leading to "one of the real success stories in the history of ISO." Similarly, when environmental problems became evident in the late 1960s, "ISO . . . established . . . new technical committees for air and water quality, and for soil preservation," and "the oil crisis in the 1970s resulted in . . . a new committee on solar energy." He noted current areas of intense discussion on "information processing, image technology, and industrial automation, and I would not be surprised if an ISO role were soon defined in the field of biotechnology."[181] Sturén praised ISO's council, which "included representatives of several countries without diplomatic relations with Israel," for overlooking Israel's international isolation when it appointed and reappointed SII's representative (also, unusually, a woman) to its Technical Board, saying, "Merit and nothing else speaks in ISO, or I should be careful to say has up to now spoken, but then I must add that there are no signs of any change."[182] Like le Maistre, Sturén was a true believer who may have exaggerated the extent to which the ideology of the standards movement was lived by all involved.

At his retirement celebration a few months later, Sturén spoke about the centrality of internationalism to the mission and success of voluntary standardization from its start, which he traced to the first formal meeting of what became the International Association of Testing Materials in Dresden in 1886 (see chapter 1), which "showed that it was about time to start working on problems that had *only* international solutions or no solutions." He then spoke more lyrically about his beliefs:

> In fact, if there is a truly philosophical element in my aspirations . . . it is this—that I believe in internationalism, and I am convinced that what we do in international standardization is a contribution to this end—but more so, that standardization, by codifying and stabilizing technological development, assists in directing technical progress to the general good.[183]

Sturén had given a similar, but somewhat more prosaic, response to a reporter's question about whether international standard setting would "inevitably lead to worldwide '1984'-style regimentation, deadening conformity, and elimination of creativity." The reporter cited the playwright John Osborne's recent sarcastic comment in a letter to the London *Times*, "How ineffably depressing it is to know that such a thing actually exists and officially functions called the International Organization for Standardization." Sturén responded, "Standards free mankind; they don't chain us. . . . the alternative to world standards is world chaos—and chaos is what the 1700 ISO committees around the world are working diligently, unsung and almost unnoticed, to avoid."[184]

The next chapter provides a closer look at the standard-setting committees, both inside and outside the ISO network, in a high-technology field throughout the second wave.

6

US Participation in International RFI/EMC Standardization, World War II to the 1980s

In the decades following World War II, many different types of standard-setting organizations made up a complex network at the national and international levels, with extensive overlap in domain and membership, as well as interaction among organizations. The container story in the previous chapter demonstrated the simplest structure, with interactions between the US national body (ASA and later ANSI) and ISO on the international level. The standards field in a particular industrial area was often more crowded and less hierarchically ordered than in that case, however, and could also include governmental bodies and hybrid standard-setting organizations as in the color TV case. This chapter explores one domain of standards activity—initially known as radio frequency interference (RFI) but since the 1960s as electromagnetic compatibility (EMC)—at the nexus of US national standards bodies and international bodies. It demonstrates the complexities of second-wave standardization in fields where technologies and their use pushed companies, governments, and engineers toward developing international standards.

In such fields, many of the same experts took part in several national and international committees in which relevant standards were being set. They developed friendships and shared knowledge, allowing leaders to emerge—arena-specific standardization entrepreneurs and diplomats. This chapter highlights one of them, an American professor of electrical engineering named Ralph M. Showers. In the larger history of private standard setting, Showers exemplifies the hundreds of standards leaders in specific fields. He had many of the qualities shared by these people and by the standardization entrepreneurs we have discussed in the earlier chapters—especially the diplomacy, personal reserve, intellectual independence, and commitment to standardization of Charles le Maistre. These were useful talents for an American in the second, internationally focused wave of standardization, when the United States was at a disadvantage in comparison to European countries that had, for geographical reasons, always been more involved in international efforts.

The opening section of this chapter introduces the field of RFI/EMC standardization and the emergence of several national and international organizations in this domain before and shortly after World War II. At the end of the war, tensions between US standardization bodies and Europe-oriented international ones immediately became evident. The subsequent story revolves around how standard setters in the United States gradually recognized the importance of international standardization and learned to participate in it effectively. From the 1960s to the 1990s, electromagnetic compatibility issues touched everything from household appliances to cars to computer equipment to space-based satellite transmissions. Achieving international standardization required reaching across the Cold War divide, as well as across different industrial constituencies. Showers, along with a few colleagues, guided the US standardizers as they struggled to integrate the United States into the system of international standardization in this domain, and he personally became a leader in the field.

European and US Standardization around RFI before the 1960s

Radio frequency interference occurs when one set of electromagnetic waves interferes with another. To many older people, the most familiar manifestation of RFI is the crackling and loss of radio and TV reception caused by electromagnetic waves (or *noise*) emitted by vacuum cleaners, passing cars, and microwave ovens. Radio transmissions can also interfere with each other. Standardization in the field of RFI began before World War II with relatively uncoordinated efforts at the international and national levels through a variety of governmental, hybrid, and private organizations. During and after the war, more consistent and coordinated efforts occurred at both levels.

Interwar Emergence of RFI Standardization

As radio broadcasting emerged in the 1920s, interference became an issue for national governments and private standard-setting bodies.[1] In 1922 when the

British Broadcasting Corporation was established, British electrical engineers immediately began publishing papers on RFI. Radio broadcasting began in Germany in 1923, and by 1924 the VDE (the society of German electrical engineers) had established a High-Frequency Committee to develop guidelines around interference. At the request of the German Ministry of Posts and Telecommunications, the VDE prepared a national RFI standard (published in 1928) for constructing and testing equipment to prevent radio wave emissions that would interfere with broadcasting.[2] In the 1930s voluntary national standard setting began to occur alongside government regulation throughout Europe, including in the UK, the Netherlands, and Poland.[3]

US standards work on interference started in the military during World War I. The Army Signal Corps created standards to prevent interference with military radio communications from radio waves emitted by tanks and other combat vehicles. The corps also used interference to jam enemy transmissions and created technologies to prevent jamming of its own transmissions.[4] In 1931, US private efforts began when the Edison Electric Institute, the National Electrical Manufacturers Association, and the Radio Manufacturers Association (the trade associations for private electrical utilities, manufacturers of electrical equipment, and makers of radios, respectively) formed the Joint Coordination Committee on Radio Reception. The committee quickly published a report providing specifications for a radio noise meter to measure interference.[5] In 1936, ASA formed the Sectional Committee on Radio-Electrical Coordination, designated C63, with the Radio Manufacturers Association as its initial sponsoring (funding and overseeing) organization, but the committee issued no standards until after the war, when C63's work becomes central to this chapter's story.[6]

International developments in voluntary, firm-dominated standard setting around RFI also began before World War II. As musical programming increased and amplifiers and loudspeakers improved in acoustical quality during the 1920s, radio interference became an increasing problem for radio manufacturers, broadcasters, and the listening public, especially in Europe. In April 1925, broadcasters from ten European countries met in Geneva to coordinate radio frequency use to avoid interference, establishing the Union internationale de radiophonie (International Radio Union, UIR).[7] In 1933, the International Electrotechnical Commission, the UIR, and other interested international organizations called a Paris meeting to discuss problems posed by radio interference. Most organizations represented agreed to establish a joint committee called the Comité international spécial des perturbations radiophoniques or the International Special Committee on Radio Interference (CISPR), a body that became (and remains) central to international standardizing activities in radio frequency interference and electrical apparatus and is, like C63, central to this chapter. Participants at the Paris meeting agreed to focus initially on establishment of a uniform "method of measurement" of interference and on "stipulation of limits to avoid difficulties in the exchange of goods and services."[8]

CISPR's first official meeting took place the following year, with seven representatives from national committees of the IEC, two representatives from the UIR, and one each from several other organizations.[9] Although it met "under the aegis of the I.E.C.," the new commission was allowed freedom to operate independently of the IEC national bodies because of the perceived urgency of the interference problem.[10] CISPR invited the International Telecommunication Union's hybrid public-private standardization body, the International Radiocommunication Consultative Committee (see chapters 3 and 5), to the 1933 meeting but the CCIR chose not to become a member, probably because a clear division of responsibility existed between the two groups. The CCIR handled interference among radio services, recommending to signatory governments how to set national laws to avoid interference, while CISPR would focus on standardizing interference measurement apparatus and setting voluntary limits for radio wave emissions from industrial and consumer equipment that interfered with broadcast reception.[11] The CCIR and CISPR would, however, regularly send observers to each other's meetings.[12]

CISPR was primarily a European committee at this point because radio interference was common across national boundaries in Europe, while countries as large as the United States faced more domestic than international interference challenges. The commission met eight times from 1934 through 1939 (figure 6.1 shows attendees at the 1937 Brussels meeting) and made several "recommendations," as it originally designated its standards (following the contemporary practice of the IEC and CCIR).[13] It progressed on establishing uniform methods of measurement but not initially on setting limits for interference.[14] Before its July 1939 Paris meeting, CISPR agreed on draft specifications for a CISPR "measuring set" (a device intended to provide uniform measurement) based on the Belgian Electrotechnical Committee's prototype apparatus and commissioned the Belgians to build several more sets; at that meeting, CISPR distributed these sets to other national commit-

Figure 6.1. CISPR met in Brussels in 1937; its work at this point focused primarily on standardizing uniform methods of measuring interference. IEC Photo File, courtesy of the IEC.

tees so they could test them in other locations.[15] Then the war intervened, and CISPR did not meet again until 1946.

During the war, most RFI standardizing efforts once again became national. Germany's VDE issued several wartime standards for suppression of radio interference.[16] In the United States, ASA's C63 committee worked closely with governmental agencies including the Federal Communications Commission and the military.[17] The Allies coordinated some standards through the wartime United Nations Standards Coordinating Committee, but when the UNSCC suggested coordinating on radio interference in May 1945, the ASA Standards Council responded that it needed to do more research before cooperating in this domain.[18] In 1946 C63 published its first American Standard, C63.1, on measuring interference in the 150 kHz to 20 MHz range; it was simply a reprint of the June 1945 joint army-navy specification JAN-I-225 with a new cover.[19] In subsequent years, C63 would issue its own measurement standards for an ever-broadening range of frequencies.

Postwar Tensions between the United States and International RFI Standardizers

International standardization efforts in radio interference reappeared after the war, and as the United States joined them, tensions emerged. CISPR's Committee of Experts met again in 1946; as before the war, the IEC convened the committee and provided meeting support.[20] Although CISPR was still predominantly European, its 1946 meeting included, for the first time, an official delegation from the United States as well as delegations from other non-European countries, such as Canada and Australia.[21]

Integrating the United States into CISPR posed challenges, both technical and cultural, as revealed in reports of its first two postwar meetings. In 1946 the European member countries reported on their progress in making international comparisons of the CISPR set, and Canadian and Australian delegates said that they were anxious to cooperate. The US delegation, however, reported that it currently measured interference with a different standard set developed by the prewar Joint Coordination Committee on Radio Reception that preceded C63. They agreed to compare it to the CISPR set but stated that "the U.S. delegation could not bind themselves to accept the international standards."[22] At the 1947 meeting, the US delegation reported that more than 2,000 of their radio noise meters were already being used in the United States, and that "the time was not considered to be ripe for a permanent specification [standard] for radio noise meters" because not enough research had yet been done. They argued that, in any case, any such specification should not address *construction* of the meter but only its *performance*. Moreover, they judged the CISPR meter unsuitable for US standardization because its frequency band was too restricted, and it was not sensitive enough. The US delegation presented a C63 statement urging CISPR to undertake comparative tests of the two noise meters and publish correlations between them, moving them toward a "universal interim standard" by making their measurements directly comparable, and the meeting chairman agreed to arrange the comparative tests.[23] Measurement methods would continue to cause disagreements between the United States and other CISPR delegations until the mid-1960s.

In 1946, delegates also discussed the relationship of CISPR standards to national laws, another area where the United States was an outlier. All the delegations reported on existing legislation on interference in their countries and whether "authorities in their countries were satisfied that the matter should continue to be dealt with by the C.I.S.P.R."[24] Some countries had or wanted detailed legislation based on CISPR recommendations (e.g., Belgium, France, Switzerland), and others stated that any legislation was necessarily very general (e.g., Canada, the UK). The US delegation insisted that any reduction of interference in the United States would be completely voluntary, based on standards established by ASA, in conjunction with the FCC. Reflecting the importance placed on CISPR's relations with both governmental and private bodies, Charles le Maistre (attending as general secretary of the IEC) proposed that IEC national committees, which were members of CISPR, include government representation and offered to work with the CCIR to establish strong liaisons with CISPR.[25]

Other issues were less divisive. Attendees at the first two postwar meetings repeatedly asserted CISPR's need to coordinate with other standardizing bodies, international and national, governmental and nongovernmental. Representatives from several international standards organizations and trade associations attended the 1946 and 1947 CISPR meetings. The meeting reports also frequently referred to the special committees of the ITU, including the CCIF (the International Long-Distance Telephone Consultative Committee) and CCIR, urging cooperation with them.[26] Expansion of the organization's scope to new domains and frequencies was considered in 1947. A Belgian delegate noted that extending "the frequency range brought television into the field," and the UK delegation expressed concern about motor car ignition systems interfering with television receivers. The US delegation contributed, describing a standard related to automobile ignition interference with television reception that the Society of Automotive Engineers (SAE) and the Radio Manufacturers Association had jointly arrived at, using the C63 standard radio noise meter.[27] The delegations agreed to expand the frequency range to include television and asked the IEC Council to change the last word of its name from Radio*phoniques* to Radio*électriques*, a change approved in 1948 and made official in 1953.[28]

CISPR's structure and processes evolved in the postwar era. In 1947 CISPR became an autonomous special committee of the IEC, with the UK's British Standards Institution as its secretariat.[29] In the late 1950s, CISPR's procedures, heretofore informal, were formalized: "In Brussels 1956 Terms of reference for CISPR were stated and a Procedure worked out for Recommendations, Reports and Study Questions with the same form as used in CCIR," an interesting influence of the hybrid public-private CCIR on CISPR.[30] In addition, CISPR adopted the IEC's six-months rule for voting by correspondence.

The intergovernmental ITU went through a similar and simultaneous process of institutional reform. After the war, the ITU's prewar international regulations for radio needed significant updating because of enormous advances in radio and radar technology. It held two coordinated 1947 conferences to update radio regulations.[31] The conferences affirmed that, to avoid interference, ITU would allocate the radio frequency spectrum and register frequency assignments, creating the International Frequency Registration Board (IFRB) as the intergovernmental body to keep the official global list of broadcast frequencies.[32] The ITU's CCIR would "undertake studies, formulate recommendations [standards], and collect and publish information on telecommunication matters," with a CCIR specialized secretariat established to plan for the meetings.[33] In 1948 CCIR convened for the first time since the war (subsequently meeting every three to four years).[34] It established 13 study groups (SGs) on topics from radio transmitters and receivers (SG I and II) to television (SG XI, see chapter 5).

The United States Finds a Modus Vivendi with International RFI/EMC Standardization in the 1960s

In the early 1960s, on both the international and national levels, multiple standards organizations with different interests operated, all committed to the consensus process but with very different cultures and practices. The ITU's hybrid public-private CCIR focused on radio interference among broadcasting sources on land and in space; CISPR developed standards to measure and minimize interference from industrial and consumer equipment to radio and television reception; and ASA's C63 Committee set and coordinated American national standards among its many member organizations, focusing on the same area as CISPR. In this section, we examine how United States and international standards organizations and people interacted in their RFI work during the 1960s. A cast of characters who worked in both C63 and CISPR—including Ralph M. Showers, Brooks H. Short, and Frederick Bauer—give us a window into the tensions between the organizations and communities that spanned them. Before describing their efforts to establish a modus vivendi between US and international RFI standardizers, we provide context around the CCIR's processes (which shaped CISPR's processes) in the early 1960s and C63's national standards activities.

The Bureaucratic Culture of CCIR International Standard Setting

At the CCIR's Ninth Plenary in Los Angeles in 1959, less than two years after the launch of Sputnik by the USSR and Explorer I by the United States, the CCIR expanded its purview to space, refocusing SG IV on "technical questions regarding systems of telecommunication with and between locations in space."[35] Space satellites made matters of radio frequency interference more inescapably international than before. The CCIR's 10th Plenary Assembly was scheduled for 1963, and in October 1962 CCIR director Ernst Metzler from Switzerland, along with several study group chairmen, documented the activities of CCIR headquarters and study groups since the 1959 Los Angeles plenary.[36] Their reports reveal the CCIR's formal, bureaucratic methods, including how it coordinated with other standards organizations, how it documented its technical work, and how it achieved consensus (including across Cold War political blocs). Records from a US delegate provide insight into how that delegation worked within this international body.

First, the CCIR actively and formally coordinated with private international standardizing bodies, particularly the IEC and CISPR. For example, a CCIR secretariat member joined an IEC special working party to update IEC specifications for testing radio transmitters, "unifying the texts of the two organizations" in this area.[37] Reciprocally, the IEC maintained a liaison with CCIR and the other ITU consultative committees, "where liaison enables us to assist, as well as to learn the limitations set by or expected from governmental requirements."[38] In 1962 the US bodies for the CCIR and IEC also appointed liaisons on the national level.[39] Coordination took place in study groups, as well. The chairman of SG I on transmitters distributed several CISPR documents to his study group and proposed assessing whether CISPR measuring methods, when finalized, should be adopted by the CCIR. The chairman of SG II on receivers referred to cooperation with CISPR and liaison with the IEC around television receivers.[40] One CCIR document (reproduced within a CISPR document) recommended that the CCIR provisionally accept CISPR's measuring methods and apparatus.[41]

Voluminous publications, reflecting its bureaucratic orientation, were a second characteristic of the CCIR's operations. The secretariat had produced five volumes of documents, each in three languages, for the 1959 plenary assembly, and another 14-100 documents (in three languages) for each of the 13 study groups' interim meetings, and the volume was increasing.[42] A stickler for correct process, Metzler, the CCIR's Swiss director, had spent 30 years in his country's governmental telegraph, telephone, and radio administrations while also serving as a delegate to the UIR and ITU.[43] In his report to the 1959 assembly, he specified the documentary structure of the CCIR's standardizing process, which provided one model for CISPR's methods. Several related and sequential document genres structured the CCIR's process, guiding study groups along a path toward recommendations.[44] A question, agreed to by the Plenary Assembly, initiated the sequence, followed by a study program describing the work a study group needed to perform to answer the question. A report of studies (carried out by study group members or outside sources) partially responded to a question. A recommendation or standard was a "statement issued when a Question has been wholly or partly answered. A Question is normally terminated by the issue of a Recommendation."[45] Finally, if the committee wished to express an opinion on nontechnical matters, a rare occurrence, it used the resolution. Delegates at the plenaries every four years voted in person (not by mail as often occurred in the IEC or CISPR) to approve new recommendations. Documents were prepared and circulated six months before meetings to allow discussion within and across delegations.

Although reports (widely used in other contexts) varied in format, the CCIR's question, study program, recommendation, and resolution genres used similar structures designed to emphasize technical rationalism and modeled on the practices of intergovernmental conferences (including those of the ITU) developed prior to the League of Nations era and continued under the UN system, of which the ITU and CCIR were now a part.[46] The format for each is schematically illustrated in a 1963 CCIR Drafting Committee document.[47] Recommendations took a standard form, with a varying number of items in the lists, as follows:

> The C.C.I.R.,
>
> CONSIDERING
>
> (a) ....
>
> (b) ....
>
> RECOMMENDS
>
> 1. ....
>
> 2. ....

The "considerings" could include references to first principles, context, reports, and existing recommendations. Questions, study programs, and the rare resolutions followed a similar format, beginning with "considerings" and followed by "DECIDES that the following question should be studied" or "DECIDES that the following studies should be carried out" or "RESOLVES."[48] This common format for the standardization-specific genres suggested that what followed the "considerings" was logically and technically necessary, thus rhetorically reinforcing belief in the CCIR's technical (rather than political) orientation.[49]

A third characteristic of CCIR operations was its consensus process. The CCIR, unlike the IEC or CISPR, had delegations approved by each country's government rather than by its technical and standards bodies, a factor that could make achieving consensus more difficult, as we saw in the failed attempt to standardize color TV (chapter 5). Yet technical contributions from member countries on both sides of the Cold War divide sometimes led to apparently noncontroversial recommendations.[50] In other cases unanimous decisions could not be reached, though the study group chairmen rarely detailed the conflicts that obviously existed.[51]

One rare case where a conflict was documented demonstrates the tensions caused when the US delegation failed to follow the CCIR's strict bureaucratic processes. A US delegate distributed a proposed draft recommendation to SG II during (rather than the required several months before) its interim meeting, clearly breaking CCIR rules and norms. The late document proposed new parameters for assessing receiver system sensitivity. The chairman of SG II circulated the US draft to CCIR members with a statement saying this change "completely altered the spirit and practical value" of an existing recommendation.[52] The US failure to respect the CCIR's process apparently led him to document the conflict, resulting in rejection of the American recommendation.

The deliberations of the US National CCIR Organization (in charge of the US delegation) further illuminate the US relationship to the CCIR. Because the United States had no telecommunications ministry, its delegation to the CCIR included experts from private firms as well as government agencies; nevertheless, the State Department retained oversight and veto power over the delegation, unlike in the IEC and CISPR.[53] The US National CCIR Organization consisted of an executive committee responsible for "the development and application of national policies within the purview of CCIR activities" and study groups corresponding to those of the CCIR. Any interested firm was invited to participate, but the Executive Committee often had trouble securing US industry interest and representation.[54] Attendance at Executive Committee meetings every few months between August 1961 and

January 1963 included a large majority representing government agencies (e.g., the State and Defense Departments and the FCC), in part due to these agencies' proximity to the State Department headquarters where the meetings took place, far from the headquarters of relevant private companies.[55] Nonetheless, when it needed funds to host CCIR interim meetings for two study groups, the Executive Committee turned to industry, emphasizing that "the Government does not operate telecommunications and, as a matter of tradition, cannot provide many funds for the activities mentioned above," leaving firms responsible.[56]

Unsurprisingly, given the government dominance, the US Executive Committee focused on national interests, in addition to technical merit. For example, when the American vice chairman of a CCIR study group was made chairman in 1961, he recommended that the United States consider "whose interests are compatible with United States interests" in deciding who to support as his successor.[57] Sometimes "US interests" were influenced by Cold War conflicts. Another American complained that the CCIR made free enterprise look bad when it requested detailed information on radio receivers and the USSR provided it but the US could not because "United States companies are reluctant to supply what is confidential, competitive information."[58] At other times, US interests were framed in strictly commercial terms, as when the US Executive Committee discussed a complaint from some American manufacturers that a Tokyo ITU-sponsored seminar had been turned into a sales event by Japanese manufacturers, and it decided to have the US representative in Geneva informally talk to ITU secretary-general Gerald C. Gross, from the United States, to prevent such problems in the future.[59]

More surprisingly, throughout the 1960s the United States and other national delegations shared technical information across Cold War lines even in SG IV's high-profile space arena. Secretary-General Gross decided that radio communication in space was a critical area for international cooperation; consequently, in 1961 he asked the USSR and the United States each to send an expert in this area to the Geneva secretariat to work together on space-related issues.[60] In these early years of space research, one US report was a "factual statement of present activity on 7 systems of communications in the communication satellite field"; one Soviet document provided a table of its delegation's proposals for allocation of frequencies in space; and at least one UK document reported actual experiments with the Telstar satellite, launched on July 10, 1962, using the Goonhilly Downs satellite dish.[61] Thus neither the CCIR's bureaucracy, the ITU's intergovernmental nature, nor the Cold War suspicions of government-dominated national CCIR organizations completely overrode the standardizers' belief that technical matters should transcend political issues, though it surely limited the extent of technical exchange.

Developing and Coordinating American Standards in C63

Next, we turn to US national standard setting in RFI through C63. Just as the international standardization field had multiple organizations characterized by different interests requiring coordination, the messy US system (see chapter 3) also included many and varied standardization bodies working in the RFI arena. US national standardization thus required standards coordination as well as development, and University of Pennsylvania electrical engineering professor Ralph M. Showers would lead the committee toward that strategy.

Showers, like Olle Sturén, was born at the end of World War I and spent his formative years on the outskirts of a large city, in this case Philadelphia rather than Stockholm, though he spent summers with extended family in Penns Creek, a small town surrounded by farms in central Pennsylvania. Showers attended high school in Havertown, Pennsylvania, graduating in 1935. His high school yearbook reads, "Ralph is a thoroughly likeable fellow. He is persevering in his manner and can be depended on to do his best. He is another one of those quiet chaps noted for deeds rather than words." He received a scholarship to attend the nearby University of Pennsylvania, where he received bachelors (1939), masters (1941), and doctorate (1950) degrees. Unlike Sturén, Showers was a distinguished engineer in his own right. World War II directly and indirectly led him to the field of electrical interference. After a brief stint at GE, in 1941 Showers was invited on short notice to return to Penn's cutting-edge Moore School of Electrical Engineering (where ENIAC, one of the first computers, would soon be developed) as a research assistant and instructor to replace a man who had been called to active duty from the reserves. Showers's research at Penn was deemed "vitally important" to the war effort, for which he received a Selective Service deferment. This research was the beginning of his lifelong interest in RFI; he investigated noise measurement, weapons systems communication, and all the related questions of improving telecommunication by controlling electrical interference. Showers joined the Penn faculty as assistant professor in 1945, and the university remained the center of his professional life until he became professor emeritus in 1989. After the war, he continued working in the field of radio interference, instrumentation, and measurement under research contracts with the US Navy, Army, and National Aeronautics and Space Administration, the largest sources of funding for this type of work at the time. He continued contracts with the military until late in his career. Showers began volunteering as a standards setter in 1948 by serving as the Institute of Radio Engineers (IRE) representative on C63. This was just a few months after Olle Sturén took his first standardization job, and both men's preoccupation with standard setting and their roles in international standards organizations began a few years after that, in the mid-1950s.[62]

In 1958, 10 years after he began his national standardization work, Showers attended his first CISPR meeting when he headed the American delegation to the Sixth CISPR Plenary in The Hague (figure 6.2 pictures Showers as part of the US delegation[63]). He wrote to his wife from that first overseas trip, "The meetings went O.K. so far as U.S.A. is concerned, we pretty much got our way with a British proposal for reorganization, but I still don't trust them," though he also claimed to enjoy socializing with them. The personal qualities described in his high school yearbook helped him develop productive relationships with other standardizers and successfully navigate the divide of different personalities, cultures, and standards environments. These meetings also evidently made him aware that US participation in CISPR was not what it should be, commenting, "There is a lot that needs to be done in the U.S.A. and I hope it can be done in the next 3 years" before the subsequent CISPR plenary meeting. He indirectly assured that some of the work would be done when, "at the chairman's suggestion, I tentatively proposed the next one in Philadelphia," putting the United States and C63 on the hook for organizing that conference.[64] From this point on, Showers was highly involved in RFI standardization on both the national and international level. The 1958 meeting was the first of scores of overseas international standardization trips that Showers made over the next 53 years. They would take him to dozens of countries across six continents; sometimes (as was also the case with Sturén) Showers was accompanied by his wife. While these trips were, from the very beginning, exhausting, they were also meaningful and exciting. Most significantly, they gave Showers a sense of accomplishment.

Showers's national base—the private and voluntary C63—had membership consisting primarily of professional and trade associations and government departments, rather than individuals or representatives of firms. In 1960, member organizations included professional associations such as the

Figure 6.2. Ralph M. Showers (to the right of his fellow US delegates) at work during a CISPR meeting, probably his first 1958 meeting in The Hague. Courtesy of the Showers family.

American Institute of Electrical Engineers and the Institute of Radio Engineers (IRE); trade associations such as the National Electrical Manufacturers Association (NEMA, C63's sponsor during the 1960s); government departments including the National Bureau of Standards and the FCC; and only two member firms—Western Union Telegraph Co. and Aeronautical Radio, Inc., which dominated US telegraph and commercial airline radio communication, respectively.[65] Finally, two members-at-large (a professor and a consulting engineer) completed the membership. The C63 Steering Committee had a membership policy favoring associations over firms; in 1960, for example, they rejected the membership application of a company that manufactured meters to measure radio noise, recommending instead that "a general letter be addressed to all known manufacturers of radio noise and field intensity meters recommending an organization of these manufacturers to thus obtain representation of their interests on Committee C63."[66] At this time, when C63 focused primarily on standardizing *measurement* of noise rather than products making noise, ASA considered C63 a scientific committee and waived the strict classification of members as producers, consumers, or general interest.[67]

Regardless of what member organization they represented, many members of C63 belonged to IRE's Professional Group on RFI in 1960, including the two chairmen of C63 during the 1960s.[68] The first was William E. Pakala, a brilliant but aging electrical engineer at Westinghouse Electric Company who had been active in the AIEE when Showers was in grade school.[69] Pakala was C63 chairman from 1955 to 1968, and represented NEMA on the committee. His immediate successor was Showers, who initially represented the IRE, then its successor, the Institute of Electrical and Electronics Engineers (IEEE).[70] Showers had chaired the IRE's Professional Group on Radio Frequency Interference in 1960–1961.

In 1962 and 1963, IRE and the AIEE went through a complicated merger to form the IEEE, a process that included merging the AIEE's standards-oriented technical committees with the IRE's education-oriented professional groups.[71] At the same time, the electrical engineers in this specialty also changed the name of their area from radio frequency interference to electromagnetic compatibility, reflecting a broader scope that included not just radio frequency *emissions*, such as those from a car or a microwave oven, but also radio frequency *immunity*, or how well protected a radio or other device was from external emissions. By June 1963, the newly merged IEEE Professional Technical Group on Electromagnetic Compatibility emerged from the process. Showers led a committee of this new group toward standardization by sponsoring an invitation-only standards workshop and proposing revisions to its constitution and bylaws "to accommodate 'standards generation'" as part of its scope.[72] In 1963, the merged IEEE created a formal standards body, the IEEE Standards Board, and, by 1964, Showers sought the EMC group's representation on the IEEE Standards Committee.[73] Later in the 1960s what became known as the IEEE EMC Group (and eventually as the IEEE EMC Society) was added to the mix of national organizations whose standards C63 coordinated. The EMC Group also became a professional home to a community of American electrical engineers, centering on Showers, who were involved in C63 and other national standardizing bodies, as well as in international bodies such as CISPR, the IEC, and even the CCIR, for the rest of the twentieth century.[74]

During the early 1960s, C63 developed important basic American EMC standards. By 1964, the committee had generated three national standards for noise measurement at different frequencies (designated C63.2, C63.3, and C63.4), in addition to C63.1, the reprinted military standard).[75] Task groups established to work on specific technical areas researched, debated, balloted, and compromised to reach consensus on standards. C63 did not use the formal, bureaucratic set of document genres used by the CCIR (as CISPR did at this time), but it regularly circulated correspondence including reports, proposed revisions of standards, and ballots on proposals, in addition to holding steering committee meetings twice a year. It also followed due process rules, tabulating and responding to each negative vote or comment on a ballot, often resulting in further correspondence and additional rounds of balloting. After a lengthy development and approval process, C63 finally published all three standards in 1964.

At this point, work on standards development went more or less dormant until 1968. With the publication of the three standards, Pakala apparently shifted his attention to the international level, and Harold A. Gauper (who chaired the C63 subcommittee developing national standards) stopped attending most C63 Steering Committee meetings.[76] Other members of the Steering Committee periodically spoke of the need to reactivate standards development, but nothing happened until 1968, when the younger and more energetic Showers took over the C63 chairmanship from Pakala. In the final meeting that year, Showers explained that he had called Gauper and discussed several areas needing attention, including reconstituting some of the subcommittee's working groups. Gauper attended the July 1969 meeting and reported that his subcommittee had met for the first time since 1965 and that it was planning to ballot a reaffirmation of C63.2, C63.3, and C63.4, since the renamed USA Standards Institute (which succeeded ASA in 1966) required that all American Standards be revised, reaffirmed, or rescinded every five years. At the next meeting (by which time the US national standards body had changed its name to ANSI), Gauper reported the successful reaffirmation of the three existing standards and commencement of new activities.[77]

Standards development was only part of C63's role in American standardization of EMC. Given the complexity of the American system, it also aspired to *coordinate* related standards of other bodies, including recommending that standards developed by member organizations be put on track to become American Standards.[78] In 1968, even with the resumption of standards development activity, standards coordination grew under Showers's leadership. He incorporated into C63 meetings reports on standards-related activities in member organizations, including NEMA, the IEEE, the FCC (which reported legislation rather than voluntary standards activity), the US Army, and the Society of Automotive Engineers.[79] At the final meeting of 1968, he distributed a summary compilation of the activity previously reported and subsequently circulated an update that listed activities of even more organizations. The IEEE provided the most detailed and extensive list of activities, including standards from five different committees or professional interest groups, including the IEEE's EMC Group, of which Showers was an active member.

Navigating Tensions between CISPR and C63 in the International Arena

Although C63 developed and coordinated American standards, international activities—especially US relations with CISPR—dominated its meetings during the 1960s. The US National Committee (USNC) for IEC oversaw the delegation to CISPR, and most of the delegates were C63 members. Within C63, first Pakala then Showers pushed US participation in CISPR because of the potential impact of CISPR decisions on trade, which was becoming increasingly international. Since many European countries incorporated CISPR standards into national legislation, US products aimed at those markets (e.g., cars and computing equipment) might have to meet CISPR standards for radio wave emissions and immunity, a worry that would grow stronger in industry during the 1970s. Thus, acquiring a voice and influence in CISPR was important but not easy. The friction demonstrated in the immediate postwar period when the United States first joined the body persisted into the 1960s, reflecting irregular US participation, technical differences, difficult personalities, and differing cultures of standardization.

To attend meetings generally based in Europe, Americans faced the greater time, cost, and difficulty of overseas travel. In 1958 when Ralph Showers led his first CISPR delegation to meetings in The Hague, he combined attending the meetings themselves (which lasted more than a week) with a little sightseeing and visiting European labs. In his long letters to his wife, who stayed at home with their children, he commented on the hard work involved, saying that "so far I don't think I have been to bed before 1 pm any night—Monday we had a conference at 7:30 & my room-mate and I wrote a report 'til 3:45 A.M." We have seen his assessment at the close that the meetings were "O.K." for the United States, and that he had tentatively proposed that the United States would host the 1961 CISPR plenary in Philadelphia.[80] This was only the beginning of Showers's long relationship with CISPR.

US participation at CISPR working group meetings was often spotty, in part reflecting the greater lead time, planning, and funding needed for them to travel to Europe than that needed for Europeans to travel within Europe. C63 complained, for example, that US working group members were informed so late about 1960 working group meetings in London that they could not arrange adequate representation, a situation they claimed had recurred periodically. Pakala, as technical advisor to the IEC's USNC for CISPR, rapidly organized a few US members who could make the trip on short notice, arranging for them to cover all the working group meetings.[81] Subsequently, alternate US representatives were named for all CISPR working groups, to increase the chance that someone would be able to attend future meetings. Funding was also an issue for American delegates not employed by firms willing to pay for them to attend.

The USNC, with the help of C63, hosted the 1961 CISPR plenary in Philadelphia, a first for the United States, and, as Showers had anticipated in 1958, the US delegation undertook considerable work to prepare for it. In its support of the USNC, C63 established committees to raise money for the meetings and to organize various aspects of the meetings and social program, with Showers chairing the Committee on Arrangements.[82] One year in advance, Pakala reminded all US delegates to CISPR working groups, "One of the best ways for U.S.A. to participate and obtain recognition in this international activity is to supply U.S. documents for the coming meetings."[83] He urged them to gather all documents that addressed CISPR study questions and submit them to the IEC six months before the CISPR plenary for distribution. Later he circulated detailed instructions about submitting documents, first to the USNC for approval, then 250 copies in English and 150 in French to the IEC Central Office, where they would be assigned a CISPR number and distributed. Preparing for and hosting the Philadelphia plenary absorbed much C63 time and energy; indeed, C63 did not meet from a few months before the meeting until 18 months after it.[84] C63 leadership considered the effort worth it, however, as US involvement in CISPR consequently increased, a sign of progress in the relationship.

In addition to participation, technical issues and personalities could also pose difficulties for US and C63 relations with CISPR. Methods of measuring interference—a source of friction immediately after World War II—continued to cause problems into the 1960s. In 1958, the USNC submitted a draft standard to CISPR Working Group 4 (WG4, Interference from Ignition Systems) incorporating the peak measurement method used in the standard American noise meter rather than the quasi-peak method used in the official CISPR set.[85] Brooks H. Short, who represented the SAE on C63 and worked for Delco Remy (the division of General Motors that built starters and alternators, important sources of radio emissions), also represented the United States on CISPR WG4. He turned out to be an unfortunate choice of American representative, a prickly man unable or unwilling to adapt to international standard-setting work. After a 1960 WG4 interim meeting, he reported to C63 that "the American delegate" (as he referred to himself) disagreed with all the other delegates about whether the CISPR measuring apparatus was appropriate for measuring emissions from automobiles and argued that the true peak measurement devices used in the United States were superior.[86]

At the next working group meeting several months later, he continued to alienate the group. He distributed a statement he had asked the SAE to create, claiming that it had determined that true peak measurements correlated better with observed interference than quasi-peak measurements did, and that it "strongly recommends that any specification covering the measurement of ignition interference must be based upon the reading of peak rather than quasi-peak quantities."[87] He gleefully reported to C63, "This document threw a bomb-shell into the meeting since all of the European countries are either using or are considering the use of the quasi-peak system of measurement." This "bomb-shell" did not elicit the desired change, however. The working group decided that because CISPR had no standard to measure peak values, WG4 must use only quasi-peak measurements. They agreed only that they "would request Subcommittee B of CISPR—'Measuring Instruments' to provide a specification covering a peak reading instrument." When such a specification was available, they "will seriously consider our request for specifications based upon peak readings."[88] Unsatisfied, Short convinced his fellow members of C63 that CISPR's directive to all working groups to use the CISPR measuring set for all measurements "was unacceptable to the US," and Pakala (in his role as USNC advisor to CISPR) wrote a letter escalating the issue to the CISPR general secretariat, "requesting" that the directive be modified to include other types of measurement equipment.[89]

Why was Short so interested in escalating the conflict, rather than in addressing its technical aspects? He seems to have been a showman who had had some disappointments as an engineer. In 1946 he convinced Delco and the FCC to back his plan to improve road safety by using radio waves to have cars signal each other about their speed and direction—an idea that anticipated aspects of today's self-driving cars. Unfortunately, in 1946 it was technically infeasible, and Short's national and local hype of the idea ended up looking a bit foolish.[90] Nonetheless, Short continued to regularly report his exploits to the local newspapers in his home town of Anderson, Indiana, one of which reported his involvement in the 1960 CISPR session under the headline, "SOLE U.S. REPRESENTATIVE," noting he would "bring unique recognition to Anderson, next week" as the only American attending a "special conference . . . where vehicular radio interference will be discussed."[91] Perhaps Short heightened the conflict, in part, to play to readers back home.

Meanwhile, a study undertaken in response to the US 1958 submission, published by the British IEE in 1961, countered Short's and the SAE's attack on CISPR's quasi-peak method. It described tests done on British, American, and German measuring sets in 1959–1960 at the Delco Remy testing lab outside Anderson, Indiana, "which were attended by representatives from American car and ignition-equipment manufacturers, measuring set manufacturers and the Federal Communications Commission, as well as three representatives from Germany and the authors from the United Kingdom," but apparently not by Short.[92] The authors declared that "the most important result of the Anderson tests was the closeness and consistency of the agreement between the American, German and British sets." Moreover, they asserted, "the tests have completely disproved" the claim that quasi-peak measurement was less consistent and reliable than peak measurement. In addition, they established a conversion factor between the two methods, allowing consistent measurements to be made with either type of set.[93] Short certainly knew about these tests, as he worked at Delco Remy where they occurred and one of the authors of the paper, Walter Nethercot, was one of the English representatives to the CISPR WG4 meeting mentioned in Short's report, but he never mentioned them to C63.[94] In his report on the 1963 WG4 meeting (his last report before withdrawing from C63 and CISPR), the irascible Short claimed that other members of WG4 ignored him, and he recommended that C63 withdraw entirely from the working group.[95] Wisely, C63 did not adopt his recommendation and named a new US representative to WG4 (Frederick Bauer from the Ford Motor Company, discussed below).

In reaction to Short's complaints, however, Pakala opened a discussion documented in the C63 meeting minutes "on the manner in which the CISPR operates and possible ways to improve US participation and technical contributions" in anticipation of the 1964 CISPR meetings.[96] This discussion focused on differences in the culture of standardizing between C63 and CISPR, and on the need for US delegates to learn how to participate in CISPR. The committee members noted that CISPR representatives from other countries often represented the government of that country and might not consider voluntary industry standards as important as their own laws or as CISPR or

CCIR standards. They discussed that "Europeans, it appears, do not pay much attention to 'late stuff.' Proposals and viewpoints have to be circulated and 'sold' well in advance of meetings." This emphasis on a formal documentation system reflects CCIR, as well as IEC, influence. Thus, the last-minute proposals that US representatives often contributed did not gain support (as with the late US proposal to the CCIR discussed earlier). Face-to-face decision making in committee meetings also differed: while US committees usually ended discussions of controversial topics by taking a vote, "in CISPR it is not unusual for a committee chairman to simply reflect on all of the opinions expressed during a discussion and make an unchallenged pronouncement that this or that is the wish of the group." They recommended that "lack of continuing representation to meetings and the often spotty attendance, limited or no contribution to study questions, and late referral of documents abroad . . . should be avoided in US relations with CISPR." This discussion ended with an important realization: "US interests dealing with CISPR activities have to 'learn' to participate."

As a result of this discussion, C63 approached participation in the 1964 CISPR meeting in Stockholm differently and succeeded in establishing a viable modus operandi that enabled the United States to participate in it more successfully going forward. Pakala appointed an ad hoc committee including all the US representatives to CISPR working groups to gather and develop US materials (reports, study questions) and positions for the 1964 CISPR meetings.[97] In addition, "since the USA prefers the peak detector, a series of tests was conducted in June 1964 with the objective of establishing a correlation factor between the peak and quasi-peak detector methods of interference measurement from vehicle ignition systems."[98] They submitted a report of those results to CISPR, well before the meetings, and it seems to have completed the process that Nethercot and his colleague began five years earlier, solidifying correlations between the two methods that would allow the United States to use peak detectors while many other countries continued to use CISPR quasi-peak detectors.

When Pakala later reported to C63 on the outcome of the 1964 CISPR Plenary Sessions, he noted "that the main objectives of the US automobile industry were achieved completely at the 1964 meeting and CISPR documents relative to ignition interference reflect this achievement."[99] This result was due in great part to Fred Bauer, Short's replacement as American representative to WG4. Bauer, who represented the Automobile Manufacturers

Association in C63, seems to have been temperamentally better suited to international standardization, enjoying (and excelling at) its diplomatic as well as technical aspects.[100] At this CISPR meeting, Bauer first met Showers, who served as a mentor to him then and in future CISPR meetings.[101]

Coming out of the 1964 CISPR meeting, C63 reaffirmed and redoubled its efforts to improve US participation in the international body. Reporting on the meeting, Pakala recommended continued "active participation to prevent adverse effects of CISPR recommendation[s] on U.S. international trade," noting that many countries adopted CISPR recommendations "for regulatory purposes."[102] He also noted that "there is a substantial and increasing interest in CISPR in U.S.A. activities in the radio interference field."

In spite of improvements, obstacles to full US involvement remained. For example, it was hard to keep US industry informed about CISPR activities, due to "a simple matter of economics that precludes duplication and distribution of the many and varied CISPR documents to all possible C63 interests."[103] To increase circulation of the few available copies of CISPR documents, C63 formed eight new task groups (parallel to the eight CISPR working groups), chaired by the US representatives to each working group and including nonmembers of C63. Each task group chairman would circulate a single copy of relevant CISPR documents within the group. Acknowledging that the United States was not well represented in the radio, television, and appliance fields, the chairmen sought out individuals from those areas to attend the 1966 Prague CISPR working group meetings.[104]

One issue that might have been expected to cause problems for US participation in CISPR, Cold War politics, turned out not to do so. US standardizers worked productively with representatives of the Soviet bloc in CISPR, even more than in the CCIR. Representatives from the USSR, Yugoslavia, and Czechoslovakia, as well as many other Western countries and Japan, were involved in the 1964 and subsequent plenary and working group CISPR meetings.[105] In his report on the 1966 CISPR WG4 meeting in Prague, Bauer spoke favorably of the location and of participation by the USSR and Yugoslavia, newly represented on WG4. On one issue, he commented, "It is worthy of note that the strongest supporters of our position were the delegates from Czechoslovakia and USSR." Moreover, he continued, "the Russian technique of radio interference measurement was described and bears a striking resemblance to U.S. practice and particularly to the proposed SAE specification using the CISPR technique."[106] In his concluding section he said that "my reception in Czechoslovakia was extremely cordial. Complete freedom of movement was possible without restrictions on photography." At his request, the group visited the Skoda Machine Works, which he reported manufactured cars at a very slow rate, for purchase only with government permission. Finally, he lauded the "thoroughly democratic" conduct of the meetings, which presumably followed normal CISPR practice.[107]

US and C63 efforts to engage with CISPR continued, and relations improved in WG4. In his report on the 1967 CISPR WG4 meeting in Oslo, Bauer noted some US advances.[108] First, after "spirited and lengthy discussion," the United States successfully pushed WG4 to extend current CISPR limits for Recommendation 18/1 to cover UHF TV and land mobile services, as proposed by the American Manufacturers Association and SAE. In addition, "CISPR has made a number of conciliatory changes for the U.S. (SAE) usage," including agreement "to reopen the entire specification for discussion if technical reasons require." He also noted increasing CISPR interest in US developments, evidenced by their request for two reports that "had come to the attention of CISPR recently through unknown channels": an SAE paper by Bauer himself on ignition suppression in automobiles and an FCC working group paper on land mobile radio service.

He again commented on positive cooperation across the Cold War divide: "A surprising development was the amount of technical work being undertaken in the Communistic [*sic*] Bloc countries, notably Czechoslovakia." He added that the Czechs had conducted research on "almost every point of discussion in the WG4 Meeting, and presented ample data on a number of the points of discussion."[109] He concluded that "industrial development in the U.S.S.R. and Czechoslavakia [*sic*] has led to mobile radio problems akin to our own. These countries were the most firm supporters of the technical proposals submitted by the U.S." Indeed, he noted that more new work was coming from the latter country than from the United States.[110] Engineers on both sides of the Cold War apparently believed that standardization should be above political rivalries.[111]

Although US relations in WG4 had improved, a few points of friction remained. Bauer noted with distaste the preponderance of government rather than industry representatives in most CISPR national delegations.[112] This government slant in representation aligned with Charles le Maistre's 1948 advice to assure that national delegations to CISPR included government representatives to improve the interactions between CISPR and the CCIR, part of the intergovernmental ITU. Nevertheless, it made CISPR itself a further hybrid between the private IEC and the hybrid public-private CCIR, while the Americans strongly preferred completely private and voluntary standard setting. Further, Bauer argued, "There is a distinct problem in the 'one-man, one-vote' principle in that small nations, without equity in the *manufacture* of motor vehicles, has [*sic*] as big a voice as the larger industrial nations who manufacture and stand behind their products."[113] For example, WG4 debated the number of vehicles to be tested for compliance with CISPR standards. Germany, the United States, and France opposed as excessive the current requirement to test six vehicles in each "Production Series"; the UK (represented in this committee by "Post Office people who of course, were not themselves responsible for conducting the tests") argued to maintain the six-vehicle testing requirement; and the Netherlands, a small "non-manufacturer of vehicles," argued for testing every single vehicle produced, much to Bauer's dismay.[114]

Bauer ended his Oslo meeting report on a positive note, saying that "this group seems extremely willing to act upon good technical information, and is desirous of obtaining it. In spite of nationalistic differences, a relatively cooperative spirit was demonstrated in numerous unofficial discussions." Finally, he reinforced the importance of longevity on the working group, since "productive participation in such meetings as WG4 is accomplished only after accumulation of a large background of knowledge" about the other delegates and the issues under discussion, which he claimed, "requires two or three years' experience to accumulate."[115]

Although by this time the US delegation (mostly members of C63) had achieved a working relationship with CISPR, the challenge of getting the United States fully and productively integrated into CISPR was still ongoing. A few months after the Oslo meeting, a C63 committee met to prepare responses to materials circulated by CISPR for the 1967 CISPR Plenary in Stresa, Italy.[116] At the preparatory meeting, chaired by Showers, the group considered 47 documents that had been distributed by the CISPR secretariat, deciding on the US response to each, from none to submission of a reply. The US delegation report on the Stresa plenary (written by Showers and three other delegates) concluded that the outcome was generally good, with genuine interest and "an excellent spirit of cooperation" toward developing international standards. Indeed, "the U.S.A. contributions were effective in helping to establish an 'international' position and in most cases found acceptance directly or indirectly in the final actions taken." Nevertheless, they thought that the United States could still participate more actively by providing additional US technical and statistical data, submitting its techniques to CISPR, comparing them to those discussed there, and trying to build on CISPR work "to establish similar or related standards in the U.S. where they do not now exist."[117] In discussion of this report, C63 recommended that the United States become even more actively involved at all stages of CISPR's standardization process and formalize this participation in CISPR through a technical advisor to the IEC USNC and other mechanisms.[118] Such a position was established by 1969, and Showers served as assistant technical advisor from 1969 to 1971, and as technical advisor from 1971 until two years before the end of his life.[119]

The US EMC Community Learns to Participate in CISPR, 1970s–1980s

During the 1970s and 1980s, the C63 committee continued to evolve its role within the United States and to lead in integrating the United States into CISPR internationally. Under the quiet but energetic leadership of Ralph Showers, who became chairman in 1968, C63 standardizers increased their emphasis on coordinating, rather than just developing, American standards. More importantly, Showers led the American standardizers closer to CISPR, ultimately becoming its first US chairman from 1979 through 1985, thus completing the process by which US standardizers "learned to participate" in this initially entirely European international body.

C63: Changes in Structure and Focus

In 1969, when the American standards body took its current name, C63 became an ANSI committee. In 1978 C63 sponsorship moved from NEMA, a trade association, to the IEEE, a professional association; the IEEE's Standards Board had developed standards on voluntary consensus principles in electrical engineering since its creation in 1963, and many C63 members participated in it through the IEEE EMC Group.[120] Figure 6.3 portrays a 1971 vision of where that group (labeled G-EMC) fit into the national and international voluntary standardization system. In 1982, in response to the Hydrolevel antitrust decision and OMB Circular A-119 (see chapter 5), ANSI reorganized and shifted from developing to only approving standards.[121] C63 then became an *ANSI-accredited* body under the oversight of the IEEE Standards Board, rather than a committee of ANSI itself.

Changes to C63 membership and policies also occurred during this period. In the 1960s, engineers from firms were typically involved only as

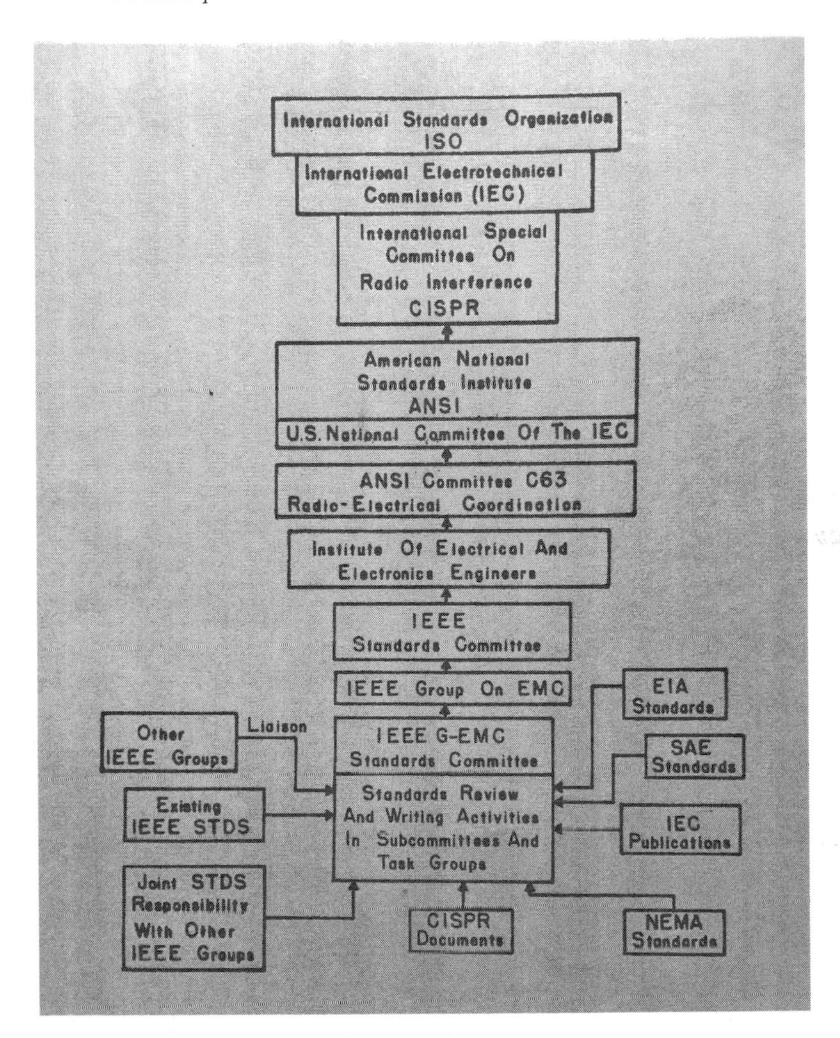

Figure 6.3. A 1971 diagram showing how the standards committee of the IEEE Group on EMC saw itself fitting into the international standardization system, showing both national and international standard-setting groups providing inputs at the bottom and its standards feeding into a hierarchical system with the IEC and ISO at the peak. From J. E. Bridges, L. W. Thomas, and W. E. Cory, "A Progress Report on G-EMC Standards Activities," attachment D of minutes of Administrative Committee Meeting of IEEE EMC Group, July 12, 1971, Philadelphia, Thomas Papers.

representatives of member trade or professional associations, or, rarely, as observers. By the late 1970s, as the committee increasingly set standards for radio wave emissions from (or susceptibility to interference of) manufactured products, more firms (e.g., Hamilton Standard, International Harvester,

Xerox, IBM) sent observers, who were not voting members of C63 but could serve on subcommittees.[122] By the mid-1990s, representatives of these and many more firms were listed as members rather than as guests.[123] In response to new ANSI policies and C63's changing focus and demographics, in 1990 its Steering Committee asked members to classify themselves into six categories—professional society, trade organization, manufacturer, government, testing lab, and general interest—"for use in determining committee balance for letter ballot purposes."[124] The old ASA designation of C63 as a scientific committee no longer held, and traditional balance rules were instituted.

During the late 1970s, external as well as internal challenges caused C63 to shift its focus from writing its own standards toward coordinating the many proliferating EMC standards activities in the messy US system. With more and more electrical and electronic technologies, issues of electromagnetic compatibility among these technologies were increasing. In 1979, Showers held a special meeting of C63 to respond to an FCC Notice of Inquiry that asked parties to identify EMC problems and engineering, consumer, equipment manufacturing, and economic issues, as well as to address the voluntary standards effort in this area. Apparently worried about possible FCC exertion of its authority to regulate in this area, Showers suggested that C63 discuss these issues and craft a careful response. This external threat brought out a blunt statement of what Showers and his fellow standardizers saw as one purpose of voluntary standard setting in the United States: "If the objective (of the voluntary standards effort) is to avoid government regulation, then we must convince the government that the voluntary standards effort will handle the problem."[125]

This statement captures one self-interested motive underlying firm participation in and support of voluntary standard setting, especially in the United States, not usually articulated this bluntly by standards entrepreneurs and leaders. American firms generally distrusted government regulation and preferred private standard setting into which they had input. As the US phase of chapter 5's color television case suggests, many manufacturers did not view FCC regulation as a good substitute for private US standard setting, in part because the government lacked the necessary technical expertise and in part because its process did not assure that the manufacturers bought into the mandated standard. Showers's statement undoubtedly also reflected what a 2011 award citation for ANSI's Elihu Thomson Electrotechnology Medal described as his "outstanding long-term commitment to the voluntary standards process."[126] Thus C63, under his leadership, decided it must submit comments to this FCC docket and coordinate its response with those of its member organizations, many of which also set relevant standards.

This apparent external threat to C63 standardization turned out not to be terribly serious because the FCC did not actually *want* to replace voluntary standards with its own regulations. In the early 1980s the US Congress passed legislation explicitly empowering the FCC to regulate electronic devices' immunity to interference, but an FCC representative clarified its position at a 1983 C63 meeting, saying that, despite pressure to regulate TV and video recorders, the FCC preferred to reference voluntary standards such as those of C63.[127] At the same meeting, C63 adopted its own Policy on Immunity Standards, which approached the issue cautiously, given the tension between market forces and federal regulation in determining whether manufacturers designed immunity into devices they produced. The policy proposed that C63 undertake a program of developing emission and immunity guidelines in its own standard C63.12 and also that it work with other organizations to *coordinate* standards in specific problem domains, an approach that apparently satisfied the FCC.[128]

Showers himself challenged C63 to move toward increased coordination of standards. In a regular meeting occurring later in 1979, C63 took up an item of new business, "coordination of the voluntary standards effort." Attendees discussed a paper Showers had written, highlighting the many organizations setting EMC-related standards and the importance of C63's role in coordinating those standards.[129] In it, he listed bodies that set standards in the EMC domain, including three international organizations (CISPR, CCIR, and IEC) and twelve US organizations (e.g., C63, IEEE, the FCC, the military, and trade associations), recalling the image of the overflowing garbage can at the picnic area evoked in chapter 3. The US organizations, most of which were members of C63, all set voluntary standards for instrumentation, techniques, and limits around electromagnetic emissions and immunity. The discussion led attendees to refocus C63 on better coordinating these efforts: "Committee C63 could offer a service to the many standards developing organizations by bringing together those working in similar areas and ensuring the coordination of effort and compatibility of the final document with other EMC related documents."[130]

A year later this new focus inspired an ambitious but unsuccessful attempt to coordinate and unify all EMC standards at the national level. In 1980, a task group of EMC standardizers led by Herbert K. Mertel, the SAE's representative on C63, met in Washington, DC, to discuss creating a single, unified national EMC standard; the group created an outline for a comprehensive core document (which would align with CISPR Publication 16 and would draw on C63.2, C63.4, and C63.12) for the proposed National EMC Standard, to be supplemented by material on specific application areas. Its purpose was "to provide a coordinated approach to areas of instrumentation, methods of measurement, emission levels and immunity levels in order to achieve electromagnetic compatibility of electronic and electrical equipment and systems."[131] The idea of taking such a coordinated approach to EMC standards and the core document outline were enthusiastically accepted at the next C63 meeting, with suggestions to build references to standards by all the US bodies mentioned in Showers's paper into the document.

By August 1981, the task force had created a bulky 200-page draft claiming to cover 80 percent of the National EMC Standard, but it would never be completed.[132] In the C63 discussion, "Dr. Showers reviewed his concept of a National EMC Standard, its purpose being not to replace existing standards written by other organizations, but to include them in the National EMC Standard by reference," an approach that could be used to shorten the draft. Showers restated his vision that "the main function of Committee C63 is that of coordination, and not to write standards." By the mid-1980s, however, the task force's overly ambitious attempt to create a comprehensive US EMC standard collapsed, in part for lack of people willing to do the time-consuming work involved (a problem that has only worsened since that time), and in part because the idea of creating a comprehensive, published EMC guide was infeasible given the frequency of updates to the various standards that would need to be reflected in it.[133] Still, Showers's vision of C63 as coordinating EMC standards remained central to its approach going forward.

Managing the Consensus Process in C63

The committee continued to develop, coordinate, and update existing standards, and Ralph Showers spent considerable time on the lengthy and painstaking correspondence required to gain consensus in C63. This process is displayed in the revision of the C63.4 standard, as reflected in Showers's own files. Recall that ASA had approved the original C63.4 standard in 1963 and published it in 1964, and the short-lived USA Standards Institute had reaffirmed it in 1969. During the 1970s, Showers moved this standard through revisions resulting in publication of ANSI C63.4-1981, and, by 1983 and 1984, the committee was revising it again to add some new materials on test sites.[134] In 1987, C63 and its oversight body, IEEE Standards, received complaints around the new proposed section and engaged in a flurry of correspondence around that and another passage, resulting in publication of a new version, C63.4-1988.[135]

C63 immediately began yet another round of revisions, this time to incorporate materials related to information technology equipment.[136] By 1990, it had presumably gone through 10 drafts, as Showers conducted a ballot on the 60-page Draft 11 of the revision of ANSI C63.4-1988; the draft received 26 affirmative votes, 5 negative votes, and 1 abstention.[137] The Ad Hoc Editing Committee (Showers and two colleagues) compiled and responded to the 5 negative votes and to suggestions made on affirmative votes, initiating another series of rewrites and votes. Draft 11 went through four more versions (D[raft]11.1-D[raft]11.4), each balloted within the next few months.[138] On the ballot for Draft 11.3, the FCC's Art Wall changed that organization's vote from negative to affirmative and expressed "my personal thanks and congratulations to you and the Committee for the effort that has gone into the development of the subject standard." The FCC took part as a member, but it was glad to have C63 do the hard work of reaching consensus. As Showers wrote in his summary report for C63 and ANSI, all on-time ballots on Draft 11.4 were affirmative, though two negative ballots came in after the deadline.[139] He explained that one of these two was withdrawn after a telephone conversation, and the other, because it was late, "can be considered for a future revision of the document." In July, Showers submitted the standard to ANSI for a public comment period and responded to new comments submitted.[140] In spite of extended correspondence about one objection, at the end of the public comment period Showers recommended publication of Draft 11.4 of the revision of ANSI C63.4-1988 with only a few editorial changes. In November 1990 he submitted it to ANSI for final approval, and in January 1991 it was approved as ANSI C63.4-1991.[141] Another round of the process would, of course, start up soon after that.

The process of achieving consensus on and updating a standard such as C63.4, which was long and covered many areas, could be extraordinarily protracted and arduous. This example highlights an aspect of the process and documentation for voluntary consensus standardizing that was not evident in CCIR member documents (since final voting in that body occurred in face-to-face plenary meetings rather than by mail) but was required for all ANSI-accredited standard-setting committees: balloting and responding to ballots. This additional set of steps constituted the due process required of ANSI-accredited standards committees. Tallying each ballot and addressing each concern—again and again and again—required patience, perseverance, and painstaking care, demonstrating the ongoing time and effort that standards committee members and especially chairmen such as Showers put into the consensus process to get results. The persevering manner mentioned in his high school yearbook entry stood him in good stead in this work.

Managing this process well required a special type of person who was devoted to the work. One of Showers's daughters began her reminiscences of her father after his death this way: "My primary childhood memory of my Dad is that he was always at work. . . . He worked 24/7 before there was 24/7. But he loved what he did." Another daughter noted, "He was very patient about hard work. The more challenging the task, the more he seemed to enjoy it, whether it was professional work, the daily chores, or a hobby." He even made time to work on standards documents when he was at the family's country cottage—between chopping wood and doing other chores. As his third daughter recalled of his final year of life, "Many doctors asked him how he wanted to spend his remaining time. He always answered, 'Do my work and chop wood.'"[142] After Showers's death in 2013, Art Wall, retired from the FCC, would characterize his friend as "a capable and gentle soul dedicated to supporting the development and use of voluntary standards for controlling radio interference."[143] His patience, dedication, and doggedness, as well as deep knowledge of his technical field, paid big dividends in national standardizing. Perhaps these qualities are important to any effective leader of standardization.

The Integration of the United States into CISPR International Standardizing

Showers would show similar qualities in his work in international standardization. In CISPR, he steadily deepened American involvement, built respect on all sides, and integrated himself and the United States into the transnational standardization community.

In the early 1970s, the US electronics industry began to recognize that it must be more competitive internationally, a realization that would grow throughout this period as it lost its preeminence in consumer electronics.[144] It also learned that US industry participation in international standard setting improved competitiveness. In 1971, the Electronic Industries Association, a trade association member of C63, circulated an article titled "U.S. Industry's Stake in International Standards Making," summarizing views expressed at a recent meeting of its International Standards Committee.[145] It highlighted the views of member Jack R. Isken (representing TRW Electronics Group) that previous US technological preeminence had lessened recently, and "the gap between the U.S. and other countries has closed dramatically"; consequently, he claimed, "The willingness of others to accept our standards has undergone a very great transition." US standards would no longer be accepted blindly, a change "in the international standards climate" that was not yet broadly recognized. In particular, Isken worried about the dominance of European countries in the IEC (and presumably the affiliated CISPR) and asserted the importance of "a U.S. effort to keep the whole IEC operation a truly international one, rather than a simple extension and cover for agreements reached between European bloc members." The Electronic Industries Association itself supported this effort by acting as the US sponsor organization for 26 technical committees and subcommittees of the IEC.

The C63 discussion of this article noted that a proposed change in IEC voting policy would require countries not planning to adopt a recommendation to "cast a negative vote with an explanation"—that is, "pressure is on for National Standards to agree with the IEC, or if not, to justify the differences that exist."[146] In December 1973 the USNC of the IEC issued a US voting policy responding to this pressure, saying that when IEC documents were circulated for voting, the USNC should vote for them if they were or could easily be aligned with US standards, but should vote against them if they conflicted with existing national standards. The policy also stated, "The USNC should include in its voting document all comments pertinent to the harmony between the IEC document and any existing U.S. standard, and insist that a summary of the basic differences be included in the preface of the document."[147] The US standards community was taking international standards in this domain increasingly seriously.

The increasing complexity and ubiquity of electronic technology during this period also led the IEC to broaden its own EMC activities. It extended its coverage beyond radio broadcasting to include lower frequency emissions and the immunity of all electrotechnical products, forming IEC Technical Committee 77, Electromagnetic Compatibility (TC77).[148] Showers was on the IEC Committee of Action on EMC Standards that recommended forming

TC77; served on TC77 from the mid-1970s to 2011, two years before his death; and served as chairman and member of the IEC's advisory committee on EMC, which helped the IEC coordinate EMC matters across CISPR, TC77, and other IEC committees that dealt with interference issues.[149]

The European Committee for Electrotechnical Standardization also entered the EMC standardization picture during this period, though it received less attention from the members of C63 than CISPR and the IEC did, especially since CENELEC looked to the IEC, including CISPR, for many of its standards. In the mid-1970s, CENELEC's adoption of CISPR's standard measurement apparatus did, however, push C63 and other US standardizing bodies to adjust if they wished to keep European markets.[150] In 1989, an EMC directive for the European Union was released, "which for the first time imposed immunity requirements on commercial products," further pressuring American companies to adjust their products and C63 to adjust its standards.[151]

Despite all these new entrants into international EMC standardization, members of C63 continued to consider CISPR the most relevant international standards body, and its US representatives (most of them members of C63), under the leadership of Showers, became increasingly oriented toward and integrated into that organization during this period. By the early 1970s, Showers was technical advisor to the IEC's USNC on CISPR (as well as on IEC TC77 and TC1 on terminology), leader of US delegations to CISPR, and chairman of C63, providing an important link between CISPR and C63.[152] His role required a great deal of international travel, with trips often lasting multiple weeks as different organizations, technical committees, and working groups scheduled meetings back-to-back to accommodate those who participated in several of them. As an academic, rather than an employee of a company with interests in the standards being set, he self-funded all his travel associated with standard setting, a very significant expense that he was willing to pay for the personal satisfaction he earned from standards work in this domain and from the community of which he was so important a part.[153]

He missed his wife Beatrice (Bea) and expressed homesickness during 1960s trips, but by the early 1970s their children were old enough that Bea began enthusiastically accompanying him on his travels, sightseeing, visiting art museums, and participating in CISPR social programs, as well as helping him carry briefcases full of paper documents.[154] Although he still worked long hours during the meetings, her participation made the trips more pleasant, and her socializing with other delegates and their wives (during mixed social events as well as the ladies programs) cemented many long-term business and personal friendships. Although CISPR committees were still all male during this period, wives, including Bea Showers (like Nalle Sturén in chapter 5) provided important social glue for the community of standardizers.

CISPR met in the United States again—this time in West Long Branch, New Jersey—in June 1973, with Ralph Showers as chairman of the Steering Committee for the meeting. By this time Bea Showers had traveled three times to international CISPR plenary or working group meetings, and she served on the Ladies Hospitality Committee for the 1973 conference, along with Fred Bauer's wife (figure 6.4).[155] The meeting itself focused primarily on

Figure 6.4. Ralph Showers's wife Beatrice (center back) was in charge of the ladies' program for the 1973 CISPR meeting in New Jersey; here she and a member of her committee, Fred Bauer's wife (far right), lead a tour of the Monmouth College campus with wives of CISPR representatives from Germany, Canada, Sweden, and Norway. Courtesy of the Showers family.

reorganizing CISPR, apparently to bring it more in line with IEC procedures (and less with CCIR ones).[156] The numbered working groups became Subcommittees A–F (to align with current IEC terminology), within which the subcommittee chairmen would designate working groups. Showers became chairman of Subcommittee A on "methods of measurement, measuring instruments, statistical methods," his area of expertise, with the United States as the secretariat. In addition, CISPR named Showers one of three new vice chairmen of CISPR, further advancing the American role in that organization.[157]

The CISPR consensus process was at least as difficult as that of C63 and the CCIR; it now resembled the IEC's more than the CCIR's, though document genres and process elements of both organizations remained. Moving from an initial idea to an international standard involved multiple steps: voting to take up a new work item (IEC genre) or to adopt a study question (CCIR genre), circulating reports (used by both the IEC and CCIR), circulating committee draft standards (IEC) to CISPR national committees for comments, and finally voting by mail under the IEC's six-months rule within CISPR as a whole on whether to make a Draft International Standard into an International Standard (the IEC's current terminology, replacing the earlier "recommendation").[158] In voting on Draft International Standards, the subcommittee secretariat conducted multiple written ballots, tabulated responses by country, and responded to each comment (IEC). The voting results were reported to the CISPR Central Office, initially relatively informally but by the 1990s on IEC forms including each country's status (participant or observer), its vote, any comments it provided, the secretariat's comments on the vote, the number and percentage of participant member countries voting in favor, against, or abstaining, and the overall result (approval or disapproval).[159] All comments and suggestions, with the secretariat's recommended responses to each, were circulated within the subcommittee and attached to the form sent to the Central Office.[160] For each section of the document, every comment was listed by country, with secretariat proposals for how to respond (e.g., suggested edits to the text). Obviously, in CISPR (as in C63) the process of reaching consensus was painstaking and time consuming.

CISPR continued to dominate C63's attention internationally, and Showers played increasingly key roles in it, as well as in other IEC committees.[161] A C63 report on the autumn 1977 CISPR interim meeting in Dubrovnik included the significant statement that at the next CISPR plenary, scheduled for 1979 in The Hague, "it is anticipated that Dr. Ralph M. Showers of the USA will be elected Chairman."[162] CISPR began as a solely European committee

Figure 6.5. Ralph Showers, chairman of CISPR (holding wine glass in center), at a reception for the 50th anniversary meeting of CISPR in Paris in 1983. Courtesy of the Showers family.

and had nine consecutive European chairmen, but in 1979 it elected its first American. In the person of Ralph Showers, the United States had finally found its place in CISPR. He served in this role from 1979 to 1985, including presiding over CISPR's 50th anniversary meeting in Paris in 1984. In that capacity, he received a congratulatory telegram from the Soviet National Committee, which extolled CISPR's contribution to "the cause of peace and creation."[163] Although an American had taken on the leadership role and the top hats of le Maistre's era had been replaced by suits and ties at the 50th anniversary (figure 6.5), CISPR was still composed entirely of male engineers.[164] If the role of gender in CISPR had not changed in 50 years, the national balance had changed, thanks in great part to the efforts of Ralph Showers, who brought the United States into it as well as encouraging other non-European participants.[165]

After his term as chairman ended in 1985, Showers remained deeply involved in CISPR (and the IEC) well into the twenty-first century, taking his last major international trip (this time without Bea, who was too ill to accompany him) to an IEC meeting in Melbourne and a CISPR meeting in Seoul in 2011 when he was 93 years old, two years before the end of his life.[166] His continued, active participation in international and national standard-setting activities long after he retired from University of Pennsylvania in 1989 earned him the sobriquet of "'the energizer bunny' since he was SO ACTIVE for SO LONG," as one colleague noted in an online comment after his death.[167] He received all of the highest awards offered to standardizers within the United States and internationally, but perhaps his greatest accomplishment was bringing the United States fully into international EMC standardization, at last.

This chapter highlights the network of organizations and community of standardizers operating in one obscure but important industrial domain from World War II through the 1980s, illustrating the move from the national to the international level during the second wave of standardization. For all their seeming obscurity, these standards were, and still are, ubiquitous and critical to our daily lives. Without the work of these standardizers, we would worry not just about microwaves and cars interfering with radios and TVs but also about today's cell phones interfering with each other and with other electronic devices in the vicinity. Without these standards, satellite signals and signals from airplanes would interfere with other technology on the ground, from phones to computers, and vice versa. Our modern technology is deeply dependent on these standards.

This case illustrates that, increasingly, international rather than national standardization was critical to the electronic age. It demonstrates how US standardizers were able to overcome their tendency to focus on the national level to the exclusion of the international level during this period, gradually learning how to participate in the international standard-setting community, led by Ralph Showers.

Showers followed a standardizing career path from the IEEE EMC Society to C63 to CISPR and the IEC.[168] After his passing, one colleague described him as "'Mr. EMC Standards' at both the national and international level."[169] Another called him "the voice of academic reason" and "the best technical and administrative titan in the EMC world."[170] Another elaborated on qualities that made him so effective and respected in standards circles:

> He was a very humble man, and never sought the limelight although his vast knowledge and experience certainly would have justified it. . . . From Ralph I learned that you did not have to have the loudest voice or a bullying attitude to get things done, you just had to have a sound reason and convince others that the reasoning was valid, and not give up if at first you did not succeed. He

> will be tremendously missed because of his ability to cut through the chaff and expose the wheat no matter how good or bad the wheat was.

In congratulating Showers for receiving the IEC's 1998 Lord Kelvin Award for outstanding contributions to global electrotechnical standardization, Fred Bauer exclaimed, "You are simply that self-effacing person who can move mountains without making waves!"[171] He shared these qualities with earlier standardization entrepreneurs like Charles le Maistre, Charles Dudley, and Olle Sturén, but Showers did what hundreds of other have probably done in different specific fields: he maintained and helped improve the committees that do the actual work of standardization, helped set their agendas and processes, shepherded specific standards through those processes, helped coordinate across organizations, and nurtured and recruited new members into an increasingly international community.

part III

THE THIRD WAVE

By the 1980s the second-wave network of international and national voluntary standard-setting organizations was well established and reasonably stable. At the end of that decade, however, the emergence of computer networking and the perceived inability of the traditional standard-setting organizations to respond to a more rapid pace of technological change around computing challenged existing standard-setting organizations. A third wave of different types of standard-setting organizations emerged initially in the digital realm. They were less international (they did not have representation or voting by nations) but ultimately more global than the second-wave organizations, setting standards for an increasingly global high-technology sector, then spreading beyond technology into social realms.

Chapter 7 focuses on the late 1980s and 1990s, relating the rise first of the free-wheeling Internet Engineering Task Force (IETF); then of another new type of organization even further from the traditional organizations—standards consortia comprised primarily or exclusively of a few like-minded firms; and finally of an organization that combined characteristics of second-wave organizations with those of IETF and consortia—the World Wide Web Consortium, or W3C. Chapter 8 provides a case study of standardization in W3C to highlight this combination of characteristics in action and to show how it both resembles and differs from the earlier model of private voluntary standard setting. In chapter 9, we turn from high tech to new realms, observing how the third wave takes the multi-stakeholder consensus model into quality management systems, then into environmental and social responsibility standards.

7

Computer Networking Ushers in a New Era in Standard Setting, the 1980s to the 2000s

As chapter 6 demonstrated, the international and national network of voluntary standard-setting organizations was complicated, with the many organizations overlapping and interacting, but by the 1980s it was well established and reasonably stable. Computers and computer networking would challenge the nature and pace of voluntary standard setting starting in the late 1980s. Initially, existing and new voluntary standardization organizations based on the traditional model of the first and second waves tried to establish standards for the new technologies, but their efforts did not meet the demand for increased speed to keep up with the new technology. De facto victory in a standards war for internet protocol standards raised a small informal body of American software engineers, the Internet Engineering Task Force (IETF), from obscurity into the major body setting standards for the increasingly important and global internet. Another new form of standard-setting organization emerged around the same time—consortia of similar multinational firms that joined forces to arrive at voluntary standards more rapidly, without going through the multi-stakeholder consensus process. The first- and second-wave international organizations, the International Electrotechnical Commission and the International Organization for Standardization, had to find ways to work with consortia to avoid becoming obsolete in this new technical field. Finally, in the 1990s, software engineer Tim Berners-Lee introduced the underpinnings of the World Wide Web and launched the World Wide Web Consortium (W3C) to set standards for it. W3C combined characteristics of earlier organizations with those of the IETF and consortia.

This chapter is the story of how this new set of organizations arose, initiating a wave of private standard-setting organizations that were neither national nor international but aspired to be *global*. Their work would have just as much influence on the global economy and everyday life as the standards created by the international network of standard-setting bodies organized under ISO.

Standard Setting for the Internet

Recall from chapter 5 that in 1960 Olle Sturén, then the head of the Swedish Standards Institute, convinced ISO to create Technical Committee 97 on Computers and Information Processing. At TC97's first meeting in 1961, the American national body (to become the American National Standards Institute in 1968) was named its secretariat and in turn delegated the secretariat duties to its own newly formed Sectional Committee X3.[1] In 1987 ISO TC97 would merge with the IEC's Committee on Information Technology Equipment, IEC TC83 (formed in 1980), to become ISO/IEC Joint Technical Committee 1 (JTC1) on Information Technology, the primary traditional voluntary standard-setting venue for issues around computer hardware today.[2] In 1968 the International Telecommunication Union's CCITT (International Telephone and Telegraph Consultative Committee, a merger of the separate telephone and telegraph standards committees mentioned in chapter 3 and elsewhere) set up a Joint Working Party on New Data Networks to work on standards for networks of digital computers.[3] Meanwhile, some newer, computer-oriented technical associations also followed a relatively traditional standards path. The pioneering Association for Computing Machinery, founded in the United States in 1947, formed standards committees and in the early 1960s collaborated with other US engineering associations such as the Institute of Radio Engineers in developing standard terminology around computing.[4] The International Federation for Information Processing (IFIP) was officially launched in 1960, growing out of the 1959 First International Conference on Information Processing. This new international body of information technology associations immediately created a Committee for the Standardization of Terminology and Symbols, chaired by a computer scientist who was active in the British Standards Institution's efforts in computer terminology and thus well acquainted with traditional standard-setting processes.[5] Bigger changes were, however, on the way.

In this section, we examine the 1980s standards war around computer networking within and beyond the primary international standardizing bodies that had profound consequences for the standard-setting world. From that conflict emerged a de facto standard that opened the way for a new and different type of global voluntary standard-setting body—the IETF.

The Internet Standards War

With the move toward connecting or internetworking computers in the 1970s and 1980s, the first challenge to traditional international standard-setting models emerged out of a standards war. Three countries—the United States, France, and the UK—were all pursuing the goal of interconnecting heterogeneous computer networks, the United States through the Defense Department's Advanced Research Projects Agency (ARPA), which was developing its own computer network, France through its national research lab's Cyclades project, and the UK through experiments in its National Physical Laboratory (NPL). The well-meaning efforts of these players to work toward international standardization through traditional standard-setting organizations such as the ITU's CCITT and ISO failed; meanwhile a new de facto standard, TCP/IP (transmission control protocol / internet protocol), emerged and catapulted the IETF—a standardization body very different from previous organizations though sharing some of their characteristics—into the primary standardizing body for the internet.[6]

ARPA funded basic research in computer science and created computing research centers at several US universities. In the late 1960s, to allow shared use of computers across sites, it created the new ARPANET project to connect these computers, placing it under Lawrence Roberts in ARPA's Information Processing Techniques Office (IPTO). Connecting computers required creating a system to transmit data from one location to another. One approach was circuit switching, used for telephone calls, which assigned a dedicated channel or circuit over which all the data would be transmitted at once. But Roberts was aware of work on a second, more robust alternative, packet switching, which broke up the communication into smaller chunks of data and routed them independently to their destination in addressed packets. It had been independently developed by both Paul Baran at the Rand Corporation in the United States and Donald Davies at NPL in the UK.[7] Packet switching would become a critical technical component of ARPANET.

To coordinate computer networking efforts within the ARPA research network, a group of researchers and graduate students at the various locations began meeting regularly, as the Network Working Group (NWG), coordinated by UCLA graduate student Steve Crocker. The more senior researchers worked on hardware development, leaving what was then considered the less prestigious software development primarily to the graduate students. In 1969, for the NWG's internal decision making, Crocker suggested using unofficial and modestly named requests for comments (RFCs) to solicit feedback and seek consensus on standard protocols and other issues. Because NWG members were working entirely within the ARPA-funded project, they did not look to standard-setting bodies such as ANSI's X3 committee or the IEC for models of standardization genres. After Crocker issued the first RFC, his fellow graduate student on the NWG, Jon Postel, took over editing such documents into the NWG RFC series (initially on paper but in electronic form as soon as ARPANET was live in the early 1970s), which quickly became the vehicle for quasi-official standard setting for ARPANET.[8]

Meanwhile, in 1972, a live demonstration of ARPANET highlighted the First International Conference of Computer Communications in Washington. Three groups of researchers dominated the conference: the ARPANET group led by Robert Kahn, director of IPTO; a team from the British NPL including the packet-switching developer Donald Davies; and the French Cyclades packet-switched network project led by Louis Pouzin of the French national research lab. At the conference they formed the International Network Working Group, or INWG, to forward work on packet-switched networks. Initially conceived as an informal body in the spirit of the NWG, the INWG named as its chair Vinton Cerf, an ARPANET researcher who had just completed his PhD from UCLA and was starting out on the faculty at Stanford.[9]

Although the NWG, working within the ARPANET community, could disregard the larger world of standards, the international INWG, interested in setting protocols and standards that would work across public and private networks in multiple countries, faced a very different situation. In particular, members of the INWG worried about efforts by the ITU's CCITT to create international standards for data networks. The CCITT was a hybrid public-private standard-setting committee, like the CCIR (the International Radiocommunication Consultative Committee) discussed in chapter 6, and its members were almost exclusively national postal, telegraph, and telephone agencies (PTTs). They focused on the telephone lines that connected the network, rather than on the computers at the nodes, and thus they sought to put control in the lines with virtual circuits. By 1972 the CCITT's Joint Working Party, which had now been formalized into CCITT Study Group VII, New Networks for Data Transmission, had approved its first recommendations on connecting public and private data networks, and it had created a study program and study questions focused almost entirely on virtual circuit connection to be addressed before its next plenary in 1976.[10]

Members of the INWG, who saw datagrams allowing connectionless control within nodes as the future of computer networking, thought they should try to redirect the CCITT effort. They decided to do so through IFIP, of which Pouzin was a member. Establishing the INWG as an IFIP working group was easy; moreover, as Cyclades collaborator Hubert Zimmerman later noted, IFIP was "a meeting place for the scientific community."[11] Although that orientation made IFIP a congenial host for the INWG, it was not at all similar to the more bureaucratic CCITT, which was dominated by PTT functionaries, not by computer scientists or even by engineers representing firms. As an IFIP working group, the INWG saw its role as contributing material on datagrams to the CCITT as well as to ISO TC97.

The INWG's efforts to have an influential voice in international standard setting over the next few years were unsuccessful. Although all INWG members supported datagrams for international networking, research was ongoing in this very new area, and they did not yet agree on what exact approach to use. At the same time, by the mid-1970s, ARPA was attempting to link its own domestic ARPANET (connecting mainframe computers in the United States) to two other experimental networks that used radio and satellite connections. In 1973 Kahn and Cerf (now working for ARPA in Washington) had started working toward an ARPA "*Inter*net" to link these various systems, the kernel of what we now simply call the internet. The researchers decided to use the newly devised and underdeveloped transmission control protocol (TCP) to link the systems. Although the ARPANET-wide switchover to TCP/IP (in 1978 TCP was split into two parts, TCP and IP or internetwork protocol) would not occur until the beginning of 1983, by 1975 those ARPA researchers who were trying to connect existing networks were already committed to further developing this new approach, while the other members of the INWG were still trying to find a datagram approach that all could agree upon.[12] This technical area was not yet mature enough for traditional voluntary standard setting, which depended on existing, tested technology.

Shortly before the 1976 CCITT Plenary, a four-person subcommittee of the INWG (including Cerf) hammered out a compromise proposal that received lukewarm support within the INWG. But, because they waited so long, the proposal went to the CCITT deliberations late (we saw in chapter 6 how the CCITT's sibling ITU committee CCIR rejected a late submission) and without having been ratified by a vote in the IFIP working group.[13] At the 1976 plenary, CCITT ignored the datagram approach entirely and approved its X.25 standard based on virtual circuits, which put functionality in the network and consequently control in the hands of the PTTs.[14] To add to the INWG's failure, even after the INWG/IFIP vote, when Cyclades, NPL, and the European Informatics Network agreed to convert to the chosen compromise system, ARPA did not follow suit.[15]

Meanwhile, ISO was launching its own international standardization effort related to internetworking, focusing not on national PTTs but on manufacturers and users of computers, and on ways to connect them. By this time companies had introduced proprietary networking systems such as IBM's System Network Architecture (SNA) and Digital Equipment Corporation's DECnet, but none allowed interconnection across proprietary systems. ISO took up that challenge in 1978 when it began its Open Systems Interconnection (OSI) effort in TC97. The CCITT's X.25 standard had been rushed to completion without consideration of many issues (e.g., datagrams), closing off options; in contrast, ISO's OSI standard took too long and left too many options open.

In 1977 several computer scientists and engineers on the British Standards Institution's computer standards committee, including some members of the INWG, requested that ISO create a new subcommittee of TC97 to develop network standards for open systems, as a way around both the PTT-dominated CCITT standard and the IBM-dominated proprietary system, which threatened to become a de facto monopoly. ISO responded by creating Subcommittee 16 (SC16), Open System Interconnection, with the United States (specifically ANSI's X3 committee) as secretariat, and Charles Bachman of Honeywell as chair. In other X3 standard-setting attempts, Bachman had already faced IBM's power to stall and prevent reaching a consensus. Before SC16's first meeting in 1978, Bachman and X3 had proposed a seven-layer reference model for open systems interconnection that strategically expanded on IBM's five-layer model for SNA in a way that created problems for IBM, while fitting well with Honeywell's proprietary networking system.[16]

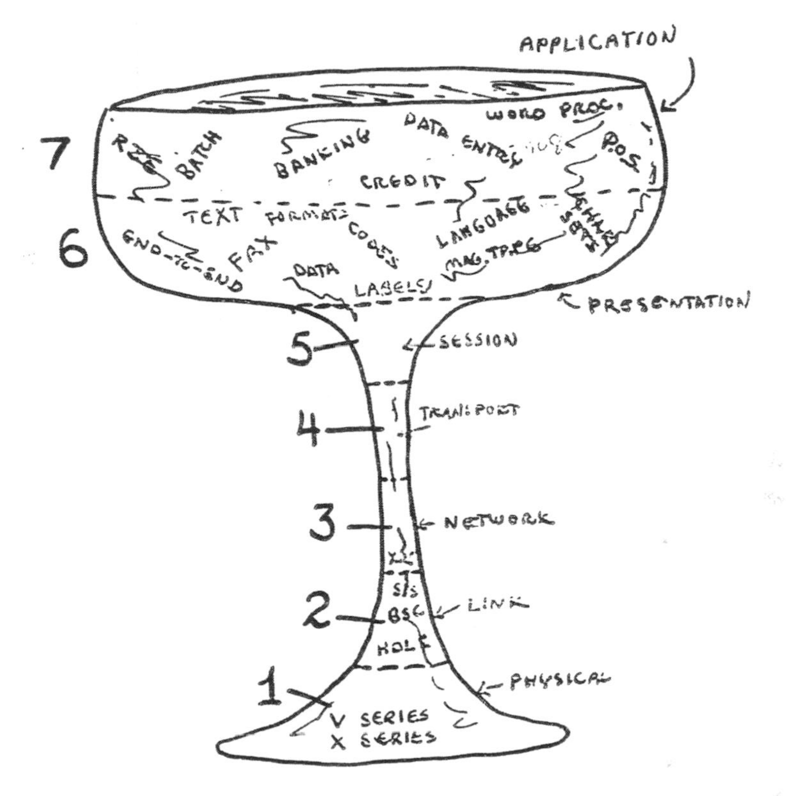

Figure 7.1. The seven-layer reference model for open system interconnection is drawn as a drink glass in 1990 committee notes for ANSI X3.T2, Data Interchange. Used by permission of Haverford College Archives.

Undoubtedly X3 and ISO SC16 had plenty of members happy to counter a potential IBM monopoly, and Bachman pushed hard for consensus on the model, trying to achieve a framework within which protocols could be adopted as standards before the PTTs or IBM could interfere too much.[17]

Despite many stakeholders as well as delays generated by IBM and others, the reference model progressed through the ISO standardization process from working draft in 1979, to draft proposal in 1980, to draft international standard in 1982, to approved international standard ISO 7498—the Basic Reference Model for Open System Interconnection, known as OSI—in 1983. The final model had seven protocol layers, from bottom to top—physical, data link, network, transport, session, presentation, and application—as portrayed by a participant in figure 7.1.[18] For such a broad-ranging and multi-stakeholder standard, five years from proposal to standard was not unusually long. But the Basic Reference Model was only a framework. Five years into the effort, the committee had barely started filling in the conceptual layers with specific standards and getting it adopted and implemented in the market.

OSI was unusual for an ISO standard in one very important respect. Typically, ISO and other traditional private standard-setting organizations waited to begin the process until a product or technology was reasonably mature, and existing, tested methods were proposed as standards. SC16, in contrast, was developing new—not standardizing existing—technology. No products had yet implemented the seven-layer system, and specific standards were still being added to each layer into the early 1990s, after oversight of the OSI standards moved from ISO TC97, SC16 into ISO/IEC JTC1 in 1989. Many stakeholders were brought on board by OSI's incorporation of multiple specific standards into those acceptable for different layers of the model (e.g., a virtual circuit switching service from CCITT's X.25 standard was incorporated into OSI's transport layer), and the committee even attempted to merge ARPANET's TCP/IP into OSI by including TCP-like and IP-like programs as standards in OSI, but internet engineers fought fiercely against any compromise with OSI, which they saw as the enemy.[19] National governments and agencies, including the US Defense Department, which had funded ARPANET with its (incompatible) TCP/IP protocols, signed on to support OSI (or a special version of OSI called Government OSI Profile, or GOSIP), but what did such a commitment mean in practice, given the many alternatives for each layer and the lack of widespread OSI implementations?

Although the OSI reference model would continue to dominate conceptually, its implementation stalled completely in the early 1990s. Carl F. Cargill, longtime director of standards at Sun Microsystems, Inc. and later standards principal at Adobe Systems, a widely published expert in computer standardization, located the problem in OSI's use of what he termed *anticipatory standards*: "Anticipatory standards are those which standardize a technology in advance of that specific technology being available as a product in any viable commercial form." Because no "underlying acceptable implementation (or even clear market definition)" existed for OSI, "the standardization effort expanded to cover nearly every possible contingency." Ultimately this approach sowed confusion: "Because of the number of options available in the standards, an OSI system provider could both comply to the OSI standards and be totally noninteroperable with another compliant system."[20]

The fate of the OSI project is illustrated by the records of an ANSI committee—X3T2, Data Interchange—working on part of one layer of OSI, kept by its long-serving secretary, Murray Freeman of Bell Labs, who joined this committee in 1986. In the beginning, dozens of members from every major computer and telecommunications company and many government departments attended, exchanging voluminous paper documentation (copies of more than 100 documents mailed to more than 100 people in 1990). Freeman constantly urged members to shift from paper to email, but with little effect until the mid-1990s. Over time, the number of participants dropped, and those who remained had increasing difficulty getting support from their employers. By 1995, too few members attended face-to-face meetings, so "motions were prepared . . . and sent by a combination of e-mail and facsimile to absent members prior to the closing plenary asking for attendance at the closing plenary or votes 'in absentia.'" The last meeting for which Freeman kept records occurred with only a handful of men in 1998, the year in which JTC1 published the one frequently updated OSI standard the committee contributed to, ISO/IEC 8824:1998, Information Technology—Abstract Syntax Notation.[21]

While Freeman was dealing with the frustrations of trying to create an anticipatory standard, the Defense Department had implemented TCP/IP across ARPANET and begun to spin off the civilian internet in 1983; TCP/IP became, de facto, the most workable and prevalent protocol for internetworking. Ironically, a paper distributed to one of the first meetings of X3T2 entitled "An Almost-Everything-Independent Data Transfer Method" had raised (at least theoretically) the possibility of such a universal standard.[22] TCP/IP had won the standards war over ISO's OSI and the CCITT's X.25, setting the stage to make the informal body that descended from ARPANET's NWG into a major new standardizing body going forward.

The Internet and IETF as a New Model for Voluntary Standard Setting

Soon after ARPANET was up and running in the early 1970s, the original Network Working Group ceased to operate, but the somewhat more formal technical oversight groups that emerged to govern it continued to use some informal processes developed by the NWG, including requests for comment (or RFCs) as the main (now electronic) document series used to standardize protocols. Throughout the 1970s, Cerf and Kahn led ongoing protocol development. To advise them, they initially established a group called the Internet Configuration Control Board, chaired by David Clark, faculty member at MIT. In the early 1980s a broader body called the IAB (initially Internet Advisory Board, then Internet Activities Board, and ultimately Internet Architecture Board), still chaired by Clark, took over this role; although it was larger than the original advisory group, it was still composed primarily of members of the ARPANET research community, from Department of Defense agencies, from government contractors like Bolt, Beranek and Newman, and from universities with Defense Department funding such as MIT. Even after the private and soon increasingly commercial internet broke off from ARPANET, the IAB continued to be dominated by a small set of expert computer scientists and was referred to as a "council of elders." The IAB further spun off protocol development and standardization activities to a new body, the Internet Engineering Task Force, which first met in January 1986.[23] Although participation at that first meeting was by invitation (mostly to government-funded researchers), the IETF quickly opened up to anyone who wanted to attend, including representatives of firms that wanted to commercialize the internet. The group, which still sets standards for the internet, shed the more autocratic aspects of ARPANET governance within the Defense Department, moving rapidly toward a more democratic model of governance.

The new model of internet standard setting was refined in what has been called its "constitutional crisis" in 1992. This story has been well told elsewhere but deserves brief summary here.[24] It was initiated by the IAB's decision to adopt OSI's ConnectionLessNetwork Protocol in place of IP (the second part of TCP/IP) to solve an anticipated shortage of internet addresses. At an IETF meeting a month later, IETF participants massively and vociferously protested the decision, also complaining to the Internet Society, a nonprofit organization newly formed to replace the Defense Department in overseeing the IAB and IETF. The IAB and key players Vinton Cerf and David Clark ultimately backed down in the face of the protest. Cerf became a hero by stripping off his jacket, vest, and shirt during his talk to reveal a T-shirt reading "IP on Everything." More significantly for internet standardization, Clark gave a rousing talk that summarized and lauded the IETF approach as follows: "We reject: kings, presidents, and voting. We believe in: rough consensus and running code."

This manifesto framed the process as bottom-up and democratic, rejecting kings and presidents—and presumably ISO committee chairs—as well as voting.[25] The IETF has had no official membership since its first meeting. Its rejection of defined membership was enshrined in the procedural and cultural guide to the IETF—"The Tao of IETF: A Guide for New Attendees of the Internet Engineering Task Force"—first published as an RFC in 1993 and maintained and periodically updated since.[26] It explained, "There is no membership in the IETF. Anyone may register for a meeting and then attend. The closest thing there is to being an IETF member is being on the IETF or Working Group mailing lists."[27] The IETF similarly rejected formal voting. Since 2001, the "Tao" has explained, "Another aspect of Working Groups that confounds many people is the fact that there is no formal voting. The general rule on disputed topics is that the Working Group has to come to 'rough consensus,' meaning that a very large majority of those who care must agree."[28] Different working groups use different (nonvoting) methods of determining rough consensus, including, for example, having advocates of different opinions hum and judging which group is louder.[29] The reference to "rough consensus" in Clark's manifesto indicated that the IETF rejected what participants saw as the bureaucratic voting rules of ISO, the CCITT, and other more traditional standard-setting organizations. Finally, "running code" referred to the IETF's rejection of anticipatory standards like OSI, instead insisting that new protocols or code had to run on multiple implementations before becoming a standard.

IETF processes resembled those of traditional voluntary standard-setting organizations in some ways. It depended on volunteers who were overwhelmingly technical experts (typically computer scientists and software engineers). It established standards based on consensus, even though it defined consensus more loosely than did the traditional organizations, and it did not require the due process of responding to every negative view. Like traditional organizations, the IETF followed defined steps toward a standard (proposed standard, draft standard, and internet standard). And it acknowledged the need for oversight and coordination of the process, so it created the Internet Engineering Steering Group (IESG), made up of the IETF chair and area directors as well as IETF participants who volunteered for it and were then chosen through an elaborate (but nonvoting) process.[30]

Despite these similarities, differences existed in several areas. IETF standards were made available free on the internet, rather than sold in paper form to support the body, as was the case for many traditional standard-setting organizations. Unsurprisingly, the IETF also embraced electronic ways of conducting its work between meetings long before the more traditional standard-setting organizations did so. Indeed, during the 1990s, at the same time that the IETF was using its electronic RFC process (modeled on the even earlier electronic RFCs in the NWG) and working group email lists to do work between meetings, OSI-related committees of ISO were either using the internet primarily to coordinate meetings or resisting electronic communication altogether (recall that it took until about 1995 for Murray Freeman's X3T2 committee to consider email a normal means for communicating). The IETF's use of email lists and electronic RFCs between meetings undoubtedly permitted some speed-up compared to other standards organizations. The ITU's CCITT worked on a four-year cycle of plenaries at which standards could be approved (or delayed for four more years), and the OSI reference model took five years from beginning to international standard in ISO. Of course, processes in these traditional venues could extend much longer and sometimes never produced standards. At least initially, the IETF proceeded somewhat more rapidly, though between 1992 and 2000 its pace slowed considerably.[31] Complaints about IETF slowness became common in the twenty-first century, with one IETF participant referring, in 2015, to "the points of ossification in the IETF process" and characterizing the IETF as "the paradigm of being old and slow and very concerned about posterity."[32] In 2011, the draft standard stage was eliminated, presumably to speed up the process.[33] More recently many standards remain at the proposed standard level but are effectively treated as internet standards.[34]

Certainly, the tone of interaction in the IETF differed from that of more traditional standard-setting organizations, probably reflecting both the 1990s and 2000s culture of computer experts and heavy use of electronic communication, as well as the younger age of participants in its early years. While the paper-based documentation of standard setting shown in chapter 6 was neutral, impersonal, and professional in tone (and no transcriptions of the actual meetings survive to indicate how face-to-face meetings differed), IETF email lists reveal a more combative, personal, and uninhibited tone of argumentation, punctuated by occasional "flame wars," a style that has been shown to be gendered.[35] And certainly the July 1992 IETF meeting, as described in personal accounts, involved uninhibited face-to-face interactions, as well. Attire at IETF meetings was decidedly different from the top hats at the 1926 IEC conference in New York (see figure 3.1) and the suits at

Figure 7.2. This cartoon of an IETF meeting comes from an instructional video, "Top Things to Know When Attending an IETF Meeting." Reproduced with permission of the Internet Society and the IETF Trust.

the 1984 CISPR 50th Anniversary meeting (see figure 6.5); "The Tao of IETF" emphasized the informality of dress at IETF meetings:

> Since attendees must wear their name tags, they must also wear shirts or blouses. Pants or skirts are also highly recommended. Seriously though, many newcomers are often embarrassed when they show up Monday morning in suits, to discover that everybody else is wearing T-shirts, jeans (shorts, if weather permits) and sandals.[36]

The IETF and internet standardization challenged the formality and courtesy that characterized earlier generations of standardizers. Although men still dominated (and the aggressive style of argumentation no doubt deterred many women from participating in) IETF standardizing, most members of the IETF would have rejected any "gentlemanly" pretensions. Figure 7.2, from a YouTube video preparing people to attend an IETF meeting for the first time, suggests both the informality of dress and the freewheeling tone.

Perhaps the most significant difference from the principles of standardization developed within the traditional organizations at the national level was the lack of a balance requirement. Anyone could join a working group email list and register for and attend a meeting. On the one hand, open participation meant that individuals from all interested stakeholder groups could, in theory, participate in internet standard setting, which served as the basis for claims about the IETF's democratic nature. Of course, a small set of expert players willing to do the work typically dominated. The IETF's openness combined with its lack of defined membership or balance rules, however, meant that a commercial firm could, if it wished, send many software engineers to participate in a working group dealing with an issue particularly important to it, thus stalling the standard or potentially even railroading its preferred solution through the process. When asked about the potential for firms to game the system in the absence of a balance requirement, a recent IETF chair, Russ Housley, said that "when they want to, [it] is fairly easy for them to cause delay, but I think it's fairly difficult for them to completely derail" the standardization process. He also noted that although it was sometimes hard to get people from all the sectors that they would like represented, "we do what we can to knock down barriers and to encourage participation from places we can identify are low." Nevertheless, "we don't look at the industry, government, academia kind of model, whereas the IEEE"—which follows the traditional balanced consensus process of a first wave organization—"uses that very heavily when they evaluate a ballot pool."[37]

Of course, even in more traditional standards organizations with balance rules, firms found ways to delay and obstruct consensus. Moreover, large multinational firms such as IBM could wield undue power in ISO international standardization efforts by getting IBM employees into the delegations of multiple countries, allowing them to combine efforts to stonewall and delay.[38] Still, the IETF's lack of a balance rule was a significant departure from the norms that evolved in voluntary standard-setting committees, potentially allowing a single player or group of players to wield greater influence.[39] Moreover, the IETF's notion of "rough consensus" allowed it to ignore disagreement if it was limited, rather than responding to and attempting to resolve every negative comment. As Housley put it, "It doesn't have to be unanimous and sometimes we tell people they're in the rough and that's kind of the rough part of rough consensus that also aligns with the golf terms."[40] That is, they were outside the consensus and without recourse.

The IETF was only the beginning in the evolution of new standard-setting organizations. Although it differed from older standard-setting organizations in several respects, it shared many of the original democratic—at least within a technocratic framework—values and principles. The proliferation of consortia for standard setting introduced a bigger departure from the principles of standardization traced in earlier chapters.

The Emergence of Standards Consortia

Andrew Updegrove, a Boston lawyer who has, by his count, set up more standards consortia than any other attorney, described the impetus for the rise of standards consortia as follows:

> Rightly or wrongly, a feeling arose in the late 1980s that the SDOs [standards developing organizations] acted too slowly to provide useful standards in the fast-paced world of technology. Further, their dedication to permitting all parties to participate in standard setting led some companies wishing to create products based on standards to feel that their needs were not being directly enough met.
>
> The result was the explosive formation, beginning in about 1987, of a myriad of unofficial groups of companies, each usually formed to create a standard to address a single commercial need.[41]

Such consortia typically comprised a set of vendors with aligned interests in creating interoperable products, who came together to create standards without first going through the time-consuming process of traditional multi-stakeholder standard setting or an expensive and time-consuming standards war.[42] In addition to standard setting, such alliances sometimes adopted a variety of other functions, ranging from research and development to implementation and marketing. Standards consortia have taken on different legal forms and different ways of operating, making a precise definition difficult, but standards scholar Tineke Egyedi uses the term to "refer to an alliance of companies and organizations financed by membership fees, the aims of which include developing publicly available, multi-party industry standards or technical specifications."[43] The membership fees in such consortia could be quite high, leading to the occasional description of them as "pay to play."[44]

Although a very few emerged earlier, Updegrove dubbed 1988 "the year of the consortium."[45] As indicated above, he saw as precipitating factors the slow speed of the traditional standard-setting organizations and firms' desire for more control than standards organizations afforded them. Carl Cargill, a standard setter who has written extensively about the standards world, served on the boards of many standards consortia, testified to Congress on standards, and served as standardization expert on panels for the Office of Technology Assessment and General Accounting Office, has argued that the key year was 1989.[46] He noted additional factors encouraging formation of consortia: the weakened positions of the ANSI-accredited X3 committee and of ISO/IEC JTC1 in the United States and internationally. The US Department of Commerce under President Ronald Reagan had challenged ANSI as the "first among equals" in US standardizing organizations, proposing a new body—Standards Committee, USA—to oversee all US standardizing bodies, including ANSI; ANSI ultimately pushed back the proposal, but to get ANSI-accredited organizations to support it in this effort, it granted them some concessions that weakened the national body. Meanwhile, ISO's image among US companies was also declining, Cargill argued, based on the OSI debacle, the emergence of ISO 9000, which was looked down upon by many US companies (see chapter 9), and the lack of recent successes in JTC1.[47] Turning to political economy rather than standards politics, historian Andrew Russell recently noted a US policy change as yet another factor: to increase American competitiveness, the National Cooperative Research Act of 1984 permitted cooperative standard setting under antitrust law as long as no price fixing or other collusive behaviors took place, opening the way to participation in standards consortia by American high-tech companies.[48]

Probably for all these reasons, in the late 1980s large information technology companies in the United States and elsewhere turned away from multi-stakeholder consensus-based standard-setting organizations and toward each other to address their standardization needs, especially around attempts to establish open, rather than proprietary, systems in many areas. They soon learned that consortia also had other advantages; most notably, in addition to producing a specification, such an alliance could also support its implementation in products, a concrete outcome that ISO, for example, did not provide.[49] Thus, some firms became more strategic in establishing such collaborations to achieve business aims.[50]

The goals of early consortia shifted as the situation shifted. In 1984 European computer companies Olivetti, Bull, ICL, Siemens, and Nixdorf founded X/Open to create an open computing environment allowing applications in various UNIX-related environments to be ported between systems, to counter the advance of IBM's SNA proprietary network. By 1990, X/Open's shareholders had increased from 5 to 21, a group that now *included* IBM, as well as AT&T, Sun, and other US firms. The so-called shareholder members were the companies that paid the most and had the most influence on the common applications environment that X/Open was creating, but ultimately other firms could buy into a user council, two vendor councils, and other membership categories at lower levels.[51] In 1996 X/Open merged with the Open Software Foundation, founded in 1988 by seven major companies (Digital Equipment

Company, Hewlett Packard, IBM, Apollo Computer, Groupe Bull, Nixdorf Computer, and Siemens AG) that wanted to create an operating system to compete *against* the UNIX operating system (controlled by AT&T and Sun Microsystems at that point).[52] The merged Open Group consortium still existed in 2017 and described itself as "leading the development of open, vendor-neutral IT standards and certifications."[53]

Like X/Open, consortia often had different types of membership, such as a secondary set of user firms paying lower fees and having no say in creating the standards, but helping to get the standards adopted (the standards developed by consortia were normally available free, though in some cases implementing them might involve paying royalties to patent holders, because the members wanted as many companies to adopt them as possible). Typically, the firms leading development of the standards shared interests defined by the coalition they created or joined. As Cargill explains, consortia have "a precondition that SDOs [standards developing organizations] do not enjoy—basically, the members of the consortium are usually like-minded and usually wish for action to occur."[54] This common condition assured that any standards they set were the consensus not of a balanced set of stakeholders but of an intentionally unbalanced set. They were typically controlled by technology producers—at least during the initial stages of standards development—though users often joined them during implementation.

Standards consortia also shared some characteristics with trade associations. Updegrove sees a convergence between trade associations and standard-setting organizations in the high-tech world. He notes, "Trade associations have existed for over 100 years to serve the general needs of an industry, providing a place for participants to share knowledge, plan promotional activities, share lobbying expenses, and otherwise further their collective interest." In today's technology sectors, however, "agreement on standards became part of the necessary strategy to enable the growth of these new sectors, and the furtherance of the collective good."[55] Recall that trade associations often included standards committees that played a role in national standard-setting bodies in the UK, Germany, the United States, and other countries throughout the twentieth century (see chapters 3 and 6). These trade associations, however, typically included a wider set of industry players than many modern consortia, which were at least initially often composed of a very small set of firms.

One very early organization exemplifies the merging of trade associations into consortia. The European Computer Manufacturers Association (ECMA) was founded in 1961 as a European trade association focused entirely on standard setting.[56] Although three companies—Compagnie des Machines Bull, IBM World Trade Europe Corporation, and International Computers and Tabulators Limited—initiated it, they invited all known European manufacturers of computing equipment to a meeting to form a new trade organization focused on setting standards. They defined explicit criteria for membership, including that they be "companies who in Europe develop, manufacture and market data processing machines or groups of machines used to process digital information for business, scientific or other similar purposes."[57] The twenty original member companies they recruited paid dues to support the organization. Because ECMA made standard setting its purpose from the start, it immediately became engaged with ISO and the IEC in discussions of standard setting around computers and became a liaison member to JTC1 from that joint committee's beginning in 1987.

As consortia captured business from the traditional standards organizations in the 1990s, their value was hotly debated. In the context of European Union standards, for example, "the general feeling is that standards consortia work more effectively, but that they have restrictive membership rules and are undemocratic."[58] That feeling accounted for the distinctions that the US OMB A-119 and the EU New Approach (see chapter 5) drew between voluntary consensus standards organizations (to use the US term) and other standard-setting bodies, especially because these regulatory actions dealt with standards that were referenced in law. Nevertheless, European standardization scholar Egyedi has argued that "dominant rhetoric underestimates the openness of most industry consortia and overestimates the democratic process in formal standards committees."[59] She bases her conclusions on studies of two consortia: ECMA and the World Wide Web Consortium (introduced in the next section). Both of these consortia were more open in membership and more closely allied with the traditional voluntary standards organizations than were many consortia formed during the 1980s and 1990s. Updegrove, too, has argued that over time "consortia also started to adopt most (but not all) of the same principles that had become normative in the traditional world of standards development."[60] Following these principles, such as openness of process, gave them legitimacy, thereby encouraging implementation of standards. Still, he noted that the traditional requirements to respond to all negative votes and the ability to appeal a decision were not followed in consortia, thus speeding up the standardization process but reducing due process.

The traditional voluntary standards organizations had to adapt to the demand for more rapid standardization and to the proliferation of consortia for setting standards. The ISO/IEC's JTC1, for example, adopted new processes for addressing both issues early on. A standard in JTC1 normally began as a new work item proposal, then became a working draft, which became a draft proposal when the committee reached consensus; only then did it become a draft international standard, to be voted on by all the national members of JTC1, and, if approved, it became an international standard.[61] The first version of JTC1 directives issued in 1989 included a fast track process to deal with demand for more rapid standardization in the information technology world; it allowed participating national members of JTC1 (e.g., ANSI) as well as organizations with A-Liaison status (a small set of very closely cooperating and trusted standards bodies including ECMA and the ITU's CCITT) to submit a standard they or other standards bodies had published to JTC1 to be voted on as a draft international standard. This accelerated process truncated several stages of the normal JTC1 standardization process, making it possible for an ECMA standard, for example, to be approved as an international standard in six months and published shortly after that.[62] Specific subcommittees of JTC1 could also give A-Liaison status to other trusted standard-setting bodies such as IEEE.[63]

Beginning in 1994, JTC1 responded to the proliferation of standards consortia by creating the new PAS (publicly available specification) process. A contemporary account described the change and the reasons behind it:

> JTC1's traditional paradigm has been criticized for several reasons. These include slow responses to dynamic market pressures, ineffective efforts to develop anticipatory standards, dogged pursuit of perfect solutions when "good enough" would be tolerable, and requests for additional benefits from organizations participating in expensive consortia.
>
> The body has responded to this criticism with unprecedented action. During its October 1994 meeting in Geneva, it approved new procedures to transpose publicly available specifications into international standards. This new standards development paradigm, called PAS, can be used during a trial period ending January 1997.[64]

JTC1's PAS process, designed to rapidly turn consortia's publicly available specifications into standards, required the originator of the specification to work with or to become an "authorized PAS submitter" (applying to JTC1 for PAS submitter status was a three-month process that was much easier than becoming an A-Liaison) to submit the specification and other required materials. Consortia such as X-Open and W3C became JTC1 PAS submitters.[65] If the publicly available specification did not conflict with any existing JTC1 standards, these materials would be distributed to national participating members of JTC1 as a draft international standard for a six-month voting period, as in the fast track process. If approved, it was published as an international standard. This process differed somewhat from the PAS processes that evolved for ISO and the IEC outside of JTC1; the consolidated ISO/IEC directives that have been issued since 1989 did not initially call specifications that followed this shortened process standards, instead just designating them as publicly available specifications, defined as "a document not fulfilling the requirements for a standard."[66] JTC1, concerned that it would otherwise become irrelevant to standardization in information technology, went further to designate them as standards.

We might view this integration of consortia into the private standard-setting system as a version of economists Farrell and Saloner's notion of the most economically efficient world, with standard setting by private committees in a world in which powerful actors (in this case firms establishing consortia) can skip the process and set a standard that many others are likely to follow.[67]

W3C: A New Type of Consortium

With its role as the principal standard-setting organization for the core technology of the web, W3C has become one of the most visible and influential new standards organizations. It is, in name and by Egyedi's definition, a consortium. Nevertheless, in many ways W3C resembles the more traditional standard-setting organizations, while in other ways it resembles neither the consortium nor the traditional model. This section examines the emergence and nature of W3C, and chapter 8 provides an example of standard setting in one W3C committee that highlights its similarities to and differences from the older model.

While most consortia were formed by firms to serve their business needs, W3C was formed by an individual in conjunction with some research universities, and with a very different goal. In the late 1980s and early 1990s, Tim Berners-Lee, a British software engineer at Geneva-based CERN, the European Organization for Nuclear Research, developed a new hypertext browsing system to allow better information sharing among scientists. In setting it up at CERN, he developed universal resource identifiers (later standardized in the IETF as universal resource locators, or URLs); hypertext markup language, or HTML, and the hypertext transfer protocol, or HTTP, the critical building blocks of the World Wide Web.[68] Wanting his innovation to be widely available and further developed, in 1993 he persuaded CERN to release his code for the web to the public, though not to provide resources to support the web's growth from a base in that institution. He worried, however, about potential balkanization in the decentralized system and knew he would have to stay involved in it to assure its growth and availability. Berners-Lee considered and rejected the option of setting up an entrepreneurial business venture, as he did not think it would prevent the balkanization he feared. Instead, he followed a path of what Andrew Russell has termed "dot-org entrepreneurship"—"an Internet-based, non-proprietary endeavor that is oriented primarily around social cooperation, not market competition," a more narrowly defined form of what we have called standardization entrepreneurship.[69] Clearly protocols would frequently need to be standardized for the web, but Berners-Lee believed that the IETF standards process he had engaged in to get the original protocols standardized took too long and required him to compromise his expectations for the web too much. Instead, Berners-Lee (figure 7.3) moved to MIT and, with the advice of Michael Dertouzos, director of MIT's Lab for Computer Science, decided to found his own standards consortium.

In 1994 W3C was established with institutional support from the academic and scientific world to provide legitimacy and from corporate and organizational members to provide funding. MIT provided the initial institutional support, followed in 1995 by INRIA (Institut national de recherche en informatique et en automatique), the French computer science organization, which became its European host. In 1996 the University of Keio in Tokyo became the Asian host, cementing W3C's global intellectual and institutional credibility.[70] To gain financial stability, it signed up dues-paying business and organizational members. From the beginning, however, Berners-Lee used W3C's dues model to foster broad membership. W3C initially charged firms with revenue of at least $50 million an annual fee of $50,000 and charged all others (smaller firms and nonprofits) one-tenth that amount. Early members included large firms such as AT&T, Deutsche Telekom, Fujitsu, and Oracle, as well as small nonprofits such as the Swedish Institute of Computer Science, and the Center for Mathematics and Computer Science in the Netherlands.[71] Since then, the sliding-fee schedule has become more granular, taking into account revenue, profit/nonprofit status, and home country (classified by

Figure 7.3. Tim Berners-Lee, inventor of the World Wide Web and founder of the World Wide Web Consortium, addresses the audience at the Bilbao Web Summit 2011. Picture by Coralie Mercier, published with her permission.

income).[72] Moreover, all member organizations, no matter what they pay, have the same voting rights, suggesting relative openness and egalitarianism in participation (though individuals and tiny organizations that could not afford the lowest rates could not be voting members). Its membership model differentiated W3C from the IETF, its fee model from traditional private standardization organizations, and its voting rights and fee structure from more typical consortia.

In process, W3C shares many characteristics with traditional standard-setting organizations. For example, its standards proceed through multiple levels of consensus that provide opportunities for members and the broader web community to express views, which are given thorough consideration. Yet, despite W3C's relative openness in process and membership and its founder's proclaimed appreciation of "rough consensus and running code," Berners-Lee did not reject kings; indeed, the rules of W3C give him, as director, ultimate authority. He has to agree to charter any new working group, for example, giving him control over new directions the web might take. He has been criticized for these broad powers by the internet community and others.[73] As he acknowledged in 1998, when tasked with his kingly role by MIT's *Technology Review*, "a lot of people, including me, believe in the 'no kings' maxim at heart and try to find ways of making organizations run well and achieve their objectives in a timely and open fashion." To reconcile this belief, he explained, "the wise king creates a parliament and civil service as soon as he can, and gets out of the loop."[74] And that is what Berners-Lee seems to have done, though he retains the formal power to intervene should he so choose.

At the turn of the twenty-first century, tensions emerged in W3C between commercial and open-source values around patent policy. Up until then, there had been no question of incorporating patents or fees in W3C recommendations (as it calls its standards).[75] Berners-Lee saw open standards as the source of the web's growth, and the internet community and the growing population of open-source programmers certainly shared that view. But by 1999, after the 1998 Supreme Court decision in *State Street Bank & Trust Co. v. Signature Financial Group, Inc.* that business processes (and software) could be patented and in the midst of the dot-com bubble, patent claims on some W3C recommendations delayed the process, and some corporate members of the consortium complained about existing or potential patent infringement.[76] Thus, W3C chartered a Patent Policy Working Group. After two years of work, it issued a working draft proposing a new W3C patent policy that attempted to balance open-source and commercial values:

> In developing a new patent policy for W3C Activities, our goal is to affirm the Web community's longstanding preference for Recommendations that can be implemented on a royalty-free (RF) basis. Where that is not possible, the new policy will provide a framework to assure maximum possible openness based on reasonable and non-discriminatory (RAND) licensing terms.[77]

This attempt at balancing the two sets of values, with its endorsement of patents under RAND terms (a common policy in standards consortia), elicited an explosive negative response from the open source community during the comment period, comparable to the IETF revolt discussed above. An expanded working group (with added open source advocates) met to consider all the comments, but it was unable to reach a consensus. It consulted with the W3C Advisory Committee, which recommended that it develop an alternative royalty-free policy to consider alongside the current draft. Four months later, after further meetings and discussions, the group issued a new patent policy recommendation, this time based on royalty-free principles. The final version, released in May 2003, allowed inclusion of non-royalty-free technology only under exceptional circumstances, but otherwise committed W3C to a royalty-free policy that pleased the open source community.

In this embrace of royalty-free policies, W3C cemented its commitment to operate in the public interest, not just in the interests of its members, distinguishing it from many standards consortia. This incident also shed light on how Berners-Lee chose to wield (or not to wield) his power in the organization. After the first-draft patent policy was issued for comment, Berners-Lee actively avoided becoming involved in the public argument, only posting to the comment list to explain why he chose not to enter the fray, in spite of "several strong calls for me to break my silence in this debate and present my personal opinion." He commented that, "of course, many people know my views on this already, as I have ranted on various media and various times." Nevertheless, he responded in terms of his larger role in the organization: "My silence arises from the fact that I value the consensus-building process at W3C. I am not (contrary to what some of the pundits might suggest! ;-) a dictator by role or nature and so prefer to wait and let the community resolve an issue."[78] To maintain legitimacy Berners-Lee refrained from weighing in publicly, though he may have expressed views privately to committee members and, if not, his well-known personal views may still have indirectly influenced the outcome.

W3C's mission as stated on its website is "to lead the World Wide Web to its full potential by developing protocols and guidelines that ensure the long-term growth of the Web."[79] The website further proclaims two design principles: (1) "Web for All," a commitment to making the web's advantages available to all people, "whatever their hardware, software, network infrastructure, native language, culture, geographical location, or physical or mental ability"; and (2) "Web on Everything," a commitment to making the web available on a wide range of devices.

Since 2012, W3C has also stated its commitment to open standards principles. In that year, five internet-related organizations involved in standard setting, either directly or indirectly—the IEEE, IAB (the internet "council of elders"), IETF, Internet Society (which organizes the IETF and publishes its journal), and W3C—joined together to establish an alliance called OpenStand. This alliance has endorsed what it calls the "modern paradigm for standards," which it identifies by the following characteristics:

- cooperation among standards organizations;
- adherence to due process, broad consensus, transparency, balance and openness in standards development;
- commitment to technical merit, interoperability, competition, innovation and benefit to humanity;
- availability of standards to all; and
- voluntary adoption.[80]

It is not coincidental that, around this time, several countries had put forward proposals to incorporate ITU-Telecommunication (formerly CCITT) recommendations around the internet into the ITU Telecommunication Treaty, thus making them mandatory for signatory states.[81] OpenStand declared the value to society of the principles of voluntary, multi-stakeholder, consensus-based standard setting, principles that stand in contrast to the ITU's intergovernmental and potentially mandatory approach.

Although OpenStand identifies its principles as the "modern paradigm for standard setting," for the most part these principles reflect and restate the principles of private voluntary standard setting developed since the turn of the twentieth century, as traced earlier, including *voluntary adoption*, commitment to *technical merit*, *cooperation among standards organizations*, *due process*, *broad consensus*, and *balance*. The IEEE and W3C both seek *broad consensus* and follow *due process* in considering all opposing opinions, while the IETF espouses at least rough consensus and has norms that are somewhat less respectful of minority opinions. Only the much older IEEE, however, has explicit rules of *balance*.[82] The IETF and W3C espouse the principle of balance in OpenStand, but both lack explicit rules to guarantee balance. *Openness* and *transparency* seem to be newer additions that were not articulated during most of the twentieth century, when industrial standards bodies were comfortable including only technical experts from established firms and organizations.[83] Although the IETF and W3C, like earlier organizations, are still, in practice, technocratic, their documents, email lists, and other electronic resources are often (but not always) open to public scrutiny.[84] *Availability* (at no charge) to all is more characteristic of the third wave, internet-related organizations, since the traditional standards organizations were typically supported by selling (printed) standards. Finally, as previous chapters show, *benefit to humanity* has traditionally been an element of the standards movement.

W3C embraces many of the traditional principles of voluntary standard setting. Although its standards do not currently qualify for US government procurement (indicating it is not officially considered a "voluntary consensus standard setting organization" as defined by OMB A-119), it is validated for PAS processes in ISO, and using that accelerated process it has submitted several specifications to ISO and gotten them approved as international standards on up-or-down votes, thus protecting them from modification.[85] W3C is very much a part of the modern landscape of standard setting, but with traditional and unique features as well. The next chapter provides a closer look at its processes.

The New Organizations and the Problem of Globalizing Standard Setting

In the final decades of the twentieth century, innovations in standard setting arose out of the networking of digital computers. In this third wave, the boundaries of standardization are also shifting, though at different rates, toward the global. The IETF, incubated in the US Defense Department, handles standards for an internet that is now global, though its progress toward global participation has been slow. Because it has no membership and no limits on participation, it can claim to be non-national, non-international, and perhaps even global, though attendees at conferences are still predominantly American and male.[86] The IETF has held three or four face-to-face meetings every year since 1986, and at the end of 2017 reached its 100th such meeting. Of these 100, 71 meetings took place in the United States or Canada. It met first in Europe in 1993 and in Asia in 2002.[87] Only since around 2010 has it met more often outside than inside the United States. W3C has, from the beginning, been more global, with institutional bases in the United States, France, and Japan. Although the first W3C conference was in the United States, at Berners-Lee's new MIT base, from 2001 to 2012, W3C's annual technical plenaries alternated between the United States and France, with Asia added into the rotation in 2013.[88] Even standardization consortia are typically formed by large global companies. These first standard-setting bodies of the third wave are not national like those of the first wave, nor strictly international (between nations) like ISO in the second wave, but aim, at least, to be global, though they have a US bias.

The standard setters involved in these bodies certainly envision their work as becoming truly global. Tim Berners-Lee is famous for his hopeful vision of a world made better through open access to information. That is why he played a role at the climax of the opening ceremony for the 2012 London Olympics. After depictions of the pandemonium and greed of Britain's Industrial Revolution, the devastation of the world wars, and the promise of the postwar welfare state, Berners-Lee appeared at a 1990s computer and typed out, "This is for everyone," which flashed on giant displays throughout the stadium. The London organizers chose to honor the software engineer for making the web free and available to everyone in the world, seeing it as the ultimate manifestation of the Olympic spirit.[89] Berners-Lee explained:

> Hopefully [the web] will make the human race work more efficiently in many, many ways. We've already seen acceleration of commerce, and the acceleration of learning. The big question is can we use it to accelerate peace? . . . If you've just been in conversation with somebody, or somebody's parents about some common interest—whether it's bird watching or global warming—you are less likely to shoot them.[90]

Figure 7.4. The Internet Society's Next Generation Leaders Program gives fellowships to technically qualified engineers and policy makers from emerging countries to attend IETF conferences; this group of fellows attended an IETF conference in Berlin in July 2016. All the fellows are identified at https://www.internetsociety.org/leadership/fellowship-to-ietf/fellows/96/, accessed August 25, 2018. Courtesy of Musa Stephen Honlue (front row center) and the Internet Society.

In an interesting twist on the rhetoric of earlier standards entrepreneurs who saw standardization as the force for peace, he saw the web as the force for peace and standardization as essential to the web.

Less-celebrated standard setters of the third wave state explicitly that truly global standard setting would be necessary to fulfill their vision. In 2009 Trond Arne Undheim, then director of standards strategy and policy at Oracle Systems, wrote that he and his high-tech-industry colleagues hoped that in the world of standardization of 2020, "The global standards process is the only game in town. . . . ISO is either revitalized or disbanded." In particular, he hoped that ISO would have dropped the "UN-like" requirement that a consensus of national bodies was needed to set a standard. And he hoped that small and medium-sized companies in every part of the world would have "a place at the table.[91]

Undheim's and Berners-Lee's visions of globalism are shared by others involved in the standardization organizations that make up the OpenStand alliance. All have been criticized for the overwhelmingly American and male character of their standard-setting committees. Nevertheless, the globalist commitments that the leaders of these organizations share have assured that they pay a little more than just lip service to the goal of expanding and diversifying the community of standardizers in the high-technology fields. For example, the Internet Society funded a program to bring underrepresented engineers to meetings of the IETF. Figure 7.4 shows a picture of its 2016 class. The contrast with the photo of the standards movement leaders of 1911 (see figure 2.4) is striking. Arguably, though, the IETF needs such a program because, unlike the traditional international standardization organizations ISO and the IEC, it has no procedures for assuring that the views from many countries will be taken into account. No matter how the third wave standardization organizations develop, standardization will undoubtedly become increasingly global simply because the industrial world is increasingly non-European and non–North American.

8

Development of the W3C WebCrypto API Standard, 2012 to 2017

Chapter 7 ended with the formation of the World Wide Web Consortium, one of the third-wave standardization organizations. This chapter provides a more detailed look at W3C's standardizing process by examining a single standards committee that operated from 2012 to 2017—the Web Cryptography Working Group (WebCrypto WG)—to develop an application programming interface (API) for web developers to use in creating secure web applications.[1] The committee's work was of critical importance to online commerce, and it touched on the deeply contentious issues of security across the internet and the World Wide Web. This chapter's case study is more micro-level than that of chapter 6. Here the focus is not on sector-wide leaders and cooperation among organizations that developed over decades. Instead, this chapter paints a portrait of a working group developing a single standard over a few years. In the process, the group struggles with issues that had already arisen in second-wave organizations in the late twentieth century, as well as with new issues that characterize this third wave of innovation in private and voluntary standard setting.

WebCrypto WG's charter described its task as "develop[ing] a Recommendation-track document that defines an API that lets developers implement secure application protocols on the level of Web applications, including message confidentiality and authentication services, by exposing trusted cryptographic primitives from the browser."[2] More simply, WebCrypto WG was tasked with developing a standard (that is what being on W3C's "recommendation track" means) for a cryptographic tool kit that would provide developers with the building blocks to create secure web applications (e.g., to create payment modules for commercial apps). The API's ultimate goal was to "promote higher security insofar as Web application developers will no longer have to create their own or use untrusted third-party libraries for cryptographic primitives [established low-level cryptographic algorithms]."[3] How much this standard should (or even could) guarantee the actual security of Web applications became a critical point of contention for the group, heightened by Edward Snowden's May 2013 revelations of global surveillance of the internet by the National Security Agency of the United States.

Below we first discuss W3C's roles and processes for recommendation track specifications (or specs) as prescribed when WebCrypto WG was launched. Then we examine the early stages of its work, including a key early decision that shaped its direction and time line; the dynamics that delayed the group's process over time; and the network of other standards organizations with which the committee interacted. Finally, we discuss a challenge also faced by many standard-setting organizations of the first and second wave toward the end of the twentieth century—the difficulty of gaining adequate participation—and the factors that motivate today's W3C standardizers. We observe a shift of participant motivation from belief in the value of standardization to belief in the value of the technology being standardized, in this case the web and the internet that underlies it.

W3C Roles and Stages

People in various defined roles shepherd a specification through standardization, and these roles are an important aspect of the process. A W3C working group has at least three critical roles: chair, W3C staff contact, and editor. When a WG is established, W3C director Tim Berners-Lee (or his delegate) designates a chair or cochairs to coordinate the group's work.[4] The chair schedules and runs audio teleconferences and occasional face-to-face meetings, monitors and encourages (or intervenes in) the online discussion on the WG's email list, and keeps the process moving forward. Committee chair was a key role common to earlier voluntary standard-setting bodies, as well. Designated W3C staff contacts support and work with the WG and chair, reminding them of deadlines, processes, and rules, and helping them to progress on schedule toward the desired outcomes. We did not see this *assigned* staff role, for example, in the ANSI-accredited C63 committee in chapter 6, but in that case the committee chair and secretary coordinated with ANSI headquarters staff.

A third role that seems common in third-wave standard-setting organizations involved with software issues, but not in the earlier organizations we have examined (where the chair and secretary handled this function), is that of editor. The chair appoints one or more editors from among the WG participants for each recommendation track specification or other publication. The editor, who may change over the course of a WG's life, plays a central role in W3C's standardizing process, creating and updating versions of a draft spec as it progresses toward a recommendation. W3C has a strong editor system, in comparison to many other software standardization groups.[5] The W3C editor (not the chair) controls all changes to what is known as the editor's draft, although the WG must vote to approve published working drafts and new versions of the spec prepared for each stage. This role demands considerable work but gives the editor significant control over the spec (as one WebCrypto WG member said, "doing the work gets you what you want"[6]). The holder of this role could change over time (e.g., when existing editors burned out or got too busy at their firms to continue), and a spec could have coeditors.

The individuals playing these roles ushered a specification on the recommendation track through a series of milestones or maturity levels enumerated in the W3C Process Document toward becoming a standard.[7] Although W3C is a third-wave organization, the milestones, as enumerated below, resemble stages seen in earlier standard-setting bodies.

- First Public Working Draft (FPWD): The WG begins by working on an editor's draft within the group, but the FPWD is the first draft that is formally published for review by other working groups and the public; its publication also triggers a call for member companies to declare any intellectual property claims that should be excluded from the royalty-free patent commitments (the default, if participants do not exclude a patent, is that they are agreeing to W3C's royalty-free commitment). After FPWD, every three months a "heartbeat publication" of a working draft is required to show progress, one of many mechanisms W3C uses to encourage forward movement.
- Last Call Working Draft: At the time the WebCrypto WG was chartered (in 2012), last call was required as a separate step (though since 2015 it has been merged with the next stage).[8] When the WG considers the draft nearly ready to advance to the next stage, candidate recommendation, it votes to issue a last call for major comments from the

public, with a deadline at least four weeks after the public announcement. For the draft to enter the next phase, the WG must show that it has addressed any issues raised during last call and has documented that the spec meets all WG requirements and has been widely reviewed.

- Candidate Recommendation (CR): When a spec enters CR, the W3C director announces its status to other W3C groups and to the public. At the CR phase, the spec is expected to be fairly stable, and the WG focuses on getting it implemented in browsers. Only features that have been labeled "at risk" in CR can be taken out at later stages without triggering a return to CR. In addition, the WG must create a test suite (a collection of tests) to assure interoperability across browsers; only features that are interoperable in at least two of the five browsers (Google Chrome, Mozilla Firefox, Microsoft Internet Explorer, Opera, and Apple Safari) may move to the next stage.
- Proposed Recommendation (PR): For a spec to move to PR, it must be "accepted by the W3C Director [or his designee] as of sufficient quality to become a W3C Recommendation." An advisory committee must also review it in this phase.
- W3C Recommendation: This is the final standard produced by the process. "A W3C Recommendation is a specification or set of guidelines or requirements that, after extensive consensus-building, has received the endorsement of W3C Members and the Director. W3C recommends the wide deployment of its Recommendations as standards for the Web."

Ideally, W3C expects a WG to develop and shepherd its draft through to recommendation in one to three years, as indicated in the WG's charter. The original WebCrypto WG charter allowed two years, from March 2012 to March 2014, to deliver WebCrypto API.[9]

So far, we have sketched the W3C roles and process as intended. Now we turn to how they actually unfolded in WebCrypto WG over time.

WebCrypto WG: Participants, Process, and Timeline

Although its original charter was created in March 2012, WebCrypto WG was officially launched two months later at an early May 2012 public phone meeting intended to encourage people to join. This meeting included discussion of the short time line to get through all the stages; the FPWD was due by the end of June, less than two months after this kickoff meeting, and the spec was due to become a recommendation by the end of March 2014, less than two years from the May launch.[10] Some meeting attendees questioned the feasibility of the time line from the start, and the charter was ultimately extended five times by January 2017, when the WebCrypto API spec finally attained recommendation status. Clearly W3C, one of the new standard-setting organizations using rapid electronic modes of communication from the beginning, still could not *guarantee* rapid consensus-based standard setting any more than earlier standards organizations could, though it tried to encourage greater speed by imposing frequent and tight deadlines.[11] A consensus-based process takes time, and the more contentious the issues are, the more time it takes.

Below, we describe the early stages of this project, including the choice of a controversial technical path and the emergence of a volunteer editor supporting that path; both would contribute to the extended timeline going forward. In the text, under our agreements with those we observed and interviewed, we refer to members of the group by their first names.[12] This convention also captures the informality of the W3C participants. T-shirts, sandals or sneakers, and casual speech were typical, in contrast, for example, to the top hats and grand speeches of the electrical engineers of the 1920s or even the suits and honorifics of the EMC meetings of the 1960s through 1980s.

Early Events and Decisions

Virginie was the WG chair from the beginning of the process. She is French and works for a digital security firm in France, making her atypical of the group as a whole, which was majority male and of US citizenship and residence.[13] She had been involved with several other industry and standard-setting bodies, including the European Telecommunications Standards Institute and a standards consortium, but had only recently become involved with W3C when her company became a member 18 months before the WebCrypto process began in 2012. She liked what she saw as the open-mindedness of the people in W3C and the democratic procedures W3C followed and was happy to be appointed chair of WebCrypto API. She saw herself as a process-oriented chair, whose role was "making sure that there is [a] fair decision based on consensus." She did not have the expertise in this particular area to drive the technical part of the task but saw that as the role of the editor(s).[14] Figure 8.1 shows Virginie at a W3C September 2014 workshop on the future of Web Cryptography in W3C.

Figure 8.1. WebCrypto WG chair, Virginie, at a meeting discussing the potential next version of WebCrypto API, September 2014 (posted in Final Report of the conference). Photo credited to and reproduced with permission of Wendy Seltzer.

The goals and scope of the new WG had emerged from a May 2011 event, "W3C Workshop on Identity in the Browser," and the initial editor's draft was based on Mozilla's DomCrypt, a publicly available JavaScript cryptography API.[15] At the public launch meeting in May 2012, the group already listed four editors representing three different browser companies (though only one, David, played a role in the life of the WG), as well as an initial editor's draft, described as a "rough draft straw man."[16] Wendy and Harry attended that public launch as staff contacts.

At the first public meeting, attendees discussed and debated what features were in scope, why certain features were designated as primary and others as secondary, and which secondary features should be addressed and when.[17] Messages posted to the WG list during May, as well as the minutes of three WG phone meetings during that month, revealed deep differences of opinion on these issues that would be ongoing (e.g., although smart cards with embedded microprocessors were declared out of scope early on, the issue continued to be raised, repeatedly but unsuccessfully, during the first year). In addition, WG members raised several other controversial issues that would periodically reappear in the debate for years afterward (e.g., Should algorithms be optional or mandatory for compliance with the spec? How much guidance should the spec give to developers on the security of various algorithms?).

The central debate of the early months was whether to create a so-called high-level API that was intended to be easier and safer for developers to use but that restricted their flexibility, or a low-level API that provided flexibility and compatibility with older systems, but that included elements that could be used poorly to make security weaker. An original WG editor and coauthor of the first, high-level editor's draft, David (representing one of the browsers), reported in a list email before the first working-group-only phone meeting on May 14 that he and another editor from a different browser organization, as well as the second editor's colleague, Ryan, had discussed the preliminary editor's draft by phone, discussing the "potential need for a parallel set of low-level interfaces." David's view was that "we should avoid this as a potential sinkhole. My initial desire to provide an 'easy-and-safe-to-use' interface is I think the point of this API." Nevertheless, he acknowledged, "A parallel, low-level interface is also desired."[18] This was the first of many discussions about whether to write a high-level API that restricted developers to "safe" cryptography or a low-level API that provided a tool box but let developers choose what cryptographic methods were appropriate for their particular applications.

In response to David, Ryan posted a more detailed description of what he meant by high-, medium-, and low-level APIs (ranging from a black box API to one that worked directly with algorithms): "I believe I enumerated them in what I believe will be the most controversial for consensus and implementations (high level) to least controversial for consensus and implementations." In debates about what high- and medium-level features to support, he argued, "it becomes a matter of opinion and potentially crypto-policy. Further, when we talk about things like 'safe to use' or 'easy', they're very hard to quantify, and reflect matters of opinion and not fact. As such, reaching consensus may be difficult." He stated his own position:

> I (as an individual) believe that for the general utility of the Web Community, we'd best be served by providing the Low Level API for applications to build on. I believe the decision about the best API for such Medium and High level APIs will be best decided by the web community at large, and will be a [*sic*] constantly evolving and adapting based on feedback, with perhaps multiple frameworks providing different alternatives.[19]

David responded to Ryan, thanking him for providing clear definitions on which to base the discussion and noting how these definitions and Ryan's arguments changed his mind to thinking a low-level API was necessary to start with, at least.[20] The conversation about level (and how each level would be used) continued in posted emails and the next few phone meetings. The decision was controversial, with disagreements even between two representatives of the same browser company. During the third WG phone meeting, David shared a draft survey he proposed to send out to WG members, asking about their planned uses of WebCrypto API and culminating in "Are you interested in a high-level, idiot-proof API or a low-level API?"[21] By early June, responses had arrived, and he shared the raw data on the list and in a phone meeting.[22] A small majority preferred a low-level API. Based on the data and discussion, the chair proposed and the group voted in favor of the following: "RESOLUTION: Start with low-level API, then focus on high-level API."[23]

This decision was a pivotal point in the WG's early history, shaping the nature of the standard that would emerge and paving the way for Ryan to become the dominant force in the WG. In mid-June, Ryan posted a "strawman proposal for the low-level API," and in a July phone meeting, David pronounced Ryan's work more appropriate for a low-level API than the original, higher-level draft.[24] As a result, Ryan's new proposal replaced the original editor's draft as the basis of WebCrypto API. At the first face-to-face meeting, on July 24, 2012, the group confirmed that the low-level API would be its priority, with a high-level API postponed to sometime in the future—a future that never arrived. The WG also added Ryan to the spec as an editor and removed one original editor who did not have time to contribute to WebCrypto.[25] Soon the other original editors' names disappeared from the editor's draft and the WebCrypto home page, and Ryan became the sole editor of the primary spec until late 2014 when he was joined by a secondary coeditor (Mark, discussed below) until January 2016; during those four years Ryan shaped the API and the conversation.[26] Coeditor Mark became the sole editor for the final year, after all major decisions had been made.

The actual time line was considerably longer than the two years envisioned in the initial charter. Figure 8.2 shows the number of email posts (left vertical axis) and of participants (right vertical axis) by month from the WG's beginning to January 2017, when the spec became a recommendation; it also indicates some key events in the group's life, including Ryan's addition as editor in July 2012 and his resignation as editor in January 2016.[27] The group reached FPWD by the end of summer 2012 (only slightly behind schedule),

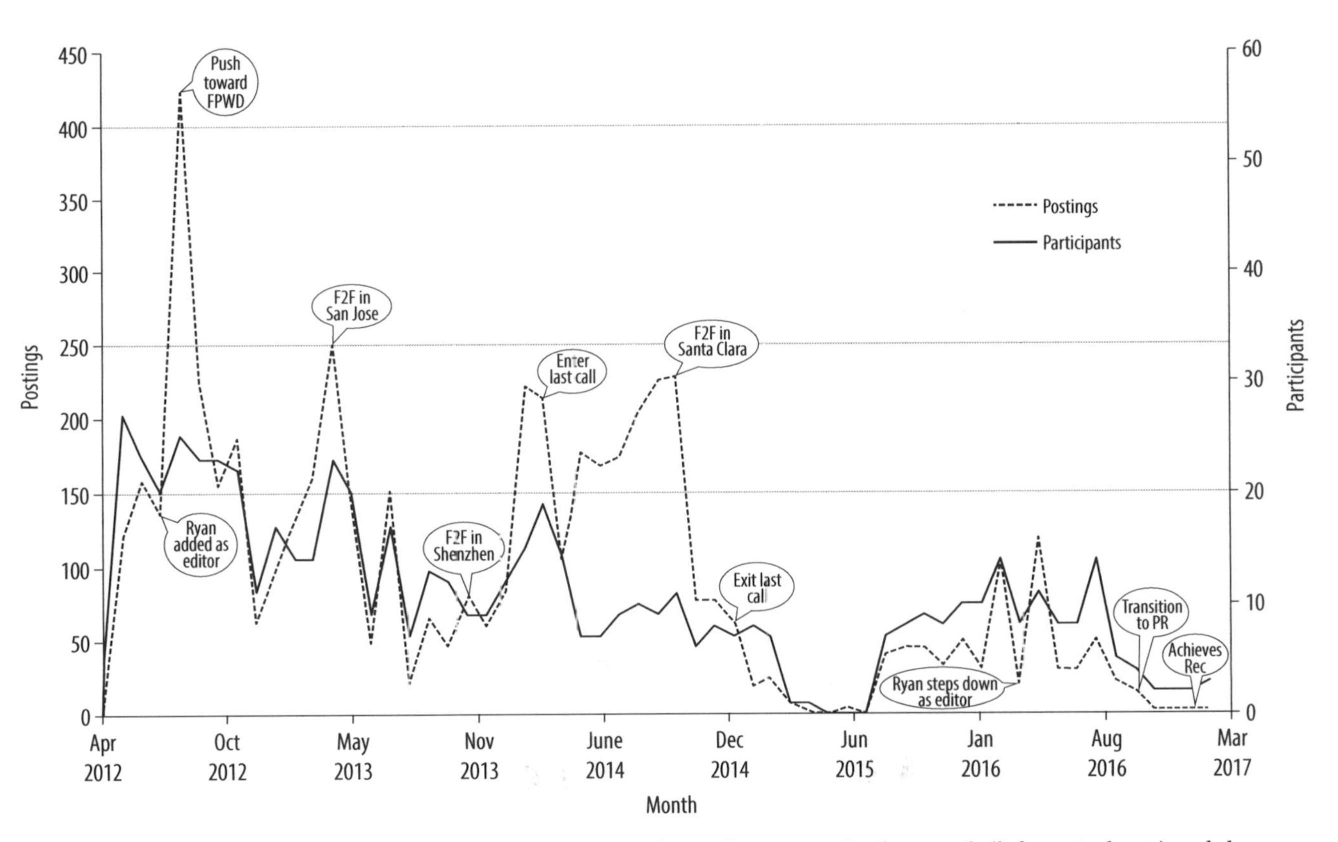

Figure 8.2. This graph shows that the number of postings on the working group list by month (left vertical axis) and the number of people making these postings (right vertical axis) varied, reflecting the stages of the standard and other events in group life, but both quantities generally declined over time.

but it did not reach the next milestone, last call, until spring 2014 and only exited last call and entered CR at the end of 2014. The spec then remained in CR for almost two years, making the transition to PR in fall 2016 and becoming a W3C recommendation in January 2017.

The next section looks at some of the factors slowing the pace and lengthening the process in this working group.

Delays and Dynamics on the Way to WebCrypto API Recommendation

Many factors contributed to WebCrypto WG's delays, including the controversial decision to create a low-level API. Issues related to level reappeared repeatedly, raised primarily by individuals outside the WG but also by those within it. For example, when the Crypto Forum Research Group, associated with the Internet Engineering Task Force, responded to a WG member's request for feedback on the FPWD, it noted that the draft used several legacy crypto algorithms that had "well-known serious weaknesses."[28] It suggested eliminating these algorithms from the standard, but, given the need for interoperability with existing systems, said it would settle for adding a section to the text describing the weaknesses and pointing toward safer choices of algorithms. In responding to the WG member who forwarded this critique, Ryan commented, "My concern is that I do not believe this draft should become a general catch-all document for algorithm-specific security concerns for every algorithm implemented." Those interested, he noted, could look to work by the Crypto Forum Research Group itself. After the caveat that "no doubt this will certainly be taken out of context," he stated his philosophy about WebCrypto API: "I'd like to try and provide as much as [*sic*] a 'value neutral' API as possible, and leave discussions about the values and merits of individual algorithms to the individual protocols and applications."[29] Ultimately, Ryan rejected even including the suggested note in the spec because he wanted it to be "clean." Similarly, he responded to a nonmember's complaint about a particular algorithm known to be cryptographically vulnerable by pointing out that a low-level API inherently includes elements that could be used unsafely, as well as safely, in any particular setting: "In seeking to provide a low-level API, it is inherent that we end up providing more than enough rope to hang oneself, and are definitely leaving a few loaded guns on the table." He argued, "So far, our hope has been that developers will be able to take these primitives, and combine them into the appropriate secure protocols and interfaces."[30] Over the next two years, such questions of protecting developers from unsafe crypto were raised again and again.[31] In an era of great concern about security (intensified by Snowden's May 2013 revelations), a low-level API providing a "value-neutral" tool kit, not protection per se, was sometimes hard to sell, and the recurrent arguments about this issue occupied list traffic and slowed the WG's pace.

The strong editor role in W3C, and Ryan himself as editor, were also factors in the WG's pace, though at different points affecting the pace differently. Initially, Ryan's willingness to jump in and devote extensive time to writing the spec and responding to all discussion on the WG list was seen as a very valuable contribution, as indicated by David's early commendation of his first strawman proposal as well as by comments from WG members in interviews. His rapid work on the spec certainly increased the group's pace at early stages, helping to get it to FPWD close to the original schedule in spite of the change of direction. But he also responded extensively to most substantive comments on the list, often engaging in prolonged and pugnacious exchanges with other members about technical (and sometimes not-so-technical) issues, slowing the process. Because he had a clear vision for what he wanted the spec to be (a value-neutral, low-level API), as well as extensive technical knowledge, he countered attempts to disagree with him vigorously and unflaggingly. On specific issues, ultimately many other WG members either became convinced that he was right or simply stopped arguing, resulting in his domination of the process—especially the online part of the process—from June 2012 through January 2016. During every month but one of that four-year period, the percentage of messages on the WG list written by him was in double digits, and ranged as high as 33 percent.[32] This high level of participation reinforced Ryan's considerable power as editor but also took time and perhaps deterred other members from participating.

When asked about whether WebCrypto WG was more or less contentious than other working groups they had served on (in W3C or other standards organizations), most members interviewed thought it was on the more contentious end of the spectrum, but not necessarily the worst. Richard, another active WG member who represented a government contractor, suggested that while WebCrypto WG was not more contentious in subject matter than one or two IETF working groups he had served on, "the elbows seem to be a little sharper" in it, pointing particularly to Ryan as having "some of the sharpest." Indeed, Ryan sharply refuted any post with which he disagreed.[33] His aggressiveness online and his ability to make others look less technically competent may have deterred participation and worn down some WG members' opposition and even willingness to participate, perhaps slowing the process over time. For example, one member said privately that he "rage-quit" the working group because Ryan did not care when he pointed out a real-world case of unsafe cryptography using one of the algorithms allowed by WebCrypto API. Figure 8.2 shows the decline in both the number of messages and the number of participants from FPWD to entering CR in fall 2014, the point when participation might be expected to decline, as it entered a period of debugging and implementation. Ryan clearly lacked the diplomatic skills of the major standardization entrepreneurs such as Charles le Maistre or of leaders in specific domains such as Ralph Showers.

An incident exemplifying the power of—and limits to—a strong and prickly W3C editor like Ryan, as well as its influence on the group's pace, occurred in November 2012. Active WG member Mark, who worked for a large application developer with a strong stake in the standard, had engaged in a long disagreement with Ryan about a capability that Mark and his company very much wanted in the spec and Ryan kept rejecting. Finally, seeking a way to get around Ryan's objections, Mark volunteered during a phone meeting to edit a separate Working Draft for Key Discovery (as the capability was later named) that the WG could move along the recommendation track in parallel to the main WebCrypto API spec.[34] Ryan agreed to this plan, but he also proposed to remove another function from the main spec that would be necessary to allow the Key Discovery capability to be used in conjunction with WebCrypto API in the future. Mark insisted that Ryan not remove that critical supporting function, and the meeting minutes did not reflect a response from Ryan.

Nevertheless, the next day Ryan removed the supporting function from the editor's draft, initiating a heated online argument about the powers of the editor. Mark strongly objected, saying, "This is absolutely no way to conduct a standardization process and will be a serious problem going forward if this capability is unilaterally removed."[35] Another WG member, who also chaired a related IETF working group, supported Mark, saying there was no consensus to remove this feature and further stating that, "as an editor (of IETF specs), I'm well aware that the editor's job is to implement the working group consensus—not to implement my own viewpoint. I assume that it must be the same in the W3C. For the process to work, that's the way that it must be." Ryan replied that W3C and IETF had different processes, and that within the W3C guidelines, "there is significantly greater flexibility afforded to authors and editors of specs," including that "editors are not required to bring everything to the list first." He argued that, since neither the supporting feature nor the additional Key Discovery spec Mark was editing had received official consensus, he was within his rights as editor to remove it until there was consensus. Mark simply responded, "We will not agree to the heartbeat publication if this feature—which was in FPWD—is removed," indicating that he and his company would register an official objection to publishing the next working draft at the three-month heartbeat required in December. At this point Virginie stepped in to halt the increasingly heated online discussion "before we enter the dangerous zone of having aggressive emails." Her intervention paused the discussion, and progress, for several days. Although Mark won this battle (the supporting feature remained in for the time being) by using W3C's due process rules, Ryan ultimately won the war, as a later section will show.

Ryan had a strong preference for discussing issues and making decisions on the email list rather than in face-to-face or especially phone meetings, which may also have slowed the process and given Ryan more control over it. Early in the group's life, Ryan proposed that decisions be made on the list rather than in meetings: "Can we just leave consensus to the mailing list, and use the phone meetings to work out details / answer questions / review the spec?" We observed and several WG members mentioned informally that Ryan's aggressiveness was less evident on phone calls and at face-to-face meetings than on the mailing list; perhaps his preference for online communication reflected his knowledge that he could wield more power on the list, with his sharp responses, than in phone or face-to-face conversation, when he seemed more constrained by norms of politeness. He argued publicly that, since some discussions of issues for consensus "have ended up spanning several phone calls, due to various participants being unable to attend phone/in-person meetings," they should simply use the mail list to reach consensus.[36]

W3C staff contact Harry responded to him that, "in general, consensus happens in meetings at W3C, not over list like IETF," since "given the amount of people on a mailing list, its [*sic*] unclear if 'not voting' is abstention or just not reading email :)."[37] He went on to explain what was considered best practice in W3C: "In general, the best format is to email relatively complete and well-thought out proposals to the list at least a week ahead of time, let the list discuss, and then make decisions at the telecon." At the face-to-face meeting in Shenzhen, China, in November 2013, Ryan again argued for achieving consensus on the mail list, this time linking his request to the WG's difficulties in finding a phone meeting time that worked for Asian members of the WG, as well as for the US and European members, a recurring problem that, along with language difficulties, marginalized four South Korean members of the WG.[38] In the wake of the meeting, Ryan succeeded in getting Virginie and the WG to institutionalize a new process for finalizing decisions online, with any votes taken in phone meetings requiring announcement on the mailing list with a two-week period for objections on the list, further slowing the process.[39]

WebCrypto WG's long dependence on one editor for the primary spec also contributed to delays when that editor had other priorities. In late 2013, at the end of the Shenzhen meeting, Ryan stated that he was very busy and decisions made at the meeting would require lots of changes; thus, he was unwilling to commit to a specific timeline that would enable them to go to last call at the time the charter indicated.[40] WG member Mark, despite his own frequent clashes with Ryan over the spec, offered to help, and Virginie said that on the next teleconference in two weeks they could discuss the schedule for making the changes and going to last call. Ryan did not attend that phone meeting, however, emailing that he was overloaded. W3C staff contact Harry noted that "being 'overloaded' generally means a co-editor is needed."[41] Harry asked Mark whether he was willing to be a coeditor, and Mark agreed if Ryan was amenable. As it turned out, the problem was not just that Ryan was burned out—he was also editor of a spec going through last call toward becoming an IETF request for comment (the equivalent of a recommendation in W3C) at the same time that WebCrypto API was going through last call at W3C, and, according to Ryan, "we had an even more spirited debate in the IETF."[42] So his engagement with WebCrypto went up and down depending on other demands (including other standardization demands) on his time and attention.

The combination of Mark and Ryan as coeditors starting in 2014 was itself an odd one. Because Mark represented a major application developer and Ryan represented a browser maker, they had very different priorities and even coding styles.[43] Indeed, they never really worked *together* over the next year. Rather, as Mark explained, "It's been time-sliced," with Mark stepping in whenever Ryan seemed too busy, then Ryan taking back over when he had time. Mark's stated motive "was really wanting to just not have this continue forever." Presumably, his company was also anxious for the spec to be completed. In any case, he was willing to take on some of the more time-consuming detailed changes, and Ryan was glad to get help with them.[44]

Another source of delay in the spec's progress was debugging and testing for interoperability. Since the API was being developed during the standardization process, it needed to be thoroughly debugged and made interoperable across at least two browsers before becoming a recommendation; otherwise, it would have been an (unacceptable) anticipatory standard. From the beginning, Virginie had stressed that the WG would need to create a test suite to check for interoperability, as a spec could not move from CR to PR without showing that every aspect of it had at least two interoperable browser implementations.[45] Moreover, implementations emerged early; at least three browsers implemented WebCrypto API during the months of debugging between entering last call (March 25, 2014) and exiting it to CR (December 10, 2014), even though CR was, in theory, the implementation stage.[46] Virginie also made multiple calls for testing, and W3C staff contact Harry offered training on W3C testing.[47] But, despite this encouragement and the existence of browser-specific tests within each of the browser companies, no one stepped forward to establish a W3C test suite or even to translate the browser-specific tests to more general ones. Without testing to determine interoperability, the spec could not enter PR. Indeed, after the transition to CR at the end of 2014, the WG went nearly dormant for much of 2015.

Starting in September 2015, staff contact Harry increased the pressure on the group to enter PR, pushing it to determine which algorithms were included in at least two implementations. At this time, for reasons not entirely clear, Ryan resisted moving to PR, pointing out many—though often small—issues still preventing complete interoperability across browsers.[48] The stalemate only gave way to forward progress after Ryan resigned as editor (leaving coeditor Mark as the sole spec editor) in January 2016, saying he no longer had time.[49] After that, Virginie recruited new representatives of several browser makers to work on tests, and they brought new energy to the process.[50] Meanwhile Mark, who was more open to negotiation and compromise than Ryan, continued to resolve remaining bugs, remove non-interoperable algorithms when necessary, and do other final editing tasks. This reenergized push and new spirit of compromise got the group to the point of requesting transition to PR by October 2016.[51] At that point the WG's work was completely over, though the spec still had to be reviewed at the highest levels of W3C before it was made a recommendation in January 2017.

This closer look at the delays in WebCrypto's time line suggests that W3C WebCrypto WG, like committees in more-traditional voluntary consensus standardization bodies, suffered in part from the long time required to reach consensus. As noted earlier, W3C has now collapsed two stages and further streamlined the process, reducing time from charter to recommendation. In addition, it has created readiness criteria to help determine when a recommendation-track effort is ready to start, cutting off some time at the beginning of the process. Still, W3C working groups face the challenges of reaching consensus among stakeholders on a complex new software standard and of assuring that people are willing to complete the testing after the original coding has been completed. Neither is the problem of timing unique to W3C among modern, software-oriented standard-setting bodies; a related IETF working group, JavaScript Object Signing and Encryption (JOSE) working group (discussed in the next section), worked on its standardization task from September 2011 to August 2016, a five-year active life. So, despite differences between the IETF and W3C models of standardization, both may face lengthy processes when struggling to reach consensus on controversial issues such as security.

Factors Shaping WebCrypto WG's Activities and Progress

In this section we take up other aspects of the W3C WebCrypto WG and related standard-setting activities that affected it. As was the case with the US and international EMC standardization committees discussed in chapter 6, W3C and WebCrypto WG existed in a network of standardization bodies and a community of standardizers. In addition, browser makers played a unique role in the web community, shaping standardization. Finally, we return to a problem that appeared in second-wave organizations, but reappears in web (and internet) standardization with a particular twist: the problem of dwindling participation, especially in the late stages of software standardization.

Networks, Communities, and Browser Makers

W3C in general, as well as WebCrypto WG in particular, existed within a network of internet-related organizations, drawing on the same community of software engineers. From the beginning, the WebCrypto charter named dependencies with three other W3C working groups and two external groups—the IETF's JOSE working group and the European Computer Manufacturer's Association Technical Committee 39.[52] The JavaScript programming language, used by JOSE WG and WebCrypto WG, was originally an ECMA standard, which accounts for the declared dependency, but WebCrypto WG did not need to interact directly with ECMA.[53] The charter required, however, a liaison with the IETF's JOSE WG, the external group with which WebCrypto overlapped and interacted most. In addition, one new and informal body not mentioned in the charter affected WebCrypto importantly but indirectly: the Web Hypertext Application Technology Working Group, or WHATWG. These two external organizations—the IETF's JOSE WG and WHATWG—were key to WebCrypto WG's organizational network.

The charter explained that WebCrypto was dependent on the IETF because it defines "robust and secure protocols for Internet functionality" that are necessary "to ensure the security and success of elements of Web Cryptography." In particular, it required WebCrypto to create an official liaison with IETF's JOSE working group, which "is producing formats relevant to the Web Cryptography Working Group."[54] Although only one official liaison with the JOSE working group was mandated, several members of WebCrypto WG were active in it, including JOSE cochairs and primary editor.[55]

After WebCrypto WG decided to create a low-level API, members discussed at length how closely related it should be to JOSE. Several members thought that WebCrypto should build on JOSE's work, while editor Ryan did not want to tie WebCrypto to JOSE. Ryan stated his position strongly: "I'd like to propose that the current low-level API specifically clarify that there is *no* specific or privileged relationship to the IETF JOSE work."[56] He explained that WebCrypto should not require the same cryptographic algorithms and identifiers being used by JOSE, since WebCrypto API "represents a generic API." Moreover, JOSE WG's products "do not represent the only consumer of this API, therefore I do not believe it makes sense to primarily design or tightly-couple this work to the JOSE WG." In the ensuing discussion, when W3C staff contact Harry reminded them that the charter required them to liaise with JOSE, Ryan argued that the connection with JOSE was based on WebCrypto WG producing a high-level API (for which he agreed that a close connection was appropriate), but not a low-level API.[57] A few others agreed with him, though the primary JOSE editor and de facto liaison, Mike, still saw some possibilities for future coordination.[58] Ryan proposed to resolve the issue minimally, by adding an explicit use case for JOSE to the use case document the WG also had to create, to "capture at least some relationship with JOSE in the WG, as reflected in the charter."[59] When Mike accepted Ryan's wording and no one else objected, consensus was declared.

The issue resurfaced six months later, in a discussion about possibly establishing a registry of algorithms outside of the spec itself. At that point one of the WG members also participating in JOSE, Richard, suggested that "another way to solve this would be to kick the algorithm ID specification over to JOSE," since he thought that "it would be a good idea in general for the two groups to use the same algorithm identifiers."[60] Ryan's response was swift and sharp: "I thought we already hashed all of these discussions out and agreed that there was consensus to NOT go with JOSE's approach." In responding to one of Richard's specific suggestions, he said, "This, to me, is unacceptable." Richard countered that "JOSE is not written in stone. It is a working draft, just like the WebCrypto spec. // Can we at least try to work together?" The online argument between the two continued for a week but ultimately subsided with no explicit change to the earlier decision not to coordinate closely with JOSE. A few months later, Richard stated in an interview that he still believed that it would be better for the two specs to be more closely coordinated and that the editors of each spec—Ryan on the WebCrypto side and Mike on the JOSE side—had each misjudged the other spec, with Ryan assuming JOSE was too high level to be of interest and Mike assuming WebCrypto was too granular or low level to be relevant.[61] Nevertheless, subsequently issues around JOSE came up only sporadically, as when Mike noted some changes that had been made to one of JOSE's algorithms, saying, "I believe that together, these changes unblock any issues for WebCrypto to directly use JWK," a cryptographic key connected to JOSE.[62] So a minimal amount of coordination still occurred between the two groups but very little considering the large overlap of people working on both.

The other group overlapping in membership with WebCrypto (and W3C more broadly) was the Web Hypertext Application Technology Working Group. WHATWG, an unusual and very informal standards group, was formed in 2004 by individuals from three browser makers—Apple, the Mozilla Foundation, and Opera Software—after a W3C workshop at which they believed that W3C was moving toward XHTML (extensible hypertext markup language) and away from HTML, abandoning the older web language even though many implementers and developers still depended on it. WHATWG defines itself on its website not as an organization but as "a community of people interested in evolving the Web through standards and tests."[63] W3C CEO Jeff Jaffe attributed its emergence to W3C's failure to listen to key implementers (primarily browser makers) at that time.[64] When WHATWG began working on HTML5, W3C decided that having two separate efforts was not good for the web, so it formed an uneasy partnership with WHATWG to work together on the new standard, the output of which was two similar but not identical versions of the HTML5 standard. The somewhat uneasy partnership has continued to the present, with W3C archiving WHATWG discussions and with WHATWG operating as a W3C community group—a structure within W3C for pre-standardization discussion of specific areas.

WHATWG follows a completely different model of standardizing from that of W3C and other types of private standards organizations. It calls its specifications "living standards," as the group updates them continuously, rather than only occasionally: "Instead of ignoring what the browsers do, we fix the spec to match what the browsers do."[65] Browser makers are central to WHATWG. Richard, who participated actively in both WebCrypto and the IETF but not in WHATWG, explained, "The WHATWG, as I understand it at least, is . . . just the browser vendors," and "I think it's notionally open, but it's really the browser vendors gathered around a single guy who writes all the specifications."[66] Initially, a single editor from a browser company, Ian Hickson, took in feedback from the broader community and then decided what to do about it—the ultimate strong editor system. Indeed, the group's initial website explicitly stated, "This is not a consensus-based approach—there's no guarantee that everyone will be happy! There is also no voting."[67] The founders of WHATWG rejected the consensus process that W3C and the private voluntary standard-setting organizations followed, which they saw as inadequately reflecting the central role of browser vendors to anything related to the web.

Although WHATWG is focused on the needs of browser vendors, its members are not browser *organizations* but *individuals* working for the browser makers (in that respect the group resembles the IETF more than W3C or a consortium). Moreover, at least one of the five major browser vendors, Microsoft, specifically chose not to get involved, even indirectly, in WHATWG during this period. The minutes of a phone meeting record a WebCrypto WG member from that firm as saying that Microsoft did not like to implement non-W3C specs; individuals from Microsoft were on the record in various forums as preferring to stick with W3C because of its clear patent policy.[68]

As an alternative venue for web standardization, WHATWG occasionally cast a shadow on WebCrypto WG's process. For example, in a contentious exchange in early 2013, two WebCrypto WG members from different browser vendors made veiled suggestions that perhaps they should turn to WHATWG if W3C could not deliver what they believed they needed. When Harry suggested creating a registry to support future growth of nonmandatory algorithms after the WG finished its work and disbanded, Ryan (also a member of WHATWG) argued that what was important was "NOT having some registry, but having industry consensus on implementing this," and that "if we believe in standards at all, we should be trying to standardize things." He went on to say, "And in that model, either there's a group to standardize things (whether it be this WG, [W3C] WebApps WG, or it requires implementers going outside W3C to get things done, such as the WHATWG), or there's not."[69] Arun, a WG member from a browser maker who was associated with WHATWG, suggested that, "instead of a registry, can we do something like active bug discussions, and keep *public [public-WebCrypto, the WG email list] active?" He noted that having an active, ongoing group to make changes "is what makes WHATWG attractive, but I understand that patent considerations might limit the merits of merely migrating there."[70] Thus, the implicit or explicit threat of doing standards work in WHATWG rather than W3C gave the browsers leverage in WebCrypto WG (and W3C more generally).

The relationship of WHATWG to W3C and WebCrypto WG also reflected a broader issue—the special role of browser vendors in W3C. As one participant who did not represent a browser vendor explained, "The other dynamic in the W3C . . . is that there's browser vendors and there's everyone else."[71] Because the five browser makers decide what to implement, they have more power in standard setting, though not always as much as they want, as reflected by the formation of WHATWG. Ryan explained that "WHATWG exists largely to get browsers on the same page and I think to a large degree recognizes that what matters is not process track documents, what matters is what browsers implement, are willing to implement, think should be implemented."[72] He thought that W3C's "consensus-based approach [was] not entirely useful," since the fee structure and processes allowed "a much greater influence, emphasis of power on non-[browsers] to influence a specification that allows them to sort of override [browser] concerns. And so the result of overriding [browser] concerns is the [browser] just ignores the specification." From his point of view, browsers should have had even more power than they had in W3C, even though their de facto power to veto by not implementing was well recognized. In contrast, W3C sees the web as existing for its billions of users, not primarily for browser vendors, and the OpenStand principles it endorsed include balance and due process, pushing back on browser control over standards.[73]

Mark's Key Discovery, which proceeded on a parallel track to WebCrypto API, encountered the de facto browser veto power. That spec was ultimately taken off the recommendation track and published as a WG note because it did not have two interoperable implementations.[74] This example also highlighted the tension between large and powerful application developers and even more powerful browser vendors. Mark—editor of Key Discovery, coeditor of WebCrypto API starting in 2014, and ultimately its sole editor after Ryan stepped down at the beginning of 2016—represented a major application developer. He had very specific goals related to his firm's needs, such as the functionality in Key Discovery, but Ryan often opposed what he wanted (many of Ryan's extended exchanges were with Mark), even though, as Mark pointed out, "the [web]site authors are supposed to be higher in the [W3C] priority of constituencies than the browser implementers."[75] Mark was more willing to compromise than Ryan, but he was persistent in pursuing his firm's goals. Thus, in the case cited earlier, Mark was able to use process rules to allow Key Discovery to be developed along a parallel track; nevertheless, he could not ultimately force the browser makers to implement that functionality. Thus, browser makers can play a powerful gatekeeping role in W3C, despite W3C's desire for balance. In this setting, browser makers seem to exert, if anything, even more power than manufacturers exercised in earlier industrial standardization, since a feature cannot be used unless it is implemented in a browser, and the application providers who depend on the browsers to deliver their services have little countervailing economic power over the browsers.

The Challenge of Dwindling Participation

Finally, this case study highlights an issue apparent by the end of the twentieth century in more traditional standards organizations, as well: dwindling participation in and enthusiasm for standard setting among engineers. WebCrypto WG undertook a large and difficult technical task and had more than 50 members, but only a small set of individuals actually did the work (as was increasingly true in more traditional voluntary standard-setting organizations, as well). In 2013 the WG officially included 55 total participants, and in early 2017 it had grown to 70 total participants.[76] Yet figure 8.2 shows that only 27 individuals contributed online in the month with the most participants, the WG's first month. Phone meeting attendance was also small, with the number of attendees occasionally reaching the 20s in early meetings but typically in single digits from the end of 2013 onward.[77] After WebCrypto API entered CR at the end of 2014, major spec development was over. Finding people to review the spec for bugs and test it for interoperability, time-consuming and unrewarding technical tasks, was particularly difficult, leading to the WG's near dormancy in 2015. Virginie and the W3C staff contacts worked hard to revive the group to get through testing in 2016, bringing in new member representatives to complete the process. As one WG member, who was a cochair of the IETF's JOSE WG (and who spoke based primarily on IETF experience), observed about this stage:

> It's easier to get people to do work to develop something than it is to get them to review something else. And so the whole question of how do you get adequate review—I think there's sort of a curve where you get a lot of people interested and then there's the final process of dotting the i's and crossing the t's and that sometimes can have a long tail on it because it's not the fun work everybody wants to do. How do you get resources to do that in a volunteer consensus-based process?[78]

Although the primary editor's aggressive approach probably contributed to the drop-off in participation in the early years, participation did not increase much during the final year, after Ryan had resigned as editor. The private standardization system depends on people willing to contribute their time to carry out demanding technical work, and firms willing to support them in doing that work. The shortage of people willing to do that work, especially in the later stages of testing and review, creates challenges for voluntary standardization.

This difficulty in recruiting active standardizers seems to reflect a long-term trend that appeared briefly in electromagnetic compatibility standardization at the end of the period covered in chapter 6. Belief in the standardization movement helped propel voluntary standardization in the early twentieth century, with new energy around international standardization in the postwar era, as embodied by Ralph Showers. But by the late twentieth and early twenty-first centuries, many engineers no longer viewed standardization as important work of which an engineer could be proud, and Showers's successor was unable to find an American engineer to be his own successor.[79] In W3C, one WebCrypto WG member explained that he did not do standards work fulltime because "that way of doing it is, I think, rightly disparaged by the people who actually build products."[80] Software engineers viewed the traditional standards organizations as particularly stodgy, but some viewed the third-wave bodies such as the IETF and W3C more favorably. As one WebCrypto WG member explained, some ANSI-accredited standards committees "are more sort of old line, sort of gray-haired people, a dozen gray-haired people sitting around the table in sort of the almost smoke-filled room of standards committees." He thought, however, that "the IETF and W3C are . . . a little bit more—I won't say necessarily democratic, but it's just a little bit more free form, a little bit more outspoken, and there is a little bit more emphasis on allowing distributed participation for example and that's kind of nice."[81] While this individual stopped short of claiming that it was democratic, WG chair Virginie was willing to embrace that term and was proud of the democracy and transparency in W3C, in which "any individual is able to speak, and is listened" to, a situation she had not found in some of her experiences with other standards bodies.[82] In face-to-face and phone meetings, she insisted that people figuratively raise their hands (by typing "q" in the internet relay chat window running during all meetings) to be recognized rather than just jumping in to make sure everyone could be heard, and she avoided back-channel communication as much as possible, preferring to communicate over the list available to all members and the public to maintain the transparency of the process.

Although the enthusiasm for standardization itself demonstrated in the earlier twentieth century was generally less evident in WebCrypto WG, a different form of movement enthusiasm motivated some of its most active participants. These members said they participated not just for business reasons (representing an employer's needs) or professional goals (reputation and resume building), but also because they believed in the importance of the internet or the web. As one WG member, who also participated in IETF standardizing, said laughingly, his work was "for the greater glory of the internet and me!" He continued, "I get a kick out of, like, making the internet a better place, and building new stuff."[83] Another WG member put it this way: "I think you kind of have to be a little bit of an altruistic idealist like that if you are working in technology anyway. I think all of us have a little bit of—we are making the world better."[84] As we saw earlier, this enthusiasm for and belief in the web and its potential led Tim Berners-Lee to found W3C in the first place. People involved in the IETF have a similar feeling about the internet; as a past chair of the IETF responded to a query about why he kept doing standards work, "It's because the internet is just a really wonderful thing. It's bringing significant changes of life to so many people and I'm excited to be a part of that."[85] So to the extent that we see standardization movement enthusiasm in third-wave standards bodies such as W3C and IETF, it seems to be more focused on the technology being standardized than on the value of standardization per se.

The standardization effort around WebCrypto API reveals both continuities and discontinuities between second-wave and third-wave standardization.

For all the proclaimed differences between new organizations such as W3C and the "stodgy" traditional organizations, their processes are remarkably similar in many ways. Like the older organizations, W3C follows a series of stages through a process from conception to final standard, with consensus necessary at various points. These stages and the consensus required also assured that even with electronic communication, the process is inherently not a rapid one. These new standards organizations, like the earlier ones, drew on communities of (in this case software) engineers who often served on technical committees of multiple standardization bodies. And both sets of communities had shortages of active standardizers by the late twentieth century.

Differences between standards bodies of the third wave and those of the earlier two waves, however, include the shift in tone and style from top hats to T-shirts, reflecting in part the new open-source and hacker cultures that emerged in tandem with computer networking. It meant that the gentlemanly pretensions of the earlier era gave way, but unfortunately not to a more gender-equal tone, since Ryan's "sharp elbows" at certain points approached flaming and, except for Virginie (who, as chair, generally contributed to process, not content), the few women on the WG accounted for very little of the communication, online or in phone meetings. The change in movement enthusiasm from believing that standardization itself created value for the world to believing that the technology being standardized created value for the world was also significant. Of course, in earlier and more recent cases, company motives were still evident, as well, as in the case of Mark's work in his company's interests. Nonetheless, the overall process and its results were influenced the most by those individual members of the committee with significant technical skills who spent the most time on the project.

Finally, the arguments around whether WebCrypto API could or should try to guarantee web security or should just give developers recognized tools with which to work in building the security themselves reminds us of what is at stake in standard setting. The issue of internet and web security is highly salient today, and related standards, including WebCrypto API, are critical parts of the infrastructure of the global economy.

9

Voluntary Standards for Quality Management and Social Responsibility since the 1980s

In 2017, private standard setters planned and celebrated multiple centenaries, including the foundation of many early national standard-setting bodies and a century of using ubiquitous standards such as A2- and A4-sized paper (which started as a German standard in 1917).[1] Over the century, the network of organizations that set private standards had grown from the International Electrotechnical Commission and the handful of national bodies at the end of World War I to an extensive network of national organizations and the first attempt at a general international body by the beginning of World War II; then, from the postwar formation of the International Organization for Standardization to the hundreds of organizations both inside and outside the nested international system gathered under the ISO by the 1980s; and, finally, since the late 1980s, to adding the new global standards bodies around information technologies. The community of engineers serving on various standard-setting committees has grown from hundreds to several hundred thousand, as large as the group of professionals working throughout the entire United Nations system.[2] And while in recent decades much of the community had lost the social movement–like zeal of the engineering standard setters a century before, the engineers who flocked to the meetings of the Internet Engineering Task Force and the World Wide Web Consortium had at least as fervent commitments to an open, global internet and World Wide Web and saw voluntary standardization as essential to achieving that end.

The most recent wave of standardization was not, however, just about information technology. It also included something fundamentally new and barely imagined a century before. The network of traditional standards organizations established through the first two waves of standardization turned to new projects that, on the surface, had nothing to do with traditional industrial standard setting. The new activities included fostering environmental and social sustainability, improving the conditions of workers in all kinds of organizations, and most of the other topics considered under the heading of corporate social responsibility (CSR). As these topics took off in the 1990s and 2000s, the community of standard setters grew to include thousands of stakeholder representatives who were not the engineers who dominated traditional standards bodies. The new members of the community extended far beyond the representatives of final consumers who had been incorporated into standard-setting committees in the mid-twentieth century. In the twenty-first century, there were representatives of labor unions, nongovernmental advocacy groups, and intergovernmental organizations, as well as many representatives of the public relations, human resources, and strategic planning departments of global corporations.

This chapter explains how these changes happened and what some of their consequences have been. We begin with the creation of ISO's standards for quality management, the ISO 9000 series (launched in the late 1980s around the time that new organizations were emerging around information technology), something that fundamentally shifted much of the attention of national standards bodies from product standards to process standards. ISO 9000 standards had far-reaching consequences for the entire network of standard-setting organizations. Most importantly, the commercial success of these standards, as well as the business of monitoring their enforcement and accrediting the monitors, all but eliminated the financial problems that had periodically plagued standard-setting bodies since the Great Depression. At the same time, competition for the ISO 9000 "business" made the entire network more complicated, less coherent, and, ironically, less focused on the social good.

The chapter then turns to a second, potentially more profound consequence of the ISO 9000 system: it became a model for people worried that economic globalization would trigger a race to the bottom in environmental, labor, and human rights standards. Beginning in the late 1990s, many political entrepreneurs—including United Nations secretary-general Kofi Annan, social investment pioneer Alice Tepper Marlin, and Davos World Economic Forum (WEF) founder Klaus Schwab—proposed new governance schemes based on process standards as ways to regulate the potential social and environmental harms of the increasingly global economy. ISO itself also stepped on this bandwagon, creating its own social and environmental standards.

ISO's relative success with selling environmental and social standards encouraged national bodies to enter the same arenas, sometimes with the support of progressive politicians and activist groups. As with standardization of information and communication technology, standard setting in the new social and environmental fields became highly contested, and the network of organizations involved became increasingly complex.

ISO 9000: Transforming the Standard-Setting System

The purpose of the ISO 9000 series of standards is, in theory, to help companies produce the highest quality goods and services. It grew out of the mid- and late-twentieth century transnational quality movement that was originally centered in the United States and Japan and that had a variety of different approaches to achieving quality (e.g., allowing any worker to stop the assembly line when any defect is discovered or looking for statistical discrepancies in charts of a company's quantifiable processes). The ISO 9000 standards emphasized setting goals, documenting the processes by which a company attempts to achieve those goals, and having independent third parties monitor the company's entire quality management system to ensure that documentation existed. ISO 9000 standards did not involve examining outputs—the products or services that a company sold—and directly assessing their quality.

The Origins of ISO 9000

There are many claims to ISO 9000's origin. Military historians remember necessary wartime process innovations (designed, for example, to keep bombs from going off before they left the factory) and trace its origin to the British and American militaries in World War II.[3] Germany's national standards organization has a claim because ISO formed the technical committee that negotiated the international standard in response to a 1977 DIN request to rationalize the many different quality standards that purchasing departments around the world were starting to impose on their suppliers.[4] Admirers of British prime minister Margaret Thatcher say that she was responsible for the success of the BSI standard that was the basis for the one that ISO eventually adopted in 1987; she supported it as a market-oriented way to push British firms to produce goods of the quality necessary to compete in global markets. If anyone else deserved credit, the Thatcher advocates said, it was the Japanese firms that produced so many high-quality goods in the first place![5]

A long-serving Swedish expert on the relevant technical committee claimed that the real trigger had been the world's largest customers: defense ministries considering purchasing whole new weapons systems, electrical power companies considering building an atomic power plant, "or when a Norwegian oil company wanted to drill [under the North Sea]." The idea of a universal standard for quality was not something pushed by "ordinary" customers; it was something that interested "strong customers who always made demands on suppliers. So it all started with those demands."[6] The man who headed ISO when the standard was negotiated, Olle Sturén, believed that the story began in 1964 when he first met one of those customers, Admiral Ralph Hennessy, controller general of the Canadian Armed Forces.[7] In 1972 Hennessy, who had become Sturén's close friend, was made executive director of the Standards Council of Canada, a new public-private corporation designed to coordinate Canada's voluntary standards activities and serve as the country's member of ISO. In 1976, Hennessy moved to ISO as vice president and chair of its agenda-setting Technical Board, the governing group that responded to DIN's request by establishing TC176, Quality Management and Quality Assurance, and recruited Canada to provide the committee's secretariat.[8]

Sturén's account of ISO 9000's origin has become semi-official. In 1999, his American successor as ISO secretary-general, Lawrence Eicher, told an audience of North Americans that the "partizan" disputes about ISO 9000's history did not really matter because, "whatever the origins, we know that ISO 9000 has become an international reference for quality requirements in business to business dealings."[9] Not surprisingly, in 2006, ISO president Masami Tanaka opened the 29th ISO General Assembly in Ottawa with the Canadian version of the story: "Among Canada's many contributions to ISO is providing the secretariats of [the] ISO technical committees" responsible for "ISO's best-known families of standards, ISO 9000 and ISO 14000 [environmental standards]." Those standards, Tanaka continued, "have been largely responsible for communicating and promoting ISO . . . to the attention of a wide international audience in business and government—and increasingly to the general public," a group that went far beyond "the managers and engineers for whom we have been proposing solutions for decades."[10]

Tanaka was certainly correct about the consequences of the standards for ISO itself, but to understand the larger picture—and to have some sense of the questions that must be raised about ISO's success at achieving "quality"—it is also worth recalling the pre-ISO history of quality management systems. In the early twentieth century, many companies experimented with new ways to use information to improve business.[11] In 1924, Walter Shewhart, a statistician working for the Western Electric Company, developed a set of linked charts to record the quality of parts at different stages of production. Shewhart's colleague, W. Edwards Deming, took these ideas to the US government during World War II and inspired creation of a new professional association of quality engineers, the American Society for Quality (ASQ), at the war's end. Deming also played an important role as a consultant to the US occupying powers in Japan. Many Japanese managers adapted Deming's ideas to create Japan's global reputation for producing quality products. Eventually, as Western fear of Japanese competition grew, Deming began to be treated as a forgotten prophet by many businessmen and government officials in Western Europe, the United States, Canada, Australia, and New Zealand.[12]

When Admiral Hennessy convinced ISO and his Canadian colleagues to follow up on DIN's 1977 request to study standards for quality management, he was familiar with Thatcher's concern about competing with the Japanese, as well as with the desire to assure "zero defects" in the manufacture of weaponry. Indeed, a government or firm might have a similar concern if it were commissioning a nuclear power plant or acquiring a floating oil derrick to drill under a nearby coastal sea. Hennessy also understood why any big purchaser, whether a global company or the government of a wealthy middle power, would also worry about being fooled into purchasing massive numbers of shoddy *everyday* items such as pencils or toilet paper. And he understood the issue raised by DIN, and by many small manufacturers around the world, about the inefficiency and potential unfairness of big purchasers imposing hundreds of different quality standards to avoid quality problems.

ISO's first quality management standard was ISO 9000:1987, Quality Management and Quality Assurance; the standards that directly updated this core standard were termed "9001," and a whole host of separate, related standards were given higher numbers in the 9000s. The core standard assumed that the operational definition of "quality" would differ from product to product or service to service. Nonetheless, the standard said, the organizational *processes* needed to produce any quality outputs involved three things that could be specified in any management system. First, those who adopted the standard had to use "customer orientation" as a central part of their own definition of quality. Second, adopters needed to document everything they did—all their processes of production and management—so that they would be able to improve their ability to provide what their customers wanted. And, third, adopters had to commit themselves to continuously improve the quality of their goods or services and the processes that produced them (including increasingly working with suppliers who were also ISO 9000 certified, a factor that encouraged the standard's spread). Adopting a customer orientation was meant to trigger the transformation of an organization's culture. As an ISO sales presentation put it, the ISO 9000 "quality management system . . . doesn't specify what the objectives relating to 'quality' or 'meeting customer needs' should be, but requires organizations to define these objectives themselves and continually improve their processes in order to reach them."[13] A quality management systems (QMS) standard was simply a standard for documenting and assessing an organization's processes, something very different from the product standards that had made up most of ISO's work before ISO 9000. Product standards specified particular outcomes—a standard product had to work in a certain way. "Management systems" standards focused on an organization's processes and culture.

ISO 9000 Certification and Auditing

While it is unclear whether the cultural transformation promised by ISO 9000 has actually taken place in the hundreds of thousands of businesses that have adopted the standards since 1987, a cultural change certainly has taken place *within* ISO and many of its member national bodies, though with some unintended consequences. Most have, themselves, adopted the ISO 9000 standards, and they now use ISO 9000's rhetoric to describe what they do. A 2002 speech by ISO interim secretary-general Christian J. Favre, gave a sense of the change. He spoke of how ISO's commitment "to satisfying the customer . . . has allowed us to maintain, including during this interim period [after the unexpected death of Lawrence Eicher], the quality of service to our members." And he went on. "Achieving customer satisfaction is required by our certification to ISO 9001: 2000 but I am glad to be able to report that it is in fact being implemented."[14] This is very different language from that used by Olle Sturén from the 1960s through the 1980s. Sturén believed that ISO should focus on the all-inclusive community of global *citizens*, not a self-defined group of *customers* created by their decision to purchase ISO's standards. The distinction between citizens and customers was a meaningful one. Not only did the new focus lead some of ISOs' critics to begin using words like "mercenary" to describe ISO and its member bodies,[15] but it also undercut the traditional argument that ISO standards, and those of its national members, were the result of processes that were, in some ways, superior even to the processes of democratic governments; that was *how* standardization served all citizens (see chapter 3). What had started out as a progressive social movement was threatening to become just another commercial operation.

The rapid shift to the new orientation to mere customers stemmed in large part from the way management system standards quickly became central to everything that ISO and its members did. The new standards brought the organizations public notice and money. ISO president Masami Tanaka was correct when he argued in 2006, the QMS boom, led by the ISO 9000 standards, had given the standard-setting organization a global public profile. For the first time, because of these standards, average people around the world had heard about ISO, no doubt a gratifying development to many standardizers. In addition, the boom not only generated sales of ISO 9000 and the many subsequent standards designed for particular sectors or individual countries; these QMS standards also greatly expanded two lucrative standards-related businesses: certification and accreditation. ISO 9000 and related standards were designed to be auditable by experts external to the companies applying the standards. Consequently, other companies were needed to *certify* compliance with the standard, new companies similar to the small number of companies such as Underwriters Laboratories that already certified that products adhered to safety or health standards. In turn, organizations similar to the regulatory bodies (public or private) that accredited those laboratories in different countries needed to accredit the companies that performed the QMS audits and certifications. National standards bodies could play both additional roles in the QMS field, and they could make money by doing so.

While it quickly became clear that the tasks of setting standards, certifying compliance with them, and accrediting the certifiers needed to be done by somewhat separate organizations to avoid conflicts of interest, the process of creating this new network of organizations transformed dramatically what was coming to be called the "business" of standardization. By 1990, just three years after the first publication of ISO 9000, Standards Australia had created a division (later a separate company) that provided ISO 9000 training and certification. The Australian division was working on mutual recognition agreements with similar divisions of the Canadian Standards Association (CSA), Great Britain's BSI (violating one of its founding principles against becoming a testing authority; see chapter 2), Underwriters Laboratories in the United States, and the national standards bodies of Japan and New Zealand. These agreements assured that ISO 9000 certifications performed in one country would be accepted in other countries.[16] In the following decade, the CSA and BSI both adopted business strategies of acquiring foreign companies involved in certifying standards—product standards as well as QMS standards—thus turning themselves into global commercial enterprises. The BSI, CSA, and Standards Australia stopped treating their names as acronyms that referred to "standards" or "standardization" and a specific country; their acronyms officially became just parts of the names of new global companies, BSI Group, CSA Group, and SAI Global.[17]

As national standards bodies were changing to get into the certification business, ISO, ANSI, and the ASQ began working on the problem of accrediting the many organizations that wanted to offer certification services. In 1986, even before the publication of the first version of ISO 9000, ISO and the IEC jointly published Guide 48, "Guidelines for Third-Party Assessment and Registration of a Supplier's Quality System." ANSI and the ASQ then became involved because of American worries that widening and deepening European economic cooperation might lead Europeans to use their QMS certification and accreditation systems to keep US companies out of the European market. The two organizations created an American accreditation body, the ANSI-ASQ National Accreditation Board (ANAB) to assure that there were American certifiers to give American companies their ISO 9000 seals of approval. Ironically, in the rapidly globalizing economy of the early 1990s, ANAB quickly became an American division of the ASQ's own, increasingly *international*, quality certification business. To overcome potential conflicts of interest within the ASQ and create an accreditation system fit for the purposes of a global economy, in 1993 ANAB and other national accreditation boards created a new organization, the International Accreditation Forum (IAF). It quickly became a global association that mutually recognizes all the auditors/certifiers accredited by any of the IAF's national members—that is, a global *accreditation* organization parallel to ISO, the global *standard-setting* organization. In 2012, ISO and the IEC issued a new standard, ISO/IEC 17065 Conformity Assessment—Requirements for Bodies Certifying Products, Processes and Services, which was used by IAF members when accrediting organizations to provide ISO 9000 audits and certifications.[18]

Those certification organizations became a much larger and more diverse set than either the standard setters or the accreditors. In 2017, ANAB listed 76 ISO 9000 series certifying agencies that it had accredited in the United States, including the national standards bodies of Israel, Jamaica, and Spain, the American Petroleum Institute, and Perry Johnson Registrars of Troy, Michigan (self-described as the "Number 1 Registrar [of ISO 9000 certifications] in North America!"). The Canadian accreditation organization, the Standards Council of Canada, listed 22 ISO 9000 certifiers that it had accredited, including the National Standards Authority of Ireland and the standards body of the Province of Quebec, as well as BSI Group America Inc. (the American part of the global company that grew out of BSI), PricewaterhouseCoopers (the second-largest of the "Big Four" global accounting firms), and UL LLC (the company that used to be known as Underwriters Laboratories Canada, the Canadian branch of the venerable American safety standards organization).[19] With the advent of ISO 9000 standards, the global network of organizations interlinked with the traditional voluntary standard setters focused on industrial products had become much more complicated and, on the surface, much less rational. National accrediting agencies were supporting foreign companies in the profitable business of certifying ISO 9000 standards. Some of those foreign companies were even the official national standard-setting bodies of other countries. And all the certifiers operated around the world. For example, ANAB reported that the Standards Institution of Israel provided ISO 9000 certification in "China, Israel, Japan, Romania, Serbia, and the United States."[20]

The confusion of the expanded global network of standards-related organizations was the first of several consequences of the staggering demand for certification of the core ISO 9000/9001 standard. Every year from 2007 to 2015, about 1 million organizations, primarily profit-seeking companies, sought such certification or recertification (which typically occurred every three years). The companies that adopted ISO 9000/9001 paid for all the associated costs. Estimates of the external costs for consultants and auditors for each ISO 9001 certification start in the tens of thousands of dollars. Thus, the global business of certifying compliance with just this one standard became very large indeed.[21]

This business also shifted geographically from its beginnings to the present. In 1993 the first global survey of certifications identified more than 45,000 organizations that received ISO 9000/9001 certifications that year. More than 60 percent of those organizations were in Britain, evidence for the hypothesis that the standard's initial success was assured by Margaret Thatcher's push for British firms to improve quality. Nevertheless, the geographic center of ISO 9001 certifications quickly shifted to larger regions where business success increasingly depended upon reaching global markets. In 2015, while 40 percent of certifications remained in Europe, an additional 40 percent were in East Asia, with China accounting for more than three-quarters of those, or about 30 percent of all organizations compliant with ISO 9001.[22]

Over time, ISO 9001 certification has also been sought by an increasing range of organization types, well beyond industrial firms. In 2006, one commentator marveled, "Who would have thought 20 years ago that financial institutions, schools, blood banks, prison systems, Buddhist temples, cruise lines, or retail chains would implement ISO 9001?" Major international relief organizations sought ISO 9001 certification, as did individual consulting astrologers.[23] Nevertheless, the most frequent users of the standard have been more prosaic organizations: construction companies and manufacturers of metals, electrical equipment, and machinery.[24]

ISO 9001's popularity, as well as widespread cynicism about its efficacy, has encouraged a great deal of research about its impact on the organizations that have adopted it. Surveys of private companies revealed that adoption of the standard led to better operating performance, but not necessarily to improved business performance; that is, adopters achieved more of their own production, quality, and cost goals than non-adopters, but that did not translate into consistently larger profits. Nevertheless, adopters oriented toward continuous improvement *before* using ISO 9000 tended to do better by most measures, including profits. Perhaps this is because these companies' earlier methods of achieving high quality focused on outcomes—quality products or services—and adherence to the management system standard simply reinforced achievement of a concrete goal. Perhaps it was also because these companies tended to choose more careful and reputable auditors who provided insightful feedback about company operations. Less-successful adopters tended to choose certifiers satisfied with a kind of rote compliance that did little to improve operations.[25] The results for these less successful adopters suggested the validity of the cynical view of ISO 9000, reflected in figure 9.1, a *Dilbert* comic strip from 1995.

Figure 9.1. This cartoon about ISO 9000 from the popular *Dilbert* comic strip (November 7, 1995) illustrates the cynical view of ISO 9000. *Dilbert* © 1995 Scott Adams. Used by permission of Andrews McMeel Syndication. All rights reserved.

So why would so many companies, at least those who hired the less reputable auditors, do something so pointless? Because, *Dilbert* aside, it was not completely pointless: it helped businesses satisfy customers, both old and new. Many companies sought certification because their largest customers, often governments or huge multinational firms, demanded it. These customers wanted to control the costs of monitoring their suppliers, while still minimizing the risk that their suppliers would provide them with shoddy goods or poor services. For companies that wanted to attract these large buyers, ISO 9000 certification provided a signal that said, "You may never have heard of us, but we are just as committed to quality as the firms you have dealt with for a long time." Thus, Chinese manufacturers and construction companies wanted to display the ISO 9000 banner on their letterheads, their business cards, and, increasingly, on flags flying outside their offices and factories to send this signal. The importance of ISO 9000 certification for companies trying to enter global markets explains why national standard-setting bodies in countries where export industries were just taking off—places like Ireland, Israel, Jamaica, and Spain, as well as China and the other rapidly industrializing countries in Asia and Africa—were happy to get into the certification business and provide the flags. Figure 9.2 shows the first private university to receive certification by the Botswana Bureau of Standards displaying such a banner.

ISO 9000 certification was not very helpful to companies that already had international reputations for high quality or for manufacturing firms in exporting countries like Japan or Germany with a general reputation for high quality. Nevertheless, even many Japanese companies sought out ISO 9000 certification, not because adherence to the standard would improve the quality of their products but because it had become a cost of doing business as

Figure 9.2. Botho University celebrates its June 2013 ISO 9001 certification from the Botswana Bureau of Standards. Reproduced with permission of Botho University, https://bothouniversity.com.

more and more companies decided to signal their commitment to customers in this way.[26]

The success of ISO 9000 led ISO to create a series of other auditable management system standards, which firms similarly displayed for all to see. In 2017, Standard Flags, a company claiming "the web's most complete selection of ISO flags," offered banners to celebrate certification with the two most recent versions of ISO 9000 (ISO 9001:2008 and ISO 9001:2015). In addition, they offered flags for the following:

- ISO 13485 Medical Devices-QMS (first published in 1996)
- ISO 14000 Environmental Management (also published in 1996 and discussed in a subsequent section)
- ISO 16949 Automotive Sector-QMS (first published in 1999 and created with the International Automotive Task Force, the global manufacturers trade association)

- ISO 22000 Food Safety Management (from 2005)
- ISO/IEC 27001 Information Security Management Systems (from 2005)
- ISO 50001 Energy Management Systems (from 2011)

Standard Flags also offered flags for non-ISO quality management systems such as AS9100 Quality Systems-Aerospace, a similar standard created for the industry by its trade association in 1999, and OHSAS 18001, Occupational Health and Safety Management Systems, an internationally applied BSI standard first set in 1999.[27]

In addition to all of these auditable standards, ISO published a number of related standards for which no flag could be raised. They were meant to provide guidance rather than to identify specific, auditable processes. For example, according to a marketing document, ISO 9004 Managing for the Sustained Success of an Organization, helped users think about how "to achieve not only satisfaction of the organization's customers, but also of all interested parties, including staff, owners, shareholders and investors, suppliers and partners, and society as a whole."[28]

Making Standard Setting a Global Business

This explosion of new management system standards had further consequences for the national standards organizations. In combination with the initial success of ISO 9000, it almost magically removed the financial problems that had plagued so many of them. The historian of Standards Australia, Winton Higgins, explains that, for this typical early national standards body, the sudden solution to the old financial problems was a somewhat mixed blessing. It certainly was a blessing in that "the old SAA [Standards Association of Australia]" had faced a "cramped, hand-to-mouth existence." Its leaders were reluctant to beg for government or industry support or follow some of the questionable practices of other cash-strapped national bodies "such as demanding an upfront fee from an interested party before embarking on a project" or "charging people to sit on technical committees," practices that undermined the norms that had governed voluntary standard setting from the start.[29] However, in the new QMS-led world, "no matter how demonstrably the organization quarantined its commercial activities from its standards development in the national interest . . . negative perceptions were hard to shake. Standards Australia looked like "a not-for-profit organization drawing heavily on the inputs of experts it did not pay" to run "a commercially successful business."[30] In 2003, Standards Australia did what many other private national standards bodies would do. It sold its publishing, consulting, and certification activities to a newly formed, profit-making private company, retaining a significant part of the new company's shares as a source of income and granting the new company a fifteen-year contract to publish Australian standards, a right for which it would pay significant royalties. "The arrangement left Standards Australia with its original and core business—standards development."[31]

While most national standards bodies could find similar ways to shield their traditional work from the newer, profit-making activity, the issue was not as simple for ANSI with its first-among-equals status within the competitive network of US standard setters. Many of these bodies also had interests in the management system business as well as their own desires to set and certify international standards. ISO's new commercial culture helped make that possible.

In the 1999, the largest and oldest of ANSI's "equals," the American Society for Testing Materials, proposed new ISO/ASTM standards that would be approved by an ISO technical committee but developed and maintained by ASTM. ASTM president James A. Thomas proclaimed, "We are proposing the creation of the strongest international standard ever . . . [providing] an international standard with unprecedented commercial power." The new hybrid ISO/ASTM standards would "increase the number of ISO standards that are used in the United States and around the world . . . another international option, an attractive, commercially practical alternative."[32] Standards Australia's business-oriented head, Ross Wraight, then serving as ISO vice president for technical management, championed the idea, arguing that ISO should include industry-based standardization bodies (in addition to national general standards bodies) among its regular members. In 2001, to cement its image as the *global* setter of standards in the materials industry, ASTM dropped "American" from its name and became ASTM International.[33] It took a number of false starts for substantive ASTM-ISO collaboration to begin, but in 2013 the two organizations published their first joint standards on additive manufacturing, more familiarly known as computerized 3D printing, used in manufacture of products or the construction of buildings.[34] The ISO-ASTM collaboration provided a model for other cooperative global standard-setting initiatives linking ISO and specific industry-based standards organizations in fields as diverse as leather tanning and the formatting of documents across different electronic media.[35]

These collaborations signaled the further disintegration of the nested structure of international, regional, and national standardization bodies that had once been the ISO ideal. Many standards users welcomed such a development. Recall from chapter 7 that in 2009 Oracle Systems' Trond Arne Undheim hoped for a world of global standardization by 2020, a world without the "UN-like" requirement that a consensus of national bodies was needed to set a standard. If turning an industry-based standards body like ASTM into a global entity and then incorporating it into ISO via partnership proved to be a step in this direction, it would be welcomed by those who shared Undheim's views. Nevertheless, he worried that the conflict among the forces driving the "messy globalization of the standards world" might be too strong for his high hopes to be realized in a single decade.[36]

ISO 14000, SA8000, and the Global Compact: New Social Standards in the 1990s

Somewhat incongruously, at the same time that private standardization was both promoting economic globalization and becoming more global, social activists concerned with the impact of globalization on the environment, on workers, and on human rights began to look to ISO's experience with management system standards as a model of how to prevent a globalization-led race to the bottom. For the activists, ISO 9000 provided a *governance model* for organizations concerned with social and environmental sustainability, one that actually promised their continuous improvement across the global economy. The relevant feature of the model had also helped make ISO 9000 certification such a successful business: as part of the "continuous improvement" that the standards demanded, ISO 9000–certified companies were expected increasingly to buy their supplies from companies that also had quality management systems in place. Adherence to this principle helped the standard spread.[37] British historian Deborah Cadbury points to a precedent for this hoped-for contagious spread of higher standards in the unusual success of nineteenth-century British and American Quaker manufacturers. They followed the highest social standards of the day in their own firms, demanded similar standards of their distributors and global suppliers, and were held accountable by an active popular press eager to point out any hypocrisy on the part of these "austere men of God" who influenced "the course of the Industrial Revolution and the [pre-Thatcher-era] commercial world."[38]

But why, in the late twentieth century, did social activists have to worry about a potential race to the bottom when governments could stop it? After all, political scientist John Ruggie argued in an influential 1982 article that the governments of the industrialized countries had done so after World War II; the rich capitalist countries created a system of free trade in the industrial products they produced, but their governments cooperated to "embed" free trade within a larger set of social norms designed to protect the gains that workers, women, and many minority groups had won during the war.[39] Initially, many analysts believed that extending these practices to the globalizing economy would be uncontroversial and relatively simple. As early as 1970, leading economists had proposed universalizing the system that Ruggie would call "embedded liberalism" by regulating transnational firms and their global supply chains.[40] In 1973, under pressure from Western Europe and the developing world, the UN created the Centre on Transnational Corporations (UNCTC), which designed a global legal code covering labor rights, anti-corruption measures, and progressive taxation,[41] but UN members debated the draft code for almost two decades, to no avail, largely due to American opposition.

Throughout the extended UN debate over regulating the new global manufacturing economy, the US Republican administrations of Gerald Ford, Ronald Reagan, and George H. W. Bush opposed any regulation of economic globalization, a bête noire of powerful conservative groups in the United States, and the Democratic administrations of Jimmy Carter and Bill Clinton had other priorities. In a 2000 *Chicago Journal of International Law* essay, John Bolton, a key conservative figure in US-UN relations under Reagan and Bush, explained the Republican reasoning: The UNCTC, Bolton said, lacked the legitimacy to establish such regulation because the majority of UN members were undemocratic governments. The International Labor Organization was no better, he argued, because it created international labor standards through committees with balanced representation of workers, employers, and governments, which was a form of the objectionable "corporatism" that Mussolini had considered "the single most essential component in fascism." Moreover, there were the suspect motives of the European Union globalists, whom Bolton considered the major proponents of the UNCTC code, who, "not content alone with transferring their own national sovereignty to Brussels . . . have also decided, in effect, to transfer some of ours [the United States'] to worldwide institutions and norms."[42] In Bolton's view, the reason governments never agreed to "re-embed" the global economy within larger social norms through regulation was that Americanists successfully blocked the self-interested supranationalist strategy of the European globalists. Indeed, George H. W. Bush fought to make Boutros Boutros-Ghali UN secretary-general in January 1992, and as new secretary-general Bush's favorite ended the two-decade-long debate over a legally binding global code for corporate responsibility, marginalized the UNCTC, and fired its head, Peter Hansen, a Danish Social Democrat who had angered many American conservatives.[43]

Yet, the UN's failure to establish a legally binding code for transnational corporations was not solely the victory of the Americanist Republicans over European globalists. From the 1970s through the 1990s, the common view of many global business leaders remained one expressed in 1974 by the Royal Dutch Shell president in early UN hearings about such a code: "The world," he argued, "is still organized nationally or locally," and, until Shell had to face global labor unions, it made little sense to think about regulating business at that level. J. A. C. Hugill, the British head of the UN Food and Agricultural Organization's Industry Cooperative Programme, added that since the "binding" of any supposedly binding code would have to be done by national governments, an internationally negotiated code would probably end up being the least common denominator among national laws, just another way to race to the bottom. It would be better, Hugill argued, to create a set of private voluntary standards to encourage companies to be on a list of the very best performers, "the kind of list that corporations should be proud to belong to."[44] These became some of the main arguments that inspired the private social responsibility standards of the 1990s.

Recall from the first three chapters that these arguments are as old as the original standardization movement. Hugill was repeating the argument that governments (and their intergovernmental organizations) are often reluctant to set necessary standards simply because constituents who are already following other standards will object to the cost of switching. The same logic suggests that any attempt to establish social or environmental standards through the legal process (treaty making, in this case), is likely, at least initially, to lead to codification of the least common denominator. Experts representing the different stakeholders involved, however, might be able to agree on higher private standards that at least some companies would adopt as the nineteenth-century Quaker manufacturers did to demonstrate their social responsibility.

Environmental Standards Based on ISO 9000: The ISO 14000 Family

Thus, in 1992, just in time for the UN's Earth Summit, BSI published the first major private environmental standard, BS 7750, Specification for Environmental Management Systems, modeled on ISO 9000/9001 and the earlier BSI quality management standards. ISO then quickly formed a technical committee to use the British standard as a template for the first of the ISO 14000 series of environmental management system standards, published in 1996. The standards required firms to identify environmental goals and targets, establish procedures and processes to achieve those goals, check and document results of these processes, and develop systems for continuous improvement. As with ISO 9000/9001, an organization could seek periodic auditing from third parties accredited to certify that it is following the standard. The standard therefore provided a way for companies to both improve their environmental performance and to signal their environmental commitments to their customers as well as to governments, environmental activists, and prospective employees.[45]

The environmental standard proved very popular, although not as popular as ISO 9000/9001. Coincidentally, about 14,000 organizations were certified as compliant with the core ISO 14001 standard in 1999, the first year for which data were collected. In 2007, more than 150,000 received certification. And, unlike ISO 9001, whose annual certifications stabilized at around 1 million after the financial crisis of 2007–2008, ISO 14001's annual certifications continued to rise, reaching more than 300,000 in 2015.[46] Nonetheless, this was still less than one-third of the annual number of ISO 9001 certifications, perhaps reflecting a weaker contagion effect of the mandate for continuous improvement. Most organizations may have been able to make many internal environmental improvements before they needed to impose ISO 14001 compliance on their suppliers.

The success of ISO 14001 generated the same kind of extensive research that followed the success of ISO 9000/9001. Ten to fifteen years after the publication of the original BSI standard on which ISO 14001 was based, some studies of the BSI and ISO standards suggested adopters polluted less than they had in the past, and less than non-adopters. Adopters also tended to comply better with national and local environmental regulations. Most adopters did so, at least in part, to signal their environmental commitments, by displaying the ISO 14001 flag to customers, government regulators, and others. Detailed case studies of nine Canadian firms, however, indicated that adoption led to "ceremonial behaviour intended to superficially show that the certified organizations conformed to the standard" but had only "an ambiguous effect on environmental management practices and performances." Nevertheless, adopters of ISO 14001 later tended to adopt additional standards from what eventually was a 35-member family of ISO 14000 standards. These fell into four categories, seven standards each for (1) monitoring and reporting emissions of greenhouse gases, (2) environmental labeling, and (3) environmental performance evaluation, and ten standards for (4) life-cycle assessment (the careful use of resources and recycling).[47] Finally, research indicated that initial fears that the standard would be popular only in Western developed countries were unfounded. As was the case with ISO 9001, the environmental standard was widely adopted in Europe (both East and West) and in East Asia, with organizations in China accounting for about one-third of all adopters.[48]

Global Social Standards outside ISO: SA 8000

At the same time that ISO 14001 was first taking off, around 1,000 corporate codes of conduct were also being newly promulgated. These included The Body Shop's Trade Not Aid initiative of 1991, the Worldwide Responsible Apparel Production labor code of 1997, and Amnesty International's Human Rights Guidelines for Companies of 1998. Most were single-company-based efforts that responded to the growing niche markets for socially responsible firms created by economic globalization, Thatcher- and Reagan-era deregulation, and fears of the growing worldwide anti-globalization movement.[49] By the late 1990s, businesses using social and environmental audits existed in many countries. For example, Brazil's Instituto Ethos established a database of model companies' socially responsible routines that it linked to a self-assessment tool to help organizations evaluate their own practices and find the best models to emulate. The institute also helped draft the corporate social responsibility standard for the Brazilian Association of Technical Standards (ABNT, Associação Brasileira de Normas Técnicas), the work for which began in the 1990s.[50]

One of the first truly global, non-ISO initiatives of the 1990s was the 1997 SA8000 labor standard, which a 2012 *New York Times* investigative report called "the most prestigious in the industry."[51] It was explicitly modeled on the ISO management systems standards and designed to reduce the confusion created by the hundreds of new private labor codes. SA8000's originator,

American labor economist Alice Tepper Marlin, explained that by 1995, "hundreds and hundreds of companies had codes of conduct for their suppliers, but the codes were widely different." Customers were confused about which codes were strong and which were just window dressing. Moreover, it was costly for factories, which typically produced for more than one brand. "Often they were supposed to comply with seven different codes. You even hear stories of companies that had 80 or 100 audits a year, each to a different code of conduct."[52]

Tepper Marlin considered herself partially responsible for this mess. To understand why requires going back to the beginning of her career. She had been one of the first women analysts on Wall Street. In 1968, her firm asked her to produce a "peace portfolio" of investments for a major religious client. That led her to create a nonprofit organization, the Council on Economic Priorities, which helped launch the socially responsible investment movement of the 1970s. Beginning in the 1980s, the council published a series of handbooks for socially and environmentally concerned consumers. In the late 1980s, when it began working on a specialized consumer guide for American students, Tepper Marlin was confronted with the problem of trying to assess the social and environmental impacts of factories in the developing world, the sources of the clothing and electronics that made up so much of what students purchased. She learned from the few American firms that already demanded high standards in their overseas suppliers' factories, such as the clothing brand company Levi Strauss. With new insight into conditions worldwide, the council began teaching other companies how to follow these leaders, fueling the explosion of company-based labor standards of the early 1990s. Its spin-off, Social Accountability International (SAI), and its SA8000 auditable global labor standard were Tepper Marlin's response to the chaos that had ensued.[53]

SA8000 followed the ISO model of a certifiable management system standard backed by an organization (SAI) that accredited companies to perform the audits and grant certifications. Most of the SAI-accredited certifiers already had experience certifying ISO 9000 and 14000. The SAI also learned how to set and maintain standards from ISO's model. The SAI created committees with what they saw as a balanced representation of stakeholders, and the committees operated by consensus. Nevertheless, both the SAI standard itself and the stakeholder process that established and maintained it differed in fundamental ways from the ISO standards and processes on which they were modeled. Perhaps most significantly, the SAI standard fo-

Figure 9.3. Social standardization entrepreneur Alice Tepper Marlin visiting an SA8000-compliant factory in Turkey in 2005. Reproduced with permission of SAI.

cused on *outcomes*, not just the documentation of management processes. SAI-certified employers had to provide a living wage, decent working conditions (including toilets, potable water, and medical attention in the event of accidents at work), and a whole range of protections found in many of the conventions and recommendations of the ILO—an intergovernmental treaty organization. Many governments have never accepted, let alone enforced, these rules. Additionally, the stakeholders relevant to setting labor standards went beyond companies that produced and purchased the goods (clothing factories in the developing world and the trademark-brand companies that marketed them to final consumers), to include representatives of labor unions, human rights organizations, development agencies, and child welfare advocates. Tepper Marlin (shown in figure 9.3 visiting a Turkish factory) recognized that most of these were people "who had not worked together before"; at best, they had sat down at the same table as antagonists. They certainly did not share a common language and culture like that provided by the engineering profession, which had facilitated the first century of voluntary standard setting. Therefore, she supplemented ISO's norms about how meetings should be conducted (e.g., with courtesy and respect) with much more explicit ground rules that she found in the practice of Search for Common Ground, a nongovernmental conflict-resolution group that had worked throughout the war zones of the early 1990s in Africa, Eastern Europe, and the Middle East.[54]

In 1998, one year after SA8000 was first published, auditors issued the first eight certifications of compliance to firms. Certifications grew to about 2,000 in 2008 and about 4,000 in 2017. Although the standard had been applied to workplaces employing more than a million people by 2017, less than 2 percent the number of organizations had adopted SA8000 than had adopted the ISO 14000 family of environmental standards.[55] SA8000 did not end the proliferation of new private labor standards. Nevertheless, the standard had been widely adopted by those firms with a history of enforcing labor standards that provided more than window dressing, and the careful design of the standard and its certification system immediately encouraged scholarly work on how the most meaningful kinds of evaluations might be done, even though few comparative studies have been completed.[56] One very harsh critic of ISO management system standards, Yale professor of architecture and globalization Keller Easterling, has emphasized what she considers the subversive potential of the SAI's model. She argues that the global activist group "mirrors the form of ISO 9000 protocols, this time precisely to piggyback on systems and habits already in place in an enormous number of organizations." Indeed, she claims that "ISO provides it with the camouflage of a relatively conservative organization" as it confronts issues of labor, the environment, and human rights. Moreover, SAI is global rather than international (like ISO or ILO), and SA8000 is not tied to national regulations. "So while the United States, for instance, may not have formally ratified [ILO] principles concerning labor protection, Social Accountability International can approach US companies individually and attempt to extract a pledge."[57]

The Global Compact: Ambitious Standards Loosely Based on the Engineers' Model

In 1997, around the time SA8000 was launched, Kofi Annan succeeded Boutros Boutros-Ghali as UN secretary-general and attempted to model his own corporate social initiative on voluntary consensus standardization. Annan was familiar with, and appreciated, the various contemporary private efforts to regulate the rapidly globalizing economy. He had known about engineering standard setting since his undergraduate days at Ghana's Kumasi College of Technology in the late 1950s, after which he studied economics in the United States and then joined the UN in 1962. Annan became attuned to the private sector when he completed his final degree, a midcareer master of science in management, at MIT's Sloan School, in 1972. When Annan returned to the UN, he cultivated his ties to global business leaders as he rose through the upper ranks of the secretariat. He regularly attended the Davos meetings of the World Economic Forum that had attracted the global economic and political elite since the early 1970s.[58] At the beginning of his second year as the UN's chief, Annan challenged the private-sector leaders at the January 1999 Davos conference with the "embedded liberalism" argument of John Ruggie, whom Annan had recruited to be the UN's director of strategic planning. The secretary-general warned, "The spread of markets outpaces the ability of societies and their political systems to adjust to them, let alone to guide the course they take. History teaches us that such an imbalance between the economic, social and political worlds can never be sustained for very long."[59] Annan proposed that transnational companies sign an agreement with the UN "to make globalization work for all the world's peoples by embedding the global market in a network of nine shared principles," which were standards derived from the Universal Declaration of Human Rights, the ILO's Declaration on Fundamental Principles and Rights at Work, and the 1992 Earth Summit's Declaration on Environment and Development. An early account of Annan's initiative written by senior staffers reported, "Immediately, enthusiastic responses from the press and from dozens of business leaders proved that the speech was a success. Letters from foreign ministers and CEOs poured into the UN, urging the Secretary-General to carry his message forward."[60]

Annan and Ruggie partnered in this and later UN attempts to promote voluntary social and environmental standards. Ruggie had risen through the ranks of Columbia's School of Public and International Affairs while Kofi Annan was nearby, rising to the top of the UN secretariat. When Annan became secretary-general, he wanted to follow the usual practice of appointing an American as a close advisor, and Ruggie was a logical choice.[61] Throughout the year after Annan's 1999 Davos speech, Ruggie joined Annan in discussions with scores of CEOs, the president of the International Confederation of Free Trade Unions, and the heads of a small group of nongovernmental organizations (Human Rights Watch, Lawyers Committee for Human Rights, Amnesty International, World Wildlife Fund for Nature, the

International Union for Conservation of Nature, the World Resources Institute, the International Institute for Environment and Development, Save the Children, the Ring Network, and Transparency International). Then, in July 2000, they officially launched the Global Compact.[62]

The compact required companies to adopt a very short standard—the original nine principles stated in three long sentences:

> Business should [1] support and respect the protection of internationally proclaimed human rights; and [2] make sure they are not complicit in human rights abuses. Business should uphold [3] the freedom of association and the effective recognition of the right to collective bargaining; [4] the elimination of all forms of forced and compulsory labor; [5] the effective abolition of child labor; and [6] eliminate discrimination in respect of employment and occupation. Business should support a [7] precautionary approach to environmental challenges; [8] undertake initiatives to promote greater environmental responsibility; and [9] encourage the development and diffusion of environmentally friendly technologies.[63]

In June 2004, a multi-stakeholder committee that advised Annan added a 10th principle: "Businesses should work against corruption in all its forms, including extortion and bribery."[64]

This standard was meant to provide guidance to companies that adopted it; no system of monitoring and certification would be provided, not even monitoring of a company's own documentation of their management practices as was required with the ISO 9000 and 14000 standards. Instead, companies that signed on were asked to make annual reports providing examples of their progress in internalizing the principles, while (to fulfill their part of the compact) UN agencies would begin holding learning forums in which company case studies of such progress would be presented and commented on by professors from leading management schools. The UN also sponsored policy dialogues among signatories on topics such as operating socially responsible firms in war zones. The secretary-general's Global Compact Office encouraged public-private partnership projects linking government agencies, nongovernmental organizations, and compact companies, including HIV/AIDS awareness programs for the employees of some firms in Africa.[65] In addition, the office immediately began working toward "the gradual convergence" of the most influential "voluntary codes of conduct," what the office referred to as the "Big Eight," which included, in addition to the Global Compact, Tepper Marlin's SA8000, the ISO 14000 series, and the ILO conventions, as well as AccountAbility 1000, the Global Reporting Initiative, the Global Sullivan Principles, and the Organisation for Economic Co-operation and Development (OECD) Guidelines for Multinationals.[66]

Most of the Big Eight, however, were not actually voluntary standards like SA8000 and ISO 14000. The Global Reporting Initiative, the only other voluntary standard, was a 1997 effort of two Boston-based nongovernmental environmental organizations and the UN Environment Programme; drawing on both ISO 14000 and SA8000, it set up a broad multi-stakeholder steering committee that established and maintained a standard for annual public reporting on environmental and social issues, often as supplements to the financial reports required of public corporations. The other five differed from voluntary standards in various ways. The ILO conventions were obligatory, not voluntary, in countries whose governments had signed and ratified them. AccountAbility 1000 was not a specific code of conduct but, rather, a standard for how companies could gain input from various stakeholders concerned with any topic that a company might specify. The OECD guidelines had first been promulgated by the Western industrialized countries in 1976 as a much weaker alternative to the UNCTC draft code. The Global Sullivan Principles were a 1999 generalization of the 1977 code of conduct adopted by foreign firms operating in South Africa that wanted to express opposition to apartheid without withdrawing their investment; in 1999, Annan and their author, American civil rights leader Leon Sullivan, reintroduced them as general principles as part of the campaign to gain signatories for the Global Compact.[67]

Tepper Marlin and some other leaders of the Big Eight were initially skeptical of Annan's Global Compact, which looked like an attempt to crowd out similar and sometimes better-designed initiatives. After all, Tepper Marlin argued, there was "no implementation measure in the Global Compact," and, unlike SA8000, the ISO standards, or the Global Reporting Initiative, the Global Compact had been created by fiat, not through a balanced multi-stakeholder process.[68] The 2002 announcement of the multi-stakeholder committee that would advise the UN Global Compact Office provided further reason for skepticism. It included ten leaders of global corporations, headed by the vice chair of Goldman Sachs, representing business, and only seven representatives of the other three groups—labor, civil society, and academia—combined. The two labor representatives were from the rich industrialized countries, and the one representative of academia was John Ruggie, who was moving to Harvard's Kennedy School after four years as Annan's right-hand man.[69]

Even though (or, perhaps, partially because) the UN failed to convince the Big Eight to merge, Tepper Marlin came to see the Global Compact Office and the SAI as "mutually supportive" because those companies that signed onto the Global Compact that had a real commitment to its principles would often turn to its office and ask, "Well, now how do we go about implementing it?" and the office would point those companies toward "other voluntary consensus-based standards that are multi-stakeholder and rigorous and consistent," including SA8000. Meanwhile, "We [SAI] often find, and Global Compact finds, that companies, let's say, implement SA8000; they then find out about the Global Compact and sign up for it."[70]

Nevertheless, the compact cannot be said to be successful, even as judged by the "mercenary" standard of the number of companies that have signed on. The compact began with a few hundred corporate signatories in 2000. In 2017, more than 9,000 companies around the world had signed on, more than twice the number that received SA8000 certification in that year,[71] but the difference may have reflected the fact that the costs of signing the compact were truly minimal. Many companies may have done so to fly a "We Support the Global Compact" flag in lieu of incurring the much higher costs of compliance with SA8000's audited performance standards. Still, the number of corporate signatories was only a small fraction of the number of companies that used ISO 14000 standards. A 2014 review of research on the compact's significance concluded that it had had little independent effect on corporate social and environmental practices, which, the authors argued, had led to "a loss of public trust" in Annan's project. The compact had become "largely dependent on the corporate sector for its very survival."[72]

Governments of many developing countries and many of the advocacy organizations that Annan recruited to the Global Compact had worried about this corporate takeover from the outset. In 2002, in response to pressure from those groups, a committee of the UN Human Rights Commission began to work on a treaty to turn the compact into intergovernmental rules that corporations would be legally obligated to uphold.[73] In 2005, Annan's penultimate year as secretary-general, he brought Ruggie back to the UN to evaluate the merits of the commission's draft norms. After extensive polling of global companies found no support for the commission's work, Ruggie delivered a pessimistic report. The firms believed that "treaty-making can be painfully slow, while the challenges of business and human rights are immediate and urgent." Moreover, they thought that "a treaty-making process now risks undermining effective shorter-term measures to raise business standards on human rights." Finally, even if governments imposed treaty obligations on companies, "serious questions remain about how they would be enforced."[74]

The companies, in Ruggie's account, had reverted to the argument that, in these fields, government standard setting would be less effective than private, voluntary efforts. Yet, the detailed responses of the organizations representing the largest group of the world's employers made it clear that many companies were also rejecting any responsibility for human rights, including those of their employees, out of hand. The International Chamber of Commerce and the International Organization of Employers jointly wrote to explain that only "the State" has a duty to uphold human rights, while "private persons," including corporations, do not. Moreover, the business associations argued, the issues discussed by the commission's draft (or, for that matter, in ILO conventions, the Global Compact, and SA8000) were not really standards. "When one tells the butcher, 'Give me a kilo of shrimp,'" they wrote, "kilo is a 'standard,'" but "an 'adequate standard of living' is not a 'standard.'" This is because, the employers' groups argued, "'Kilo' has a concrete, agreed-upon meaning by which one can judge the appropriateness of the butcher's conduct. . . . But 'an adequate standard of living' is a vague abstraction that contains no criteria for making objective evaluations."[75] Of course, there were many examples of how to set and meet meaningful standards of this sort, but that was not the real problem. The business groups simply objected to employers being required to take on such responsibilities.

ISO 26000 and ISEAL: The Twenty-First Century Social Standards

Not surprisingly, given the vehemence of this opposition to government enforcement of social and environmental standards, private efforts to regulate the global economy continued to dominate in the first decades of the twenty-first century. Two of the more consequential private initiatives were the ISO standard-setting processes that led to the publication of ISO 26000 Social Responsibility (2010); and the ISEAL (International Social and Environmental Accreditation and Labeling) Alliance's work to produce its three Codes of Good Practice for Setting (2004), Assessing the Impacts of (2010), and Assuring Compliance with Social and Environmental Standards (2012). ISO used an unprecedentedly large and inclusive (if widely criticized) stakeholder group to create ISO 26000 as a comprehensive social responsibility standard (albeit one that could not be certified), a kind of fully fleshed-out version of the Global Compact, while ISEAL's codes offered coherent guidance to those setting more specific environmental and social standards that would be subject to effective monitoring.

Expanding the ISO Stakeholder Consensus Process: ISO 26000

In April 2001, partially in response to the promulgation of SA8000 and the Global Compact, ISO's governing council formed an advisory group to discuss creating a corporate social responsibility standard that would build on ISO's experience with management system standards. In 2004, after considerable debate, the advisory group abandoned the idea of creating a management system standard and began to concentrate, instead, on a comprehensive guidance standard that would not require a new system of audits. Group members from small- and medium-sized enterprises (SMEs) in the developing world said they simply could not afford such audits, and a few big global firms with little interest in corporate social responsibility would accept nothing more onerous than a guidance standard. Moreover, some labor groups and the ILO worried about major firms whitewashing themselves with certifications from less-reputable auditors, something that would be much less costly than living up to their existing legal and contractual obligations. As Kristina Sandberg, the Swedish secretary of the advisory group, explained shortly afterward, "ISO tries to seek for the consensus and keep the work process going; therefore the 'no third party certification' of [the] standard was the way out."[76]

In January 2005, ISO members formally approved a work program to draft a broadly applicable social responsibility standard for all organizations (not just profit-making companies). The standard would cover human rights, labor rights, the environment, anti-corruption measures, and questions of transparency and public reporting. The work program listed fourteen existing ISO or national standards as possible models, along with 80 other "standards or initiatives" of intergovernmental organizations, national governments, private nongovernmental bodies, and even individual companies.[77] Instead of a technical committee, ISO formed a working group whose secretariat was jointly provided by a major standards body from a highly industrialized country, the Swedish Standards Institute, and one from the developing world, Brazil's ABNT. ISO asked national members who wanted to take part in the committee's work to appoint a delegation that included at least one person from each of six categories of stakeholders: industry, government, labor, consumers, nongovernmental organizations, and others (often academics) to represent the global interest. In addition, the secretariat strove for balance across other dimensions including gender, ethnicity, language spoken, developed versus developing world, and different sizes of organizations (from global companies to SMEs).

The committees and subcommittees of the working group met over five years in a series of eight convention-like face-to-face meetings and communicated almost continuously online, where more than 25,000 written comments on the draft standard were posted by committee members and addressed by subcommittee chairs and the standard's editors. By the last meeting, the number of official participants had grown to more than 450.[78] The entire proceedings were more like the meetings of the large and sometimes unruly IETF than the meetings of traditional ISO technical committees. The working group maintained an open website that grew to tens of thousands of web pages. ISO continued to maintain the site for many years, perhaps to interest researchers and other standard setters in studying the unusually complex and, arguably, quite successful, process. This was certainly one of the goals of the former working group leaders from SIS, Kristina Sandberg and Staffan Söderberg, who created the website ISO26000.info to disseminate information about the standard, using funds they received as winners of the Swedish Association of Graduate Engineers 2016 Environmental Prize for their outstanding work in managing such a complex and inclusive process.[79]

Five to ten years earlier, while the working group was still in session, such high praise for the process was much harder to find. The rapid succession of critical face-to-face meetings (Brazil in 2005, Thailand and Portugal in 2006, Australia and Austria in 2007, Chile in 2008, Canada in 2009, and Denmark in 2010) made it seem unlikely that the views of workers and of smaller companies would be adequately represented. Then there was ISO's "UN-like" character (of having national delegations) that, in a different context, so annoyed standards users in the high-tech industries. ISO's member standard-setting bodies served as gatekeepers. If a company, a social movement organization, or even another group involved in setting global standards wanted to participate, it had to be accepted as part of its country's delegation, and some key national standards bodies, including America's ANSI, were reluctant to open the door too wide. In addition, many national delegations operated by the rule that no one could speak for the delegation except its head; a national group consensus had to be reached first. Alice Tepper Marlin recalled how delegates from some nongovernmental organizations were, therefore, not allowed to express their own views because "virtually all [national] standards bodies" were "dominated by industry." At the same time, delegates of the national standards bodies and industry representatives from socially responsible companies indicated that they felt silenced by the political decision not to pursue the certifiable standard that most of them wanted. Finally, there were pressures from governments. At the 2008 meeting in Santiago, Chile, the governments of China, Belarus, and the United States began arguing that final approval of the standard should be delayed until all its possible ramifications could be considered; these governments worried about the way ISO standards tended to become part of official regulations, especially through the World Trade Organization principle of using "relevant international standards" as a basis for any regulations that might influence international trade.[80] These governments failed to prevent the draft standard from being approved and published in 2010.

Despite the various concerns raised about the process that produced the standard, it sold very well. One year after its publication, "ISO 26000 was selling better than ISO 14001 and was third after ISO 9001 and ISO 31000 (Risk Management)."[81] The standard proved popular in China and other rising manufacturing regions and among SMEs around the world, perhaps because many companies in those regions and SMEs everywhere considered the standard and its sponsor, ISO, to be preferable to social and environmental standards promulgated by other groups that had allowed even less effective participation in the process of standard setting. A 2008 global survey of SMEs indicated that the "ISO branding" of the standard added legitimacy to the social responsibility agenda. Many SMEs and their associations had some knowledge of, and respect for, ISO. As an interviewee from the Indian Chamber of Commerce put it, "If ISO issues a standard on these aspects, we will have to take it more seriously."[82]

Moreover, as had been the case with ISO 9000 and 14000, the social responsibility standard quickly generated additional related ISO activities, though not yet in the form of a 26000 series. In 2012, ISO published its first certifiable social responsibility management system standard, ISO 20121, Sustainable Events. ISO 37001, Anti-bribery Management Systems, followed in 2016. In 2013, a committee under a joint ABNT and AFNOR secretariat produced a guidance standard for sustainable procurement, ISO 20400. And in 2016, at the request of SIS, ISO initiated an international workshop agreement to create a system for training organizations how to use ISO 26000 in coordination with various management system standards.[83] All of these efforts have moved the ISO 26000 project closer to becoming a general governance system for socially responsible organizations based on the ISO 9000 model. Indeed, the SAI's senior director for standards and impacts, Rochelle Zaid, stated similar expectations a few months before ISO 26000 was first published: "What they will do is issue a guidance document. . . . What I'm sure they're thinking is . . . the next go around when they redo this standard, . . . which is a five-year period of time, it will be a full-fledged certification standard." Moreover, she expected national standards bodies to come up with ways to certify in this area: "They will evolve a whole pattern and . . . turn it into a certifiable standard country by country."[84] In this context, Zaid and Tepper Marlin were disappointed that ISO 26000, like the Global Compact before it, failed to make specific reference "to not only SA8000 but other major, respectable [and already certifiable] CSR standards like [those of] the ISEAL membership," despite its many explicit links to ISO 14000 and 9000.[85]

ISEAL: An Alternative to the ISO for Social Standards

ISEAL, which began at the same time as ISO 26000, was an alternative, non-ISO system for setting voluntary social and environmental standards. It was explicitly global rather than international, avoiding the need for consensus among different national bodies, and focused on maintaining and improving the consensus process that ISO's member bodies had developed over the previous century. In 2002, Tepper Marlin's SAI joined seven other standard-setting and certification organizations to form the International Social and Environmental Accreditation and Labeling Alliance; as was common at the time, it quickly dropped its full name and became simply the ISEAL Alliance. The original group included organizations that maintained standards for sustainable forestry and fishing, as well as for organic and fair-trade agriculture.

ISEAL first created a 2004 Standard Setting Code that defined all the stakeholders involved in the subject area of a particular standard, including everyone who would be affected by the standard, not just the companies that might adopt it and those companies' customers. Like ISO and the national standards bodies that were its members, the ISEAL code sought balance among the various stakeholders. Tepper Marlin considered the new code "very important because it is an alternate to ISO. . . . [F]ull members of ISEAL have pledged, and after a certain period of time they have to demonstrate compliance with this code." She considered it not just an alternative to ISO 26000 but a better option, since "it borrows from ISO, but adds a lot of things," mostly about how to forge effective and inclusive multi-stakeholder alliances.[86]

Tepper Marlin and her deputy, Zaid, clearly believed that the original ISEAL members understood the needs of weaker stakeholders better than ISO's leadership did. For example, Zaid explained that, in the ISO 26000 negotiations, ISEAL members were part of "the groups really supporting the SMEs' point of view," while the ISO leaders only paid lip service to the difficulties faced by smaller businesses in even attending and taking part in the standard-setting process.[87] ISEAL members believed that they avoided similar mistakes by effectively including all stakeholders in the process from the beginning. Nevertheless, while the earliest versions of the ISEAL code made a broad commitment to the balanced representation of all stakeholders in any standard-setting committee, clear definitions of what constituted a "reasonable balance of stakeholders" appeared only in a draft of the fifth revision of the code in 2014; in a footnote, it soberly remarked, "A balance of interests in stakeholder participation cannot be ensured but the standard-setting organization should make efforts to engage all those stakeholder groups identified."[88]

In fact, ISEAL's most important innovation may not have been its hortatory commitment to balanced stakeholder standard setting but, rather, its continuous and somewhat dogged practice of explicitly trying to improve not only its rules of inclusive standard setting, but also a second set of rules about compliance monitoring, and even a third set of rules about assessing the actual impact of members' work on social and environmental sustainability. In doing so, ISEAL moved away from a one-size-fits-all version of certifiable social and environmental standards to one more concerned with specific contexts. In 2014, the International Institute for Sustainable Development (IISD), an activist organization sympathetic to ISEAL, explained the perceived benefits of this approach as better maximizing "sustainable development outcomes," linking "sustainable consumption with sustainable production," and helping "a standards system more accurately internalize the costs of sustainable production."[89]

In addition to offering a range of noncompeting, context-specific, certifiable social and environmental standards, ISEAL tried to break ISO's dominance over the conformity assessment and accreditation businesses. ISEAL members believed that their codes for standard setting, conformity assessment, and accreditation were higher than those supported by ISO through its standard-setting norms, its standard for assessment (ISO/IEC 17065 Conformity Assessment), and its alliance with the IAF, the association of the world's main accreditation agencies.

As of 2017, ISEAL's challenge to ISO had not succeeded. ISEAL had not unified and upgraded the hundreds of non-ISO private social and environmental standards that had been promulgated since the 1990s. ISEAL had only 23 full members from among the many hundreds of providers of private social and environmental standards. Moreover, in its 2014 study, the IISD investigated the small subset of the most effective (in terms of actual outcomes) of about 400 private environmental standard setters, only about half of which were working toward compliance with ISEAL's codes, while 75 percent used the ISO standard for conformity assessment. The IISD emphasized that both ISEAL and ISO followed the same governance model. It was the ISO 9000 model that emphasized monitoring, a commitment to continuous improvement, and eventually requiring similar improvements from all of an organization's suppliers.[90]

In 2017, ISO 26000 and the standards that have evolved from it were, by themselves, much more widely adopted and publicly recognized than all non-ISO social and environmental standards. In part, this may be because ISEAL members faced a problem of achieving global inclusiveness. It was, however, a different one from that faced by the OpenStand organizations, as discussed in chapter 7. While ISEAL members made reasonable claims that they had better ways of assuring that more stakeholders are represented in their standard-setting processes than ISO did, throughout the developing world, governments, firms, and social movement organizations tended to consider ISO standards more legitimate because they originated from a network of organizations with a demonstrable history of (at least) *international* inclusiveness, while the ISEAL organizations were much newer and almost all started as philanthropic projects of groups in the global North.[91] Of course, it is not at all clear that adoption of ISO 26000 had led to more socially responsible behavior because there is no requirement that management processes, let alone outcomes, be monitored. Much of the effort may simply have been a matter of public relations, albeit one that created a lucrative business for ISO and its member bodies.

Stakeholder Standard-Setting Initiatives in Response to the Global Financial Crisis

In the years immediately after the global financial crisis of 2007–2008, the booming ISO 9000 standards industry stabilized at a very high level, but the originally much smaller businesses of environmental and social standards continued their rapid growth. Just the job of trying to keep up with all the new initiatives was, itself, becoming a major business opportunity. In 2009, the Ethical Corporation Institute published what it described as "your guide to selecting the right multi-stakeholder initiative for your company," priced at $1,000.[92] There were practical reasons for company interest in publications of this sort. John Ruggie noted that conditions had changed markedly since the early UN debates on a code of conduct for transnational corporations. Forty years later, public expectations of business had changed, and, even without new intergovernmental agreements on transnational business regulation, it was possible, in theory, for *individuals*, including corporate leaders, to be held accountable for gross violations of international human rights norms by the International Criminal Court. As a result, companies themselves were demanding greater harmonization of CSR standards as well as more explicit means for verifying compliance.[93] Nevertheless, Ruggie worried that there was little evidence that private standards would, by themselves, produce anything more than limited changes, especially vis-à-vis the conditions of workers in global supply chains. Therefore, in his role as an advisor to the secretary-general, he began calling on businesses and the UN to work toward harmonizing social standards and finding more effective means of reporting on ways to improve conditions, although he did not go so far as to say that governments needed to be called on to enforce those standards.[94]

The weakness of solely relying on even what activists considered the best of the private standards was illustrated by a series of horrendous accidents in factories along various global supply chains, including a 2012 fire in a Karachi factory that killed 262 people. Three weeks earlier, it had been certified as SA8000 compliant. Of course, the SA8000 standard was not a fire code, but it expected companies to follow one, and the standard itself was quickly amended to be much more specific about what was required. Even so, the tragedy uncovered deeply troubling issues of responsibility for the certification and accreditation businesses that were not so easy to deal with. The factory had received subsidies from the Pakistani government to hire an Italian company to carry out the certification audit, but the subsidy was contingent on the factory receiving certification. The Italian company had been accredited by the SAI, although Alice Tepper Marlin, when she heard about the arrangement, had warned them not to accept payments generated by the conditional government subsidy. In addition, the company issued SA8000 certification after "checkup meetings in the Middle East or by telephone" without ever sending any of its staff to Karachi." Separately a European clothing maker hired a local inspector to visit the factory three times in the two years before the fire. A *New York Times* investigation reported, "Each time he found a locked fire exit—as in the fatal 1911 Triangle shirtwaist factory fire in New York—minimum wage violations and other serious problems."[95]

In the immediate aftermath of the disaster, a leading scholar of private social and environmental standard setting, Khalid Nadvi, proposed that SA8000 and other voluntary standards be replaced by a government-run system developed in consultation with industry and the ILO, but neither the Pakistani government nor Pakistani business leaders showed any interest in doing so. As Scott Cooper, ANSI's vice president for policy and government affairs, grimly pointed out, even the most hardened and dedicated trade union leaders and legislators on the left were coming to recognize that, in the global climate of the early twenty-first century, voluntary standards were probably the only "practicable solutions" to the massive regulatory gaps in the global economy. In 2015, Cooper worked with US labor and civil rights icon, Representative John Lewis of Georgia, to set up a series of Capitol Hill events about ISO initiatives on worker safety, transparency, toy safety, and the like, all in the hope that Congress would get behind these efforts to "address some of the compelling voids in global public goods."[96]

Unlike John Ruggie, Cooper was skeptical that the UN would be able to step into the void left by industry and national governments. It was sad, Cooper argued, but, despite its tripartite structure, the ILO had never found a way to represent the interests of the world's vast majority of nonunionized workers or even of the small number of truly socially responsible companies that want everyone to live up to the standards they have adopted. ISO was far from perfect, but the experience it had gained in negotiating and administering the series of management system standards from ISO 9000 onward suggests that the new, certifiable ISO 45001 Worker Safety draft standard, when approved, might prove more effective than any of the large set of ILO standards that governments and industry both ignored.[97]

In the years after the global financial crisis, when Ruggie began searching for ways to make the UN system a more central player in the UN-private sector cooperation that he had helped initiate, other close collaborators with Kofi Annan promoted an even more private-sector-focused approach to global standards in this arena than either the Global Compact or the ISO approach advocated by Cooper. In 2009, the World Economic Forum's Klaus Schwab recruited Mark Malloch Brown to colead a massive two-year multi-stakeholder process aimed at redesigning global governance. Malloch Brown had been deputy UN secretary-general under Annan and served in British prime minister Gordon Brown's short-lived cabinet at the height of the financial crisis. Schwab called the project "the Global Redesign Initiative." It aimed not only to avoid future economic crises but also to ensure environmental sustainability, curb violent conflict, and assure human development.

Schwab was an economist and engineer who, as a young faculty member at the University of Geneva in the early 1970s, began hosting the annual January meetings of the world's economic and political elite in Davos, Switzerland, where Annan had proposed the Global Compact in 1999. Schwab had long believed in voluntary standards and multi-stakeholder decision making, albeit his had always been a corporation-centered and top-down view of multi-stakeholder decision making. In his major book, a 1971 monograph sponsored by the German Mechanical Engineering Industry Association, he and his coauthor, Hein Kroos, argued that corporations must act as trustees for all stakeholders, including "shareholders (owners), creditors, customers, the national economy, government and society, suppliers, and collaborators."[98] In 1971, Schwab focused on private companies as ultimate decision makers and as trustees for his list of stakeholders. Arguably, his focus had remained there throughout his many years as head of the World Economic Forum.

WEF's Global Redesign Initiative involved meetings among hundreds of public and private leaders in ISO 26000–sized private "multi-stakeholder forums," whose members were largely drawn from the attendees of Schwab's annual Davos meetings.[99] Not only did the membership of these forums differ from the more broadly representative ISO 26000 Working Group, but the drafting of the outcome document was also quite different. A group of a dozen leading academics (typically working in pairs) produced the chapters of a 600-page book that described the proposed future system of global governance. They did so, they said, on the basis of input they received from the consensus-oriented, private-sector-dominated multi-stakeholder forums. They were not, however, bound by the consensus, nor did they have any due process rules to assure that all points of view were heard.

Not surprisingly, what the authors proposed was a system of global governance based on standards created by new consensus-oriented, private-sector-dominated multi-stakeholder committees. They would include a global systemic financial risk watchdog and a "structured public-private process to identify and encourage the replication of model national labor migration policies."[100] Harris Gleckman, a retired UN official who had been deeply involved with intergovernmental attempts to regulate globalization for 40 years, argued that, while the Global Redesign Initiative built on two decades of voluntary social and environmental standard setting, especially the work of some ISEAL members, the initiative would, unrealistically, "leave governance to self-selected and potentially self-interested elite bodies."[101]

Schwab's project may have been the most ambitious attempt to extend some of the norms of voluntary standard setting to the governance of global social and environmental problems, but it also may prove to be a minor footnote in the longer history of private standard setting. In 2017, Google Scholar reported only a handful of studies that had cited the WEF's signature 600-page report, compared to about 18,000 studies citing ISO 26000, and Google reported a total of only about 1,000 web pages that contained the report's title, compared to about a half-million web pages that referred to the ISO standard.

The New Standards and Their Implications

The rise of voluntary standards for quality assurance, environmental protection, and social responsibility certainly further complicated the already messy world of standardization, but it also reinforced many of the trends we have seen throughout the history of private voluntary standard setting. As with the new information and communication standards, a social-movement-like zeal pushed forward the quality management, environmental, and social responsibility standards—in this case, the zeal of quality engineers, environmentalists, and human rights activists, rather than that of internet and World Wide Web enthusiasts. To the extent that any of the work in these new fields was consequential, it was work done by those activists who were also committed to the balanced, multi-stakeholder process. In this respect, the ISEAL Alliance member organizations were very much like the IETF and W3C, newcomers committed to improving upon the processes that had long been used by the traditional private standard-setting organizations, while the Global Compact and the Global Redesign Initiative lacked the commitment to a true balanced consensus process. ISO 9000, 14000, and 26000, as well as the certifications and auditing that followed from them, suggest that the older standard-setting bodies also initiated some innovation in the consensus process. They certainly extended it to new, more social domains and, in the process, involved more parties, but they remained biased toward business and respectful of *internationalism* rather than *globalism*, remaining committed to achieving consensus among ISO's national member bodies.

In 2000, two leading analysts of global business regulation, John Braithwaite and Peter Drahos, disparaged ISO's work in these fields as "mercenary." To them, the organization appeared concerned only with "generating income through the global promulgation of a regulatory model." That is to say, ISO and the traditional standard-setting organizations viewed their entry into the social and environmental fields simply as a new business, a way to sell more standards. This form of "model-mongering," Braithwaite and Drahos argued, could be distinguished from the "missionary" activity of socially committed men and women such as Kofi Annan and Alice Tepper Marlin, people who gained no material advantage from their advocacy of extending the engineers' standardizing methods to social and environmental regulation.[102]

None of the "missionaries" promoting this new kind of standard, however, proved as successful as ISO did, at least in terms of the number of companies that adopted their standards. In 2017, fewer than 4,000 companies had adopted Tepper Marlin's SA8000 global labor standard, even though (or perhaps because) it was highly regarded by global labor activists, while Annan's Global Compact and Schwab's Global Redesign Initiative remained mired in controversy. In contrast, from the beginning, ISO's social and environmental standards proved popular with major companies and were also treated as legitimate by many environmental groups, labor unions, and activists, especially in the developing world. Nonetheless, in terms of impact on the actual problems that the new standards were supposed to regulate, it is not clear how much difference either the missionaries or the mercenaries had made.

ISO's involvement in these new fields was very much connected to branding and securing business; it was much less about the social-movement-like mission to improve the world that motivated the first wave of private standard setters or those who pushed for the internationalization of private standard setting after World War II, especially in the Olle Sturén era. Alice Tepper Marlin was a missionary standards entrepreneur in the mold of Sturén, Charles le Maistre, or Tim Berners-Lee, but it seems doubtful that the organizations she helped create have the impact on the daily lives of people everywhere that the organizations created by those three figures continue to have. For example, even most socially responsible consumers would be hard pressed to find many objects affected by SA8000 or other ISEAL standards in their homes—perhaps a few pieces of clothing, bowls made from sustainably harvested wood from Indonesia, or an occasional bottle of Italian wine from a specific large vintner. Yet the same people could not go a minute without seeing or touching something shaped by the organizations associated with le Maistre or Sturén or Berners-Lee—everything electrical is influenced by committees of the IEC, almost every object manufactured outside the United States over the past 40 years (and that includes the majority of objects in American homes) is connected to at least one ISO standard. And, of course, whenever we interact with our phones, computers, smart speakers, and many newer appliances whose inner workings we can barely understand, we are in the world created by organizations such as Berners-Lee's W3C.

If Tepper Marlin brought the missionary zeal of standard setting to the social and environmental realms, albeit with what are (so far) modest results, Annan, Ruggie, Schwab, and Malloch Brown should really be considered less missionaries than world-weary idealists who recognized the unlikelihood of effective governmental action to deal with the negative environmental and social consequences of globalization, but who held out more than a slight hope that the engineers' twentieth-century model of private standard setting might help embed the new global economy in a set of higher moral values. ISO and many of the national standard-setting bodies include many people of both these types working on the new standards.

Nevertheless, most of the people involved with these new kinds of standards who worked in or with the old network of organizations did not view standards for product quality, environmental sustainability, or social justice in this way. To them, these standards were a business. It did not particularly matter if sometimes the business was incoherent, with a disconnect between what the standards promised and what they delivered. It even did not matter that much of the old process of reaching consensus among technical experts mostly in the same profession (engineering) was sorely tried (or entirely abandoned) when the real stakeholders involved were people with deeply conflicting interests—workers and owners, environmental activists and citizens of regions dependent on dirty industries, companies from the long-privileged global North and those from the rising South. What mattered more was that, as a result of this new business, the old organizations were still thriving. This view was not necessarily mercenary, but it certainly was one that would have made the leaders of the first and second wave of standardization—as well as the internet standardizers of the third wave—uncomfortable.

Still, until all the world's most powerful governments commit to regulating the harms of economic globalization, for better or worse, the work done on these issues by ISO and by private global standard-setting organizations in the ISEAL Alliance may be the most effective alternative.[103]

Conclusion

Since the 1880s, private consensus-based standard setting has become an important part of global life. This type of standard setting originated in the professionalization of engineering that followed industrialization, and it developed through the three waves of institutional innovation we have discussed. Today it appears to be at a new inflection point—perhaps the peak of the third wave, perhaps something more significant. Some, like the leaders of the Global Redesign Initiative, see private consensus-based standard setting becoming the main way to solve the world's most pressing problems. Others, including the initiative's loudest critics, see it as an insidious authoritarian threat. It is probably neither. It is just a relatively new, influential, understudied, and sometimes misunderstood social field, something that cannot be understood by treating it either as a realm governed solely by the logic of the market or as simply a struggle for power. And it is a field whose institutions are currently challenged by many of the same things that challenge institutions in many other fields—its aging population, legacies of sexism and Eurocentrism, and the rise of China—as well as by something particular to standardization, the waning of belief in its power to make a better world.

The story of the standardization movement and the lives of the people involved provides surprising insight into the global history of the twentieth century, even if from an idiosyncratic perspective. Historian Mark Mazower treats the standardizers' early history as illustrative of an ultimately ill-fated, yet endearing, late nineteenth-century belief that science was an always progressive "unifier" of humankind.[1] Yet perhaps the ideas that energized the original standardization movement should not be dismissed so lightly. For the average person, standardization has been a good thing; as an understated Swedish Standards Institute pamphlet put it, modern standardization has "helped the world work just a little better"[2] and the traditional private standard-setting organizations have proven much more successful at producing relatively effective standards than either the market or governments.

Of course, the voluntary consensus process has not proven to be the panacea that its most exuberant champions have imagined. Recent voluntary standards for the environment, labor, and human rights have been, at best, a partial success, and they seem to be most effective working in parallel with strong governmental regulation. Moreover, even in its original field of industrial standard setting, it did not prevent two world wars, nor did it end the Great Depression. The process has also failed to produce some of the standards that most of us would want. The International Electrotechnical Commission makes a deliberately ironic acknowledgment of this fact to the few tourists who find their way to its modest Geneva headquarters: IEC-logoed universal electrical adapters are available for purchase, right next to the IEC umbrellas and desktop IEC flags. The world's dozens of combinations of incompatible electrical plugs and outlets are, after all, the most familiar failure of international electrical standardization (the result of national standards already having been in place when the international effort began, creating extensive and locally embedded installed bases in each country). Some observers argue that failures of the process are becoming the norm in fields where technological change is so rapid that a drawn-out search for consensus simply cannot keep up.

Others see private standard setting as a threat to democratic order and perhaps even to human development. Serious theorists raise important broad questions about the ways in which obscure committees of technocrats may wield social power through standard setting.[3] Certainly it is difficult, for example, not to be a little suspicious of the interests that would be served by the World Economic Forum's proposals to replace much of the current system of global governance with consensus-oriented expert bodies in which major firms, but not other stakeholders, are always directly represented. Yet, based on what we know about today's relationship among governments, the market, and standard setting, it is clear that nothing as ambitious as the forum's proposals will grow out of the existing system. Private voluntary standard setting influences other social realms but does not control them. The struggles of the electromagnetic compatibility standardizers or the World Wide Web Consortium WebCrypto working group to get through lengthy processes with multiple checks and balances illustrate that pressures are not simply based on economic interests (e.g., when different employees of the same browser maker supported different standardization paths, as we observed in the early stages of WebCrypto WG), and that personalities can affect the process (e.g., the change when Frederick Bauer replaced Brooks Short as US representative to a CISPR committee (the special IEC committee dealing with electromagnetic compatibility), or when Mark replaced Ryan as editor in WebCrypto WG). Only the persistence of people who care deeply about the process (e.g., Ralph Showers in CISPR, Virginie in WebCrypto) or the outcome (e.g., Mark in WebCrypto, Showers in CISPR) or both provides the energy to push through the multiyear standardization process. In examining the process up close, we see the goodwill as well as the cynical power moves.

Neither a panacea nor a threat, consensus-based standard setting can still steer technological development onto a particular path, as the Internet Engineering Task Force and W3C have done with the internet and World Wide Web. In other cases, the work of technical committees can help the work of governments and even make that work possible by quickly establishing effective standards that governments can cite in regulations and treaties. Many early industrial safety standards served this purpose, and ISO's recent work on measuring carbon dioxide emissions is likely to serve the same purpose in regulations to combat climate change. Finally, the process can provide, or withhold, the legitimacy that comes from the consensus of experts endorsed by traditional democratic bodies.

As we complete this book, we believe that this realm is undergoing some inevitable changes. The third wave of institutional innovation seems to have crested, leaving us with an increasingly complex network of organizations and a larger and less unified community of standardizers. The second-wave network of organizations around the International Organization for Standardization was, in conception, hierarchically ordered, with ISO (and IEC) at the peak. The third wave gave us the addition of a new set of standard-setting bodies around the internet and web that are in theory global but in fact not fully so. In addition, traditional voluntary consensus methods moved out into process standards and then into the social and environmental arenas, in which engineers are the minority, if they are involved at all. Where does the private standard-setting system go from here?

First, what happens to this type of standard setting in traditional rather than new industrial arenas? Don Heirman (who followed Ralph Showers as head of the American National Standards Institute's C63 committee and then of CISPR) and Dan Hoolihan (current head of C63) wonder whether there will be a next generation of devoted standardizers. Hoolihan laments the lack of "younger and female engineers," while Heirman puts it more personally: "The biggest thing is to find someone that wants to replace me. No one wants to do it."[4] And those who want to be involved with standardization tend to be lower in their organizations than those in the earlier waves, when the Siemens family took part in the IEC's formation and regular work, and when Keith Tantlinger, engineer and executive at Sea-Land at a critical point was later involved in ANSI and ISO container standardizing. While we saw that energy came from belief in internet and web technology or in social justice in third-wave organizations like the IETF, W3C, and the ISEAL Alliance, in older and less trendy areas like electromagnetic compatibility it is hard to get people interested in standard setting. At least one leading standard setter has suggested that only in China are young engineers getting involved in standardization, a trend that, if true, could have large negative implications for remaining manufacturing industries in the United States and Europe.[5] However, China's leading standards guru, Wang Ping, argues that his country is no different. Wang is an engineer who began working with the predecessor of China's current national standards body in 1989; in the 1990s he was involved in both China's and ISO's effort to standardize computer-aided design and engineering data exchange. After 2000, he helped lead the Chinese national body's effort to build good relations with other major national standards bodies, especially those of the United States, Germany, the United Kingdom, France, Canada, and Australia. At the same time, he began studying and publishing on standardization history, proposed many of the major reforms of the Chinese standardization system that have been adopted since 2000, and introduced major scholars of standardization to a Chinese audience through his writing. Wang argues that, in all countries, the standardizers in older technological fields tend to be older, and young people are attracted to standard setting in cutting-edge industries. Moreover, he believes that in the information technology field, at least, leaders of Chinese companies may start to get involved simply because, given "the character of interoperability standards" in the field "and the emergence of standards-essential patents, such standards have become strategic measures." They are "a key element of industrial competitiveness."[6]

For new organizations of the third wave, the issues are somewhat different. In the IETF, while a technology-based movement around the internet, rather than around standardization per se, provides energy, leadership in that organization is still dominated by "males from Western countries," creating much talk about the need to diversify the participants and leaders in an organization that prides itself on being democratic.[7] This effort is difficult because the direct and sometimes aggressive style of IETF meetings and electronic exchanges may make it difficult for individuals from countries with a much less direct communication style or a collectivist culture to participate. Similarly, many women, even if they are from the United States or Europe, may also be put off by this style. So broadening participation in this organization as the older generation of IETF standardizers retires will be an important and difficult challenge. Adopting the explicit rules for listening and responding to others employed by Tepper Marlin's Social Accountability International and other ISEAL members would help with this problem, but that would require a dramatic cultural change. In addition, while ISO and the IEC have representation by country, encouraging more national diversity, the IETF has no such mechanism. Nonetheless, as we have seen, the IETF (working with the Internet Society) has one of the few well-funded programs devoted to bringing women, younger engineers, and people from the developing world into the standardization process. W3C has member companies and organizations from countries throughout the world, but we observed that time zone differences, as well as language and cultural differences, made it difficult for several Korean members of WebCrypto WG to participate in teleconference meetings. In the long run, these new technology-based standards bodies need to find ways to incorporate more participants from around the world for reasons of legitimacy and practical reality, as work in software moves from the developed to the developing world.

Some think that all this may change as China takes on a leading role in global standardization, but Wang Ping cautions against expecting rapid, or perhaps even significant, change. After all, it was only with Deng Xiaoping's economic opening and China's entry into ISO at the end of the 1970s that the country's government-controlled system of standardization began to rely on technical committees of stakeholders, and only much more recently that Chinese standardization organizations have taken off.[8] Yes, China will "become one of the big players in standardization in the future, together with the United States, the EU countries, Japan, and Korea," but ISO and the IEC will continue to be the main players in traditional industries and probably in the new environmental and social fields as well; ISO and the IEC (along with the International Telecommunication Union's hybrid standardization committees), will also still play a role in the new industries, but the IETF, W3C, and the like will be more important. "China cannot do standards without those standardization organizations."[9]

Whether in traditional industrial areas, new high-tech areas, or social and environmental areas, national standardization is fading in importance, and issues of how to set legitimate standards for an increasingly global world remain to be solved, even if the most important group of players is as small as Wang suggests (i.e., firms and government agencies from the United States, the EU, China, Japan, and South Korea; we would probably add Brazil, India, and some other countries with large economies). On the one hand, US standardizers complain about European bloc voting in ISO: "There are 20-some countries in the European Union, and each one gets a vote. . . . [T]he gross domestic product of the European Union, it's pretty close to the United States. They get 25 votes and we get one vote."[10] We also heard complaints about cases where many developing countries voted against developed countries that manufactured the items for which standards were set. On the other hand, the ISO and IEC one-nation/one-vote system is an important reason why those bodies have an unusual degree of legitimacy in less industrialized countries.[11] The question of stakeholder balance at all levels will no doubt continue to be an issue for discussion in the world of standardization, with no simple answer acceptable for all situations.

Equally important, and much less widely discussed, is the question of the power of different stakeholders *outside* the standard-setting process itself. Because everyone working on the web must use one of a handful of browsers, the browser manufacturers are always more equal than others taking part in web standard-setting committees; because every global company wants access to the American, Chinese, and European markets, the views of the US government, the Communist Party of China, and the European Commission always matter in standard setting, even when they are not sitting at the table. Standard setting may have improved life for almost everyone, and by supporting globalization it even may have helped lessen the global economic inequities existing a century or more ago, but the community of standard setters and the network of their organizations work alongside the other powerful institutions of the modern global economy; they do not replace them.[12]

essay on primary sources

This section contains a descriptive note on the primary sources used in this study. Comments on secondary sources are in relevant notes at the end of the book.

Detailed archival materials that show how the process of private standardizing works in practice are surprisingly difficult to find. Standard-setting organizations such as ANSI and ISO initially told us that they had only central office administrative records—that any records of actual standard-setting activities would be held by the organizations that served as secretariats for a given standardization effort. If ANSI served as a secretariat for an ISO committee, ANSI would in turn ask one of its member organizations (e.g., the IEEE) to serve as secretariat. The IEEE might then turn to a particular group within it (e.g., the IEEE EMC Society or ANSI C63 committee) to take this role. These groups did not keep records either. Thus, the complicated organizational networks meant that records were typically not saved in a centralized way but had to be tracked down from individual chairs of technical committees. We have tried to guide the records we discovered this way into repositories that will make them more accessible in the future.

For the pre–World War II period, we depend heavily on proceedings, such as those of ASTM; an array of engineering journals in which leading standardizers published, such as *Electrical Review* and *Notice sur la Société des Ingénieurs Civils de France*; and other contemporary publications. In addition, several engineering bodies and standards bodies have created online repositories of scanned original documents, accessible from their websites. For its 100th anniversary, the IEC assembled some of its key documents and made them available on its website under "IEC History," http://www.iec.ch/about/history/. We were also able to consult its central paper records at its headquarters in Geneva. The IEC also holds some historical records of CISPR.

The ITU has an excellent timeline and collection of formative documents at its "History of ITU Portal," http://www.itu.int/en/history/Pages/Home.aspx, as well as much more comprehensive archives in Geneva. The Swiss Federal Archives in Bern includes some materials on the secretary of the interwar ISA, as do the central records of ISO in Geneva, which also include the records of the World War II Allied agencies and the conferences that founded ISO. For national standards organizations, although ANSI maintains no archives, we were able to obtain access to meeting minutes of the governing committees of the AESC, ASA, USA Standards Institute, and ANSI at ANSI headquarters in New York. Some of the early documents of the British national body survive in the London archives of ICE. Although we were unable to locate Charles le Maistre's papers, some of his correspondence survives in the papers of Elihu Thomson at the American Philosophical Society in Philadelphia and those of Viscount William Douglas Weir at the University of Glasgow.

For the postwar period up to the late 1980s, we located a variety of primary sources, many from unusual sources. Engineering and standardization publications were again useful. Although it lacks an archive, ISO has a small collection of its early documents. Materials about Olle Sturén's work with ISO were obtained directly from his sons, Lars and Lolo Sturén, who plan to deposit them with the Swedish Standards Institute. Materials on radio frequency interference and electromagnetic compatibility standardization for chapter 6 were obtained primarily from a series of individuals interested in preserving the history of the EMC area as its older generation dies. The family of Ralph Showers initially gave us direct access to the Showers Papers, which are now deposited at Hagley Museum and Library. Similarly, the Thomas Papers originally came from a family member, by way of Dan Hoolihan, and are being deposited at Hagley Museum and Library. Don Heirman, past chair of CISPR and C63, has scanned many of his records to preserve them; although we obtained access to them in scanned form directly from him, the records are now available in MSA 291, Donald N. Heirman Papers, Karnes Archives and Special Collections, Purdue University. Finally, the Donald MacQuivey Papers, in the Department of Special Collections and University Archives, Stanford University Libraries, provide a complete and useful set of documents from the ITU's CCIR standardizing in the 1960s. We supplemented these collections of documents with interviews with Hoolihan and Heirman, both of which have been deposited at Hagley Museum and Library. And at a late stage of book revision, Janet Showers Patterson, daughter of Ralph Showers, scanned and sent us some additional family papers.

For the most recent period, starting at the end of the 1980s, many of our sources are available online. IETF and W3C electronic message lists and meeting minutes from most standardizing efforts are publicly available through their websites (https://www.ietf.org/ and https://www.w3.org/). We supplemented these with interviews of leaders of both organizations, as well as of individual members of the W3C WebCrypto WG. Similarly, ISO maintains extensive information about its process standards (ISO 9000, 14000, etc.) on the websites of the relevant technical committees. ISEAL maintains a significant archive of historical records at https://www.isealalliance.org/online-community/resources, some of which may only be accessed by participants in ISEAL's work. And http://iso26000.info provides archival access to most of the electronic records of the development of the ISO 26000 standard. We supplemented these with interviews with many of the principals involved in the negotiation and maintenance of the standards discussed in chapter 9.

notes

Introduction

1. The definition used by the International Organization for Standardization in 2017, "We're ISO: We Develop and Publish International Standards," https://www.iso.org/standards.html, accessed January 3, 2017.

2. Robert Tavernor, *Smoot's Ear: The Measure of Humanity* (New Haven, CT: Yale University Press, 2007), 56, 60, 106, 119, quotation from 131.

3. A good summary of the history of the act appears in US Metric Association, "History of the United States Metric Board," http://www.us-metric.org/history-of-the-united-states-metric-board/, accessed January 2, 2017.

4. Hendrick Spruyt, "The Supply and Demand of Governance in Standard-Setting: Insights from the Past," *Journal of European Public Policy* 8, no. 3 (2001): 371.

5. Joseph Farrell and Garth Saloner, "Coordination through Committees and Markets," *RAND Journal of Economics* 19, no. 2 (1988): 235–52.

6. See Michael A. Cusumano, Yiorgos Mylonadis, and Richard S. Rosenbloom, "Strategic Maneuvering and Mass-Market Dynamics: The Triumph of VHS over Beta," *Business History Review* 66, no. 1 (1992): 51–94.

7. Jean-Christope Graz, "Hybrids and Regulation in the Global Economy," *Competition and Change* 10, no. 2 (2006): 230–45.

8. Sidney Webb and Beatrice Webb, *A Constitution for a Socialist Commonwealth of Great Britain* (London: Longmans, Green, 1920), 56; Mary Parker Follett, *The New State: Group Organization and the Solution to Popular Government* (London: Longmans, Green, 1918), 344–60, and see chapter 3.

9. See chapter 9.

10. Mark Mazower, *Governing the World: The Rise and Fall of an Idea 1815 to the Present* (New York: Penguin Books, 2012), 102.

11. ISO's website reports, "Because 'International Organization for Standardization' would have different acronyms in different languages (IOS in English, OIN in French for Organisation internationale de normalisation), our founders decided to give it the short form ISO. ISO is derived from the Greek isos, meaning equal. Whatever the country, whatever the language, we are always ISO." "About ISO," ISO, accessed July 3, 2017, https://www.iso.org/about-us.html). However, in 1997 the Swiss delegate to ISO's founding conference in London wrote, "Recently I read that the name 'ISO' was chosen because 'iso-' is a Greek term meaning 'equal.' There was no mention of that in London!" Willy Kuert, "The Founding of ISO," in *Friendship Among Equals: Recollections from ISO's First Fifty Years*, ed. Jack Latimer (Geneva: ISO, 1997), 20.

12. Joseph Schumpeter, *Theory of Economic Development* (New York: Oxford University Press, 1934); JoAnne Yates and Craig N. Murphy, "Charles le Maistre: Entrepreneur in International Standardization," *Entreprise et Histoires* 51 (2008): 10–27. Andrew L. Russell introduced a similar concept for a much more recent era in "Dot-org Entrepreneurship: Weaving a Web of Trust," *Entreprise et Histoires* 51 (2008): 44–56.

13. Olle Sturén, "Collaboration in International Standardization between Industrialized and Developing Countries" (speech given at the German Institute for Standardization [DIN], Bonn, November 1981), 4. Sturén Papers.

14. E.g., Steven W. Usselman, *Regulating Railroad Innovation: Business, Technology, and Politics in America, 1840–1920* (Cambridge: Cambridge University Press, 2002) for railroads; and, most recently, Andrew L. Russell, *Open Standards and the Digital Age: History, Ideology, and Networks* (Cambridge: Cambridge University Press, 2014) for information and communication technology.

15. E.g., Alain Durand, *AFNOR: 80 Années d'Histoire* (Paris: AFNOR Éditions, 2008); and Winton Higgins, *Engine of Change: Standards Australia Since 1922* (Blackheath, AU: Brandl & Schlesinger, 2005), both official histories. Also see our (nonofficial) historical study of ISO: Craig N. Murphy and JoAnne Yates, *The International Organization for Standardization: Global Governance through Voluntary Consensus* (London: Routledge Press, 2009).

16. E.g., Christopher Spencer and Paul Temple, "Standards, Learning, and Growth in Britain: 1901–2009," *Economic History Review* 69, no. 2 (2015): 627–52; Tineke Egyedi and Jaroslav Spirco, "Standards in Transitions: Catalyzing Infrastructure Change," *Futures* 43, no. 9 (2011): 947–60; and, notably, Philip Scranton and Patrick Fridenson, *Reimagining Business History* (Baltimore: Johns Hopkins University Press, 2013).

17. E.g., Aseem Prakash and Matthew Potoski, *The Voluntary Environmentalists: Green Clubs, ISO 14001, and Voluntary Environmental Regulations* (Cambridge: Cambridge University Press, 2006); and Richard M. Locke, *The Promise and Limits of Private Power: Promoting Labor Standards in the Global Economy* (Cambridge: Cambridge University Press, 2013); for global markets, see Tim Büthe and Walter Mattli, *The New Global Rulers: The Privatization of Regulation in the World Economy* (Princeton, NJ: Princeton University Press, 2013).

18. Lawrence Busch, *Standards: Recipes for Reality* (Cambridge, MA: MIT Press, 2011); Martha Lampland and Susan Leigh Starr, eds., *Standards and Their Stories: How Quantifying, Classifying, and Formalizing Practices Shape Everyday Life* (Ithaca, NY: Cornell University Press, 2008); and David Singh Grewal, *Network Power: The Social Dynamics of Globalization* (New Haven, CT: Yale University Press, 2008).

19. Thomas G. Weiss and Rorden Wilkinson, "Global Governance to the Rescue: Saving International Relations?," *Global Governance* 20, no. 1 (2014): 19–36.

20. Russell, "Dot-org Entrepreneurship."

21. Grewal, *Network Power*, 4.

22. The classic study of this distinction remains Berenice A. Carroll, "Peace Research: The Cult of Power," *Journal of Conflict Resolution* 16, no. 4 (1972): 585–616.

23. Paul A. David and Shane Greenstein, "The Economics of Compatibility Standards: An Introduction to Recent Research," *Economics of Innovation and New Technology* 1, nos. 1/2 (1990): 3–41; Shane Greenstein, *How the Internet Became Commercial: Innovation, Privatization, and the Birth of a New Network* (Princeton, NJ: Princeton University Press, 2015).

24. Alejandro M. Peña, "Governing Differentiation: On Standardisation as Political Steering," *European Journal of International Relations* 21, no. 1 (2015): 52–75. Nils Brunsson makes a somewhat similar argument but foregrounds questions about the conditions under

which standardization has been successfully used as an alternative to hierarchical organizations, normative communities, or the market, Nils Brunsson, "Organizations, Markets, and Standardization," in *A World of Standards*, Nils Brunsson and Bengt Jacobsson, eds. (Oxford: Oxford University Press, 2000), 31–35. For a discussion of the entire range of sociological approaches to standardization, see Stefan Timmermans and Steven Epstein, "A World of Standards but Not a Standard World: Toward a Sociology of Standards and Standardization," *Annual Review of Sociology* 36 (2010): 69–89.

Chapter 1 · Engineering Professionalization and Private Standard Setting for Industry before 1900

1. For the evolution and spread of the metric system see Robert Tavernor, *Smoot's Ear: The Measurement of Humanity* (New Haven, CT: Yale University Press, 2007); Robert P. Crease, *World in the Balance: The Historic Quest for an Absolute System of Measurement* (New York: W. W. Norton, 2011); Ken Alder, "A Revolution to Measure: The Political Economy of the Metric System in France," in *The Values of Precision*, ed. M. Norton Wise (Princeton, NJ: Princeton University Press, 1995).

2. Crease, *World in the Balance*, 29.

3. Tavernor, *Smoot's Ear*, 133–34; 145–46; Crease, *World in the Balance*, 134–37. Crease notes that the organization actually first met briefly in 1870, then adjourned because of the war that had broken out between France and Prussia, reconvening in 1872.

4. Crease, *World in the Balance*, 133; "History of the IGU," International Geographical Union, accessed June 12, 2017, http://igu-online.org/about-us/history/.

5. John Noble Wilford, *The Mapmakers: The Story of the Great Pioneers in Cartography—from Antiquity to the Space Age* (New York: Alfred A. Knopf, 1981), 220–21.

6. *International Conference Held at Washington for the Purpose of Fixing a Prime Meridian: Protocols of the Proceedings* (Washington, DC: Gibson Bros., 1884), 6. Countries represented at the conference were Austria-Hungary, Brazil, Chile, Colombia, Costa Rica, Denmark, France, Germany, Great Britain, Guatemala, Hawaii, Italy, Japan, Liberia, Mexico, Netherlands, Paraguay, Russia, San Domingo, Salvador, Spain, Sweden, Switzerland, Turkey, Venezuela, and the United States.

7. *Prime Meridian*, 20.

8. *Prime Meridian*, 199–200. The votes were slightly different on the resolutions concerning related matters, including time and the universal day.

9. Bruce J. Hunt, "The Ohm Is Where the Art Is: British Telegraph Engineers and the Development of Electrical Standards," *Osiris* 9, no. 1 (1994): 48–63. This account draws heavily on this source, and the quotation is from page 55.

10. See Hunt, "The Ohm," 57–61, for the story of the committee's formation and activities. See also Larry Randles Lagerstrom, "Constructing Uniformity: The Standardization of International Electromagnetic Measures, 1860–1912" (PhD diss., University of California, Berkley, 1992), 15–27.

11. Lagerstrom, "Constructing Uniformity," 17, 27–28; Hunt, "The Ohm," 57–61.

12. Hunt, "The Ohm," 57–61; Lagerstrom, "Constructing Uniformity," 27–48.

13. Lagerstrom, "Constructing Uniformity," 49–81.

14. See, e.g., Bob Reinalda, *Routledge History of International Organizations: From 1815 to the Present Day* (London: Routledge, 2009), 85–89; and, earlier, F. S. L. Lyons, *Internationalism in Europe, 1815–1914* (Leyden: A. W. Sythoff, 1963), 39. More recently Robert Mark Spaulding has argued that the earlier Central Commission for the Navigation of the Rhine should be seen as the first organization with most of the characteristics of modern intergovernmental organizations and as the source of many of the practices of the ITU and later agencies. See his "The Central Commission for the Navigation of the Rhine (CCNR) and

European Media, 1815–1848," in *International Organizations and the Media in the Nineteenth and Twentieth Centuries*, ed. Jonas Brendebach, Martin Herzer, and Heidi J. S. Tworek (London: Routledge, 2018), chapter 2. Reinalda (*Routledge History*, 28–32) and others acknowledge the influence of the commission on later practices of major intergovernmental organizations, but they contrast the commission's regional mandate with the intended universal focus of the ITU and later public international unions.

The American political scientist Paul S. Reinsch coined the phrase "*public* international unions" to distinguish international associations whose members were primarily governmental bodies—administrative agencies such as the national telegraph and postal administrations—from the much larger number of strictly private international associations that were established in the second half of the nineteenth century. See Paul S. Reinsch, *The Public International Unions, Their Work and Organization: A Study in International Administrative Law* (Boston: Ginn, 1911), 4.

15. Craig N. Murphy, *International Organization and Industrial Change: Global Governance since 1850* (New York: Oxford University Press, 1994), 86.

16. "Overview of ITU's History," ITU, accessed June 12, 2017, http://www.itu.int/en/history/Pages/ITUsHistory.aspx; International Telecommunication Union, "CCITT/ITU-T 1956–2006, 50 Years of Excellence," ITU, July 20, 2006, accessed October 11, 2012, http://www.itu.int/ITU-T/50/docs/ITU-T_50.pdf; George Arthur Codding Jr., *The International Telecommunication Union: An Experiment in International Cooperation* (Leiden: Brill, 1952; repr., New York: Arno Press, 1972), 13–23.

17. Codding, *International Telecommunication Union*, 38–79.

18. It was originally established as the General Postal Union in 1874 but in 1878 changed its name to the Universal Postal Union. Lyons, *Internationalism in Europe*, 45. See also Codding, *International Telecommunication Union*, 52.

19. Lyons, *Internationalism in Europe*, 42–51.

20. Lyons, *Internationalism in Europe*, 40–43.

21. "Our History," Institution of Civil Engineers, accessed October 26, 2012, https://www.ice.org.uk/about-ice/our-history.

22. M. A.-C. Benoit-DuPortail, "Notice sur la Société des Ingénieurs Civils de France," *Annuaire—Société des ingénieurs civils de France, 1903* (Paris: Société des ingénieurs civils de France, 1903), 8–16.

23. For the early history of the Franklin Institute, see Bruce Sinclair, *Philadelphia's Philosopher Mechanics: A History of the Franklin Institute, 1824–1865* (Baltimore: Johns Hopkins University Press, 1974).

24. Edwin T. Layton Jr., *The Revolt of the Engineers: Social Responsibility and the American Engineering Profession* (Baltimore: Johns Hopkins University Press, 1986), 28–29.

25. Kees Gispen, *New Profession, Old Order: Engineers and German Society, 1815–1914* (Cambridge: Cambridge University Press, 1989), 45–46.

26. Tanabe Sakuro, "Progress of Engineering in Japan," in *World Engineering Congress Tokyo 1929 Proceedings 2: General Problems Concerning Engineering* (Tokyo: Kogakki, 1931), 130.

27. Rollo Appleyard, *The History of the Institution of Electrical Engineers (1871–1931)* (London: IEE, 1939), 38; "The Royal Institution of Naval Architects and Its Work—1860–1960—a Brief Historical Note," Royal Institute of Naval Architects, accessed August 25, 2017, http://www.rina.org.uk/Historical.

28. Appleyard, *Electrical Engineers*; "A History of the Institution of Engineering and Technology," Institution of Engineering and Technology, accessed June 22, 2017, http://www.theiet.org/resources/library/archives/research/guides-iet.cfm. The Society of Telegraph Engineers initially had membership requirements based on age and years of experi-

ence as a telegraph engineer, but it gradually broadened its focus to include other uses of electricity.

29. Peter Lundgreen, "Engineering Education in Europe and the U.S.A., 1750–1930: The Rise to Dominance of School Culture and the Engineering Professions," *Annals of Science* 47, no. 1 (1990): 73–74.

30. "History of the Institution of Mechanical Engineers," Institution of Mechanical Engineers, accessed June 12, 2017, http://www.imeche.org/about-us/imeche-engineering-history/institution-and-engineering-history; Bruce Sinclair, *The Centennial History of the American Society of Mechanical Engineers, 1880–1980* (Toronto: University of Toronto Press for ASME, 1980), 22–24.

31. Master Car-Builders Association, *History and Early Reports of the Master Car-Builders' Association* (New York: Martin B. Brown, 1885).

32. Layton, *The Revolt*. For a nuanced reassessment of this argument about American engineers in the early years of the twentieth century, see also Peter Meiksins, "The 'Revolt of the Engineers' Reconsidered," *Technology and Culture* 29, no. 2 (April 1988), 219–46. For a sociohistorical discussion of engineers' push for professional identity in the United States, see Yehouda Shenhav, *Manufacturing Rationality: The Engineering Foundations of the Managerial Revolution* (Oxford: Oxford University Press, 1999), especially chapter 1.

33. "Accueil: Historique," Union Technique de l'Électricité, accessed June 12, 2017, http://ute-asso.fr; J. Mikoletzky, "Vom Elektrotechnischen Verein in Wien zum Ősterreichischen Verband fűr Elektrotechnik—125 Jahre OVE," *Elektrotechnik & Informationstechnik* 125, no. 5 (2008): 147–52.

34. A. Michal McMahon, *The Making of a Profession: A Century of Electrical Engineering in America* (New York: IEEE Press, 1984), 28–29.

35. McMahon, *Making of a Profession*.

36. Appleyard, *Electrical Engineers*.

37. "Overview," IEEJ, accessed July 27, 2017, http://www.iee.jp/?page_id=1547; "Association Suisse des Électriciens," *Dictionnaire Historique de la Suisse*, accessed July 11, 2017, http://www.hls-dhs-dss.ch/textes/f/F16470.php.

38. Michael Schanz and Frank Dittmann, "The History of the VDE: A Development from a German Technical Society to an Association Acting as an All-Inclusive Platform for Electrotechnology," VDE, accessed August 25, 2018, https://www.vde.com/resource/blob/815276/193aa7d90b1675cac22ff63833bb3f18/history of the vde data.pdf, 1. Between the Berlin ETV's founding in 1879 under the chairmanship of Werner von Siemens and the founding of the VDE in 1893, Siemens and his firm were at the forefront of electrotechnology in Germany. Unfortunately, Siemens died just before the establishment of the VDE.

39. "Our Mission," Federazione Italiana di Elettrotecnica, Elettronica, Automazione, Informatica e Telecomunicazioni, accessed August 25, 2018, https://www.aeit.it/aeit/r02/struttura/pagedin.php?cod=chi_siamo.

40. Bruce Sinclair, *Early Research at the Franklin Institute: The Investigation into the Causes of Steam Boiler Explosions, 1830–1837* (Philadelphia: Franklin Institute, 1966), 6–7, quote on page 2.

41. Bruce Sinclair, "At the Turn of a Screw: William Sellers, the Franklin Institute, and a Standard American Thread," *Technology and Culture* 10, no.1 (1969): 27–31; Sinclair, *Early Research*, 4, 9, 11.

42. *VDI-Rechtlininien-Katalog/VDI Standards Catalogue* (Berlin: Beuth, 2015), 8.

43. As cited in A. E. Musson, "Joseph Whitworth and the Growth of Mass-Production Engineering," *Business History* 17, no. 2 (1975): 109–49.

44. Musson, "Joseph Whitworth," 122.

45. Sinclair, "Turn of a Screw."

46. Philip Scranton, *Endless Novelty: Specialty Production and American Industrialization, 1865–1925* (Princeton, NJ: Princeton University Press, 1997), 63–64. Scranton emphasizes the power and cost implications of standardization, and the resistance to it. Neither he nor Sinclair, "Turn of a Screw," provides empirical data about the extent to which Sellers's screws were adopted.

47. On ASCE's professional orientation, see Layton, *The Revolt*, 29–30. On the ASCE's domination by railroad engineers, see Steven W. Usselman, *Regulating Railroad Innovation: Business, Technology, and Politics in America, 1840–1920* (Cambridge: Cambridge University Press, 2002), 217–18. The quoted phrase describing the study comes from three articles by Ashbel Welch (cited by Usselman), all entitled "On the Form, Weight, Manufacture, and Life of Rails," published in the *Transactions of the ASCE*, vol. 3 (June 10, 1874), 87–110); vol. 4 (May 5, 1875), 136–41; and vol. 5 (June 15, 1876), 327–29. These articles reported the results of the committee's survey and deliberation, also described in this paragraph.

48. Layton, *The Revolt*, 29, 48n21.

49. Usselman, *Regulating Railroad Innovation*, 219–20.

50. Usselman, *Regulating Railroad Innovation*, 215–16, 220–21.

51. Unless otherwise indicated, biographical details come from Edgar Marburg, "Biographical Sketch," in *Memorial Volume Commemorative of the Life and Work of Charles Benjamin Dudley, Ph.D.*, ed. American Society for Testing Materials (Philadelphia: American Society for Testing Materials, [1910]), 11–42.

52. Yale University, "Obituary Record of Graduates of Yale University Deceased during the Academical Year Ending in June, 1910, Including the Record of a Few Who Died Previously, Hitherto Unreported," no. 10 of the Fifth Printed Series and no. 69 of the Whole Record (Presented at the Meeting of the Alumni, June 21, 1910), accessed August 21, 2017, http://mssa.library.yale.edu/obituary_record/1859_1924/1909-10.pdf; "ACS President: Charles B. Dudley (1842–1909)," American Chemical Society, accessed July 17, 2017, https://www.acs.org/content/acs/en/about/president/acspresidents/charles-dudley.html.

53. For a complete list of his memberships, see "Statistical Data" and "Offices and Honors Held by Charles B. Dudley," in ASTM, *Memorial Volume Commemorative of Dudley*, 113–15. The American Institute of Chemical Engineering was not established until 1908, a year before Dudley died, so it is not clear whether he would have associated himself with that professional association. He was a member of the scientific association for chemistry in the United States, the American Chemical Society and, from 1886 to 1898, president.

54. See bibliography of Charles B. Dudley, PhD, in ASTM, *Memorial Volume Commemorative of Dudley*, 116–17, for a complete list of his publications. The pair of papers he published in *Transactions of the AIME*, vol. 7, 1878, "The Chemical Composition and Physical Properties of Steel Rails" (172–201) and "Does the Wearing Power of Steel Rails Increase with the Hardness of the Steel" (202–205), started up a controversy about what made steel wear better. He later abandoned his claim in those papers, but they played an important role in initiating research in this area. For discussion of this episode, see Marburg, "Biographical Sketch," 22–24; and Usselman, *Regulating Railroad Innovation*, 221–223.

55. Quoted in Marburg, "Biographical Sketch," 23, source not identified.

56. William R. Jones in comments to the final report of the ASCE committee on wheel wear, *Transactions of the ASCE* 21 (July-December 1889): 279–280, as cited in Usselman, *Regulating Railroad Innovation*, 232.

57. Usselman, *Regulating Railroad Innovation*, 230–32.

58. *Transactions of the ASCE* 28 (1893): 425–44, quote is from 426.

59. Robert W. Hunt, "Specification for Steel Rails of Heavy Sections Manufactured West of the Alleghenies," *Transactions of the AIME* 25 (February-October 1895): 654; subsequent quote from same page.

60. Historian Steven Usselman has also noted that "the rail sections and the guidelines for manufacture contained in these committee reports rapidly acquired the status of industry-wide standards." Usselman, *Regulating Railroad Innovation*, 232.

61. Samuel Haber, *Authority and Honor in the American Professions, 1750–1900* (Chicago: University of Chicago Press, 1991), 294–300.

62. This paragraph and the next are based on Sinclair, *Centennial History*, 46–60. For another treatment of the debate about whether ASME should participate in standardization as part of its professional project, see Shenhav, *Manufacturing Rationality*, 58–64.

63. JoAnne Yates, *Control through Communication: The Rise of System in American Management* (Baltimore: Johns Hopkins University Press, 1989), 9–15. Scientific management focuses primarily on the shop floor, while systematic management targets the entire organization, from head office to shop floor and mailroom. Advocates of systematization such as Henry Metcalfe and scientific management leader Frederick Winslow Taylor both published in the *Transactions of the ASME*. See, for example, Henry Metcalfe, "The Shop-Order System of Accounts," *Transactions of the ASME* 7 (May 1886): 440; and Frederick W. Taylor, "A Piece Rate System Being a Step towards Partial Solution of the Labor Problem," *Transactions of the ASME* 16 (1895): 856–903.

64. Sinclair, *Centennial History*, 55.

65. Robert A. Brady, *The Rationalization Movement in German Industry: A Study in the Evolution of Economic Planning* (Berkeley: University of California Press, 1933), 148.

66. Walter Mattli, *The Logic of Regional Integration: Europe and Beyond* (Cambridge: Cambridge University Press, 1999), 121–28, provides a careful evaluation of the degree to which moves toward German unification supported industrialization, concluding that the German Customs Union and the eventual adoption of a single currency in 1857 played critical roles in establishing the larger market area that enabled industrialization. In a 1981 study of German engineers, two British scholars reflecting on Albert Speer as an example of the responsibility of German engineers for Nazism note, in apparent horror, "In the German scheme of things an architect is an honorary engineer!" Stanley Hutton and Peter Lawrence, *German Engineers: An Anatomy of a Profession* (Oxford: Clarendon Press, 1981), 86.

67. Gispen, *New Profession*, 53–54.

68. Gispen, *New Profession*, 118.

69. Quoted in Gispen, *New Profession*, 117.

70. Gispen, *New Profession*, 119. Gispen quotes the director of a Berlin machine-tool manufacturer describing the US industry as a "shining example."

71. Brady, *Rationalization Movement*, 149.

72. Wilfred Hessler and Alex Inklaar, *An Introduction to Standards and Standardization* (Berlin: Deutsches Institut für Normung, 1998), 31.

73. Brady, *Rationalization Movement*, 151; Hessler and Inklaar, *Introduction to Standards*, 32; Robert A. Brady, *Industrial Standardization* (New York: National Industrial Conference Board, 1929), 116.

74. Appleyard, *Electrical Engineers*, 321–23.

75. By 1974 it was well embedded in the law, according to Michael Neidle, *Electrical Installations and Regulations* (London: Macmillan Press, 1974), 2–3.

76. Schanz and Dittmann, "History of the VDE," 2.

77. Schanz and Dittmann, "History of the VDE," 4. As with the broader VDI engineering association, the VDE began with safety standards.

78. McMahon, *Making of a Profession*, 29.

79. McMahon, *Making of a Profession*, 83–84; "The Standardization of Generators, Motors and Transformers (A Topical Discussion)," *AIEE Transactions* 15 (1898): 3–22; the following quotations are from pages 3–4.

80. "Standardization of Generators," 14 (Crocker quote), 20 (Steinmetz quote).

81. "Standardization of Generators," 5. For Kennelly's affiliation, see McMahon, *Making of a Profession*, 85.

82. McMahon, *Making of a Profession*, 85.

83. "Report of the Committee on Standardization [Accepted by the INSTITUTE, June 26th, 1899.]," *AIEE Transactions* 16 (1899): 255–68.

84. "Report of the Committee on Standardization," 255.

85. "Standardization Rules of the American Institute of Electrical Engineers," *AIEE Transactions* 35, part 2 (1916): 1551–662. This article begins with a section on the history of the AIEE standardization rules (1551–55), then includes the standards themselves.

86. Brady, *Industrial Standardization*, 71n3.

87. On iron and steel, see Usselman, *Regulating Railroad Innovation*; and Thomas J. Misa, *A Nation of Steel* (Baltimore: John Hopkins University Press, 1999); on concrete, see Amy E. Slaton, *Reinforced Concrete and the Modernization of American Building, 1900–1930* (Baltimore: Johns Hopkins University Press, 2001).

88. Stephen P. Timoshenko, *History of Strength of Materials, with a Brief Account of the History of Theory of Elasticity and Theory of Structures* (New York: Dover, [1953] 1983), 279–81. Another source suggests that the Munich laboratory was founded even earlier, in 1868, but that these laboratories were all rudimentary until the late 1870s, when they were upgraded. Gispen, *New Profession*, 154–56.

89. This description of the founding of IATM draws on Mansfield Merriman, "The Work of the International Association for Testing Materials (IATM)," (chairman's address, 2nd Annual Meeting of the American Section, August 15–16, 1899), in *Proceedings of the ASTM* 1, no. 4 (1899): 17–25; and American Section of the IATM, *History, Laws, Committees and List of Members* (Philadelphia: Office of Secretary of American Section, 1899), 7–8.

90. J. Bauschinger, in his introduction to "Resolutions of the Conventions Held at Munich, Dresden, Berlin and Vienna—for the Purpose of Adopting Uniform Methods for Testing Construction Materials with Regard to Their Mechanical Properties," trans. O. M. Carter and E. A. Gieseler for the US War Department (Washington: Government Printing Office, 1896), 7. Although the translation into English is dated 1896, the original must have been written in 1893, shortly after the Vienna conference, as Bauschinger died later in that same year (Timoshenko, *History of Strength of Materials*, 301).

91. Merriman, "Work of the IATM," 18.

92. Bauschinger, introduction to "Resolutions of the Conventions," 7–8, states that one permanent committee was established at each of the first four conferences, but Merriman, "Work of the IATM," 19, says that "at the Vienna convention of 1893 there had been appointed 20 committees on technical subjects, and reports from many of these were presented at the Zurich congress of 1895."

93. Bauschinger, introduction to "Resolutions of the Conventions," 7.

94. Bauschinger, introduction to "Resolutions of the Conventions," 8.

95. "Memorial Notices of Members Deceased During the Year: Johann Bauschinger, Honorary Member," *Transactions of the ASME* 15, no. 605 (1894): 1184–88, quotation is from 1187.

96. "Memorial Notices: Johann Bauschinger, Honorary Member," *Transactions of the ASME*, 1187.

97. As quoted by Merriman, "Work of the IATM," 18–19.

98. Timoshenko, *History of Strength of Materials*, 281, 283, 301.

99. Membership was 1,169 in 1897, up from 493 in 1895. By 1898 it was 1,488, with 387 members from Germany, 315 from Russia, 158 from Austria, 83 each from England and Switzerland, 68 each from the US and Sweden, 66 from France, 48 from Holland, 42 from

Norway, 39 from Denmark, 36 from Spain, 35 from Italy, and 60 more members from nine additional countries (Merriman, "Work of the IATM," 19).

100. "Minutes, Meeting of the Executive Committee, American Section, IATM," *Proceedings of the ASTM* 1, no. 7 (1900): 76.

101. As summarized in Mansfield Merriman, "Address of the Chairman to the Third Annual Meeting of the American Section" (speech given on October 25, 1900), *Proceedings of the ASTM* (March 1901): 20.

102. American Section of the IATM, *History, Laws, Committees and List of Members*, 4, 8, 22–23.

103. American Section of the IATM, *History, Laws, Committees and List of Members*, 14.

104. The correspondence is reproduced in its entirely in *Proceedings of the ASTM* 1, no. 19 (September 1900): 154–69.

105. IATM president Tetmajer to American Section chairman Mansfield Merriman, June 17, 1899, reproduced in *Proceedings of the ASTM* (September 1900): 154. Captain Carter was undoubtedly from the US Army or Navy, both of which had officers at the Zurich meeting, along with a representative from the ASME (Merriman, "Work of the IATM," 19).

106. For Henning's role at the Congresses, see Sinclair, *Centennial History*, 58.

107. Tetmajer to Merriman on June 17, 1899, in *Proceedings of the ASTM* (September 1900): 154. Merriman's reply, dated July 3, 1899, is on page 155.

108. Merriman, "Work of the IATM," 24.

109. *Proceedings of the ASTM* (September 1900): 154–69.

110. *Proceedings of the ASTM* 1, no. 21 (March 1901): 185.

111. Merriman ("Work of the IATM," 22) states the number as 21; in a letter responding to Merriman's request to increase the American membership, Tetmajer puts the number at 26. Ludwig von Tetmajer to Mansfield Merriman, July 9, 1900, reproduced in *Proceedings of the ASTM* (March 1901): 198–99.

112. Tetmajer to Merriman on July 9, 1900, 198–99.

113. Merriman, "Address of the Chairman," 189.

114. Merriman, "Address of the Chairman," 190.

115. General Secretary Ernest Reitler to the Members of IATM, Vienna, March 15, 1915, reproduced in "Annual Report of the Executive Committee," *Proceedings of the ASTM* 15 (1915): 77.

116. *Proceedings of the ASTM* 19 (1919): title page. For comparison, see earlier volumes.

117. Guilliam H. Clamer, "Standardization" (annual address by the president), *Proceedings of the ASTM* 19 (1919): 91. In the 1920 proceedings, ASTM's Executive Committee referred to IATM as "the now defunct International Association for Testing Materials" and discussed a report from Henry M. Howe about his conversations in Great Britain and France of a possible new organization ("Annual Report of the Executive Committee," *Proceedings of the ASTM* 20 (1920): 91).

118. *ASTM Year Book* (Philadelphia: ASTM, August 1928), 26.

119. *ASTM Bulletin*, no. 29 (November 28, 1927): 2–5. The constitution is on page 2.

120. This new body held a congress in Zurich in 1931, but the cost of publishing the proceedings of that congress, combined with a fall in subscriptions because of the Great Depression, left it in considerable debt, and it did not meet again until the organization had emerged from the debt in 1937. See, for example, "International Association for Testing Materials," a short item in *Nature* (December 31, 1932): 1008–9. Also see H. J. Gough (president of IATM), Statement on behalf of the Permanent Committee, in IATM, *London Congress: April 19–24, 1937* (London: International Association for Testing Materials, 1937), xxviii–xxxii.

121. Usselman, *Regulating Railroad Innovation*, 233–34.

122. Usselman, *Regulating Railroad Innovation*, 236–37.

123. The members originally named in 1898 by the International Council are identified in the Report of the Executive Committee on the Membership of Committee No. 1, in "Minutes of the Third Annual Meeting of the American Section, Oct. 25–27, 1900," *Proceedings of the ASTM* 1, no. 21 (1901): 197. Their affiliations can be found on the longer membership list of Committee No. 1 of the American Section, in *Proceedings of the ASTM* 1, no. 18 (1900): 142. They included representatives of three iron and steel companies (Pencoyd Iron Works, Carnegie Steel Co., and Pennsylvania Steel Co.) and two individuals not from producer firms (the Franklin Institute, represented by its secretary, William Wahl, and William R. Webster, a consulting and inspecting engineer).

124. "Minutes, Meeting of the Executive Committee," 76. The report on Committee No. 1 notes that five producer companies had requested to be added to it, and that request was granted, with the further resolution that "as it is policy of the Association that its Technical Committees should be nearly equally divided between producers and consumers, that Committee No. 1 be requested to name five engineers not directly associated with manufacturing, to balance the five firms above mentioned." Webster responded to criticisms of bias toward steel makers during his *Progress Report of Committee No. 1* at the October 1900 meeting of the American Section of the IATM (*Proceedings of the ASTM* 1, no. 20, 177) and went on to invite more input from members of ASCE, ASME, AREMWA, and other associations on their draft recommendations, to help respond to accusations of being a producer committee. Usselman (*Regulating Railroad Innovation*, 235–39) certainly sees this forerunner of the ASTM as being slanted toward steel manufacturers, but he acknowledges that once Dudley became president of the newly independent ASTM in 1902, this bias disappeared.

125. "Minutes of the Executive Committee Meeting, June 25, 1898," *Proceedings of the ASTM* 1, no. 1 (1899): 5. Dudley was not able to attend the meeting, so someone else opened the discussion.

126. Robert W. Hunt, "Charles B. Dudley—a Personal Tribute," ASTM, *Memorial Volume Commemorative of Dudley*, 79–80.

127. Usselman, *Regulating Railroad Innovation*, 235–39; Misa, *Nation of Steel*, 145–55.

128. "Extract from the Engineering News," December 23, 1909, reproduced in ASTM, *Memorial Volume Commemorative of Dudley*, 102.

129. "Extract from the Railway and Engineering Review," December 25, 1909, reproduced in ASTM, *Memorial Volume Commemorative of Dudley*, 103.

130. John Wesley Hanson, *Progress of the 19th Century: A Panoramic Review of the Inventions and Discoveries of the Past Hundred Years* (Toronto: J. L. Nichols, 1900), 331–32, discussing a series of reports of the US supervising inspector of steamboats.

131. Maurice Yeates, *The North American City*, 5th ed. (New York: Longman, 1998), 99. He argues that these two innovations made the modern economy possible in Canada and the northern and western US.

132. Lists of the relevant electrical and materials standards set by 1907 can be found in H. F. Chadwick, "Standardization," *Electrical Journal* 9, no. 4 (April 1912): 15–22.

Chapter 2 · Organizing Private Standard Setting within and across Borders, 1900 to World War I

1. Robert C. McWilliam, "The First British Standards: Specifications and Tests Published by the Engineering Standards Committee, 1903–18," *Transactions of the Newcomen Society* 75 (2005): 262. This paragraph also draws on accounts in McWilliam, "The Evolution of British Standards" (PhD thesis, University of Reading, 2002); and in McWilliam, *BSI: The First Hundred Years, 1901–2001: A Century of Achievement* (London: Institute of Civil Engineers, 2001).

2. McWilliam, *Hundred Years*, 11. The following quote comes from minutes of the council's meetings.

3. McWilliam, "The Evolution of British Standards," 67.

4. "John Wolfe Barry, K.C.B, L.L.D, F.R.S., Past-President, 1836–1918," *Minutes of the Proceedings of the Institution of Civil Engineers* 206, part 2 (1918): 350–57; quote is from 355. A photograph of the Bauschinger plaque in the Munich subway can be found at Peter Specht, "Johann Bauschinger Plaque at Munich Subway Station Garching-Forschungszentrum (U6)," accessed July 21, 2017, https://www.flickr.com/photos/woodpeckar/3013745288/.

5. Douglas Fox, "Presidential Address," *Minutes of Proceedings* (n.p.: Institution of Civil Engineers, November 7, 1899), 7. Further quotes from this address in this paragraph and the next are drawn from pp. 15–17.

6. In the late 1880s, when Andrew Carnegie decided to move into the structural steel market, he competed based on the strength of the company's products, as determined by the results of extensive experimentation to identify the strongest shapes. The (internally) standardized shapes were documented through its superior sales handbook, with detailed notes, diagrams, and other instructional materials that made the handbook "a virtual textbook on structural steel." It thus came to define de facto standards used not just by buyers of Carnegie steel but also by buyers from other steel firms. Thomas J. Misa, *A Nation of Steel* (Baltimore: John Hopkins University Press, 1999), 70–74, quote on 74.

7. "The Atbara Bridge," *Railway Age* 29, no. 2 (January 12, 1900): 38–39.

8. Quoted in BSI, *Fifty Years of British Standards, 1901–1951* (n.p.: British Standards Institution, 1951), 25, 27. Note that in this case the manufacturer, rather than being worried about standards creating debilitating price competition [Philip Scranton's argument in *Endless Novelty: Specialty Production and American Industrialization, 1865–1925* (Princeton, NJ: Princeton University Press, 1997), 63–64], is worried about the country's manufacturers being unable to satisfy the too varied requests of customers for want of standards and thus losing the business entirely.

9. BSI, *Fifty Years of British Standards*, 27; McWilliam, "The Evolution of British Standards," 67.

10. Minutes of Council, ICE, Book 16 (1901–1903), January 22, 1901, item 25, quoted by McWilliam, "The Evolution of British Standards," 68.

11. Minutes of Council, ICE, quoted by McWilliam, "The Evolution of British Standards," 70.

12. This paragraph is based on Minutes of Council, ICE, quoted by McWilliam, "The Evolution of British Standards," 70–74; and McWilliam, *Hundred Years*, 17–27.

13. W. Noble Twelvetrees, "The Engineering Standards Committee and Its Work," *Public Works* 2 (November 15, 1903): 88.

14. McWilliam, *Hundred Years*, 22; McWilliam, "The Evolution of British Standards," 73–74.

15. BSI, *Fifty Years of British Standards*, 28.

16. The founders did not include an organization of chemical engineers, as the Institution of Chemical Engineers was not constituted until 1922; only the Society of the Chemical Industry existed at the founding of the ESC, and it was not constituted as a true professional association ("Our Origins", IChemE, accessed June 27 2017, http://www.icheme.org/about_us/origins_of_icheme.aspx).

17. McWilliam, "The Evolution of British Standards," 83.

18. McWilliam, "The First British Standards," 267.

19. BSI, *Fifty Years of British Standards*, 28–29.

20. McWilliam, "The Evolution of British Standards," 75.

21. McWilliam, *Hundred Years*, 22.

22. McWilliam, "The First British Standards," 269–79. Out of the first 81 standards issued before the ESC became independent of ICE, 3 were informative standards.

23. McWilliam, "The First British Standards," 266–67. The number 14 reflects an additional sectional committee that McWilliam lists in this article but not in his other publications: Railway Rolling Stock Underframes, formed in 1901 (table 1, 267). The table shows that this committee issued no standards independently but only in concert with other sectional committees, seemingly operating more as a subcommittee of existing committees than as its own sectional committee.

24. McWilliam, "The First British Standards," 267.

25. Twelvetrees, "Standards Committee," 96.

26. BSI, *Fifty Years of British Standards*, 31.

27. Quoted by McWilliam, *Hundred Years*, 33.

28. McWilliam, "The Evolution of British Standards," 85–86.

29. H. F. Chadwick, "Standardization," *Electric Journal* 9, no. 4 (April 1913): 320.

30. The eight principles are quoted by McWilliam, *Hundred Years*, 26. For a fuller form, see Sir John Wolfe Barry, "The Standardization of Engineering Materials and Its Influence on the Prosperity of the Country" ("James Forrest" Lecture, n.p., 1917), 10, in the archives of the Institute for Civil Engineering, ICE BSI 3.1-33 (1).

31. Alexander Blackie William Kennedy, "Physical Experiment in Relation to Engineering," "The 'James Forrest' Lecture," May 7, 1896, *Minutes of the Proceedings of the Institution of Civil Engineers* 126 (1896): 321.

32. Anonymous, "The Engineering Standards Committee," *Builder* 85 (July 11, 1903): 31.

33. McWilliam, "The Evolution of British Standards," 52.

34. This chapter focuses on its initial development, but for a look at how its institutional structure and procedures have developed up to the present day, see Tim Büthe, "Engineering Uncontestedness? The Origins and Institutional Development of the International Electrotechnical Commission (IEC)," *Business and Politics* 12, no. 3 (2010): 1–62.

35. "Elihu Thomson," Engineering and Technology History Wiki, accessed June 27, 2017, http://www.ieeeghn.org/wiki/index.php/Elihu_Thomson; Jeanne Erdmann, "The Appointment of a Representative Commission," IEC History, accessed June 27, 2017, http://www.iec.ch/about/history/beginning/commission.htm. For his career at General Electric, see W. Bernard Carlson, *Innovation as a Social Process: Elihu Thomson and the Rise of General Electric, 1870–1900* (Cambridge: Cambridge University Press, 1991).

36. Quoted in L. Ruppert, *History of the International Electrotechnical Commission* (Geneva: Bureau Central de la Commission Electrotechnique Internationale, [1954]), accessed June 27, 2017, http://www.iec.ch/about/history/documents/pdf/IEC%20History%201906-1956.pdf.

37. Mark Frary, "Colonel Crompton: The King of Electricity," IEC History, accessed June 27, 2017, http://www.iec.ch/about/history/beginning/colonel_crompton.htm; Alexander Russell, "Rookes Evelyn Bell Crompton: 1845–1940," *Obituary Notices of Fellows of the Royal Society* 3, no. 9 (January 1941): 401.

38. Russell, "Rookes Evelyn Bell Crompton," 395–97; this obituary includes the claim of winning the Crimean War Medal and Sebastopol clasp, as does Crompton's autobiography. R. E. B. Crompton, *Reminiscences* (London: Constable & Co. Ltd, 1928), 15. See, in contrast, Frary, "Colonel Crompton." Crompton was forced out of leadership in the company after he returned from the Boer War and the company faced a financial crisis; he severed all ties with it in 1912 (Crompton, *Reminiscences*, 206–8).

39. Crompton, *Reminiscences*, 195–200.

40. Crompton, *Reminiscences*, 141–42. McWilliams discusses this standard with no reference to Crompton and attributes the lab work to the National Physical Laboratory rather than to Crompton's work in his own lab (McWilliam, "The Evolution of British Standards," 121).

41. IEC, *Report of Preliminary Meeting Held at the Hotel Cecil, London, on Tuesday and Wednesday, June 26th and 27th 1906* (London: IEC, 1906), 44, available on the IEC History website, accessed September 2, 2018, http://www.iec.ch/about/history/documents/pdf/IEC_Founding_Meeting_Report_1906.pdf.

42. Mark Frary, "In the Beginning: The Founding of the IEC," IEC History, accessed June 27, 2017, http://www.iec.ch/about/history/beginning/founding_iec.htm.

43. Ruppert, "History of the International Electrotechnical Commission," 1–2; see also IEC, *Report of Preliminary Meeting*, 48.

44. Crompton, *Reminiscences*, 205; Ruppert, "History of the International Electrotechnical Commission," 2.

45. André Lange, "Charles le Maistre: His Work, the IEC," in *The 1st Charles le Maistre Memorial Lecture* (Geneva: International Electrotechnical Commission, 1955), 8.

46. IEC, *Report of Preliminary Meeting*, 3–5.

47. IEC, *Report of Preliminary Meeting*, 14.

48. IEC, *Report of Preliminary Meeting*, 9–10.

49. IEC, *Report of Preliminary Meeting*, 10; next quote, 14.

50. Douglas Howland, "Telegraph Technology and Administrative Internationalism in the Nineteenth Century," in *The Global Politics of Science and Technology: Concepts from International Relations and Other Disciplines*, ed. Maximilian Mayer, Mariana Carpes, and Ruth Knoblich (Heidelberg: Springer, 2014), 186–89.

51. IEC, *Report of Preliminary Meeting*, 14.

52. IEC, *Report of Preliminary Meeting*, 34.

53. IEC, *Transactions of the Council Held in October 1908* (London: IEC, 1909), 44–48, IEC Records, Geneva. The French delegation wanted to retain the requirement of unanimity to enable it to protect small French companies, but Crompton stated that Britain's experience "during the past six years on Committees dealing with electrical standards had conclusively shown the practical impossibility of obtaining unanimity on even uncontroversial points, and that, therefore, a substantial majority safeguarding the general interests concerned appeared to be the only reasonable method of assuring success." Ultimately, Crompton's view won out.

54. IEC, *Report of Preliminary Meeting*, 10.

55. Kees Gispen, *New Profession, Old Order: Engineers and German Society, 1815–1914* (Cambridge: Cambridge University Press, 1989), 322.

56. IEC, *Report of Preliminary Meeting*, 10.

57. IEC, *Report of Preliminary Meeting*, 20.

58. "Ichisuke Fujioka: A Wizard with Electricity," Toshiba Science Museum, accessed August 27, 2018, http://toshiba-mirai-kagakukan.jp/en/learn/history/toshiba_history/spirit/ichisuke_fujioka/index.htm.

59. IEC, *Report of Preliminary Meeting*, 30.

60. IEC, *Report of Preliminary Meeting*, 52.

61. IEC, *Report of Preliminary Meeting*, 48.

62. Ruppert, "History of the International Electrotechnical Commission," 2.

63. Quoted in Ruppert, "History of the International Electrotechnical Commission," 2.

64. IEC, *Publication 7: Second Annual Report to 31st December 1910* (London: IEC, March 1911), 12, IEC Records.

65. IEC, *Publication 3: List of Members* (London: IEC, March 1910), IEC Records.

66. IEC, *Publication 7*, 12.

67. IEC, *Publication 7*, 2–3.

68. American Philosophical Society, Philadelphia, Elihu Thomson Papers / Ms. Coll. 74 / Series / Le Maistre, C / International Electrotechnical Commission / 1908–1916, courtesy of the American Philosophical Society.

69. Le Maistre to Stratton, Washington, May 3, 1909; the letter to Thomson quoted in the next sentence is le Maistre to Thomson, May 11, 1909, in Elihu Thomson Papers.

70. The three passages are from these three letters, respectively: le Maistre to Thomson, January 10, 1910, 5; le Maistre to Thomson, January 17, 1910; le Maistre to Thomson, February 3,1910, in Elihu Thomson Papers.

71. Le Maistre to Thomson, January 10, 1910, 1, Elihu Thomson Papers.

72. Le Maistre to Thomson, February 24 1911, Elihu Thomson Papers.

73. Le Maistre to Thomson, October 26, 2010; the following quote comes from le Maistre to Thomson, February 24, 1911, Elihu Thomson Papers.

74. IEC, *Publication 24: Fourth Annual Report* (London: IEC, August 1913), 24, IEC Records.

75. Clifford B. Le Page, "Twenty-Five Years—the American Standards Association (Origins)," *Industrial Standardization* 14, no. 12 (1943): 317–22. See also Comfort A. Adams, "National Standards Movement—Its Evolution and Future," *National Standards in a Modern Economy*, ed. Dickson Reck (New York: Harper & Brothers, 1956), 23–24; Andrew L. Russell, *Open Standards and the Digital Age: History, Ideology, and Networks* (Cambridge: Cambridge University Press, 2014), 62–63.

76. Russell, *Open Standards*, 63.

77. Adams, "National Standards Movement," 23–24.

78. "Standardization Rules of the American Institute of Electrical Engineers," *AIEE Transactions* 35, part 2 (1916): 1555, says that le Maistre's visit to New York was to help in the revision of the AIEE rules. See also Russell, *Open Standards*, 50–51. For Robertson's death, see McWilliam, *Hundred Years*, 79; Grace's Guide to British Industrial History, accessed January 15, 2018, https://www.gracesguide.co.uk/Leslie_Stephen_Robertson.

79. Like the British ESC, the AESC also lacked a founding professional association of chemical engineers. The American Institute of Chemical Engineers (AIChE) was founded in 1908 ("Governance," AIChE, accessed June 27, 2017, https://www.aiche.org/about/governance), so it could have been included in the AESC but was not, whether because the AESC's founders were imitating the ESC or because they thought that the AIChE was not yet mature enough to join the other, older associations, or for other reasons. Nevertheless, the ASTM included many chemical engineers as well as other types of engineers, so the field was represented.

80. Minutes of the AESC meeting, May 4, 1918, records in the headquarters of the American National Standards Institute [subsequently referred to as ANSI Records New York City. The next several quotes are from the minutes of this first meeting, until otherwise stated.

81. Minutes of AESC Meeting, October 19, 1918, ANSI Records; Adams, "National Standards Movement," 25; see also Russell, *Open Standards*, 65–66.

82. Adams, "National Standards Movement," 21–30; P. G. Agnew, "International Standardization: The Four Stages of Industrial Standardization—National and International Bodies—Examples of Accomplishment—Information the Basis of Co-operation," *American Machinist* 57, no. 17 (October 26, 1922): 633–38, quote from 634.

83. Agnew, "International Standardization," 634–35.

84. Adams, "National Standards Movement," 25.

85. Minutes of AESC Meeting, January 17, 1919, Minutes, ANSI Records.

86. From 1917 to 1926 it was officially Normenausschuss der deutschen Industrie (NADI). From 1926 to 1975 it was officially Deutscher Normenausschuss with the now-confusing acronym DNA. "History of DIN," DIN, accessed September 1, 2018, https://www.din.de/en/din-and-our-partners/din-e-v/history. DIN has informally been used to refer to the organization in many contexts since the organization's beginning. German standards were called *Deutsche Industrie-Normen* and were marked with a distinctive trademark with the letters DIN between two bold lines, see figure 3.3.

87. Robert A. Brady, *The Rationalization Movement in German Industry: A Study in the Evolution of Economic Planning* (Berkeley: University of California Press, 1933), 151; Wilfred Hessler and Alex Inklaar, *An Introduction to Standards* (Berlin: Droz, 1998), 32; Brady, *Industrial Standardization* (New York: National Industrial Conference Board, 1929), 116; and see Waldemar Hellmich and Ernst Huhn, *Was will Taylor? Die arbeitssparende Betriebsführung und kritische Bemerkungen über das "Taylorsystem"* (Berlin: VDI, 1919).

88. Brady, *Industrial Standardization*, 116.

89. Brady, *Industrial Standardization*, 25–26.

90. Brady, *Industrial Standardization*, 26–27.

91. Siglinde Kaiser, "Standardization Strategy" (paper presented at the Fourth International Workshop on Conformity Assessment, Rio de Janerio, December 12, 2008), accessed July 10, 2017, http://docslide.us/documents/din-deutsches-institut-fuer-normung-e-v-2008-din-e-v-standardization.html, slide 11. Kaiser's presentation was based on research by DIN's archivist, Peter Anthony; quotation from Hellmich provided by Kaiser and Anthony from a speech in the DIN archives that he gave on the 10th anniversary of the organization, S. Kaiser email to C. Murphy, October 6, 2016.

92. Brady, *The Rationalization Movement*, 26.

Chapter 3 · A Community and a Movement, World War I to the Great Depression

1. Willy Kuert, "The Founding of ISO," in *Friendship Among Equals: Recollections from ISO's First Fifty Years*, ed. Jack Latimer (Geneva: ISO, 1997), 16; Clayton H. Sharp, "Discussion on Standardization," *AIEE Transactions* 35, no. 1 (1916): 491.

2. Kuert, "The Founding of ISO,"17.

3. Andrew L. Russell makes similar claims about American Paul Gough Agnew, but Agnew had far less international influence than le Maistre. See Andrew L. Russell, *Open Standards and the Digital Age: History, Ideology, and Networks* (Cambridge: Cambridge University Press, 2014), 68.

4. This estimate is based on the "non-technical" description of the history of BESA that Charles le Maistre drafted in 1927 as part of a campaign to obtain a Royal Charter: "Over 2000 Engineers and business men throughout the country give their time and experience to this national work without fee or recompense." Enclosure in a letter from le Maistre to Weir, November 4, 1927, Papers of Viscount Weir, University Archives, Glasgow University, collection 96, box 15, page 1. Two other national standards bodies are regularly described as of the same size as BESA in the late 1920s—those in Germany and the United States (e.g., Robert A. Brady, *Industrial Standardization* [New York: National Industrial Conference Board, 1929], 103–21). If we assume that, combined, IEC and the 16 remaining national standards bodies in 1927 involved as many people in technical committees as the big three, we reach an estimate of 12,000. We found no evidence that any women were standardizers during this interwar period.

5. Marie-Laure Djelic and Sigrid Quack, "Transnational Communities and Governance," in *Transnational Communities: Shaping Global Economic Governance* (Cambridge: Cambridge University Press, 2010), 28; and Marie-Laure Djelic and Sigrid Quack, "The Power of 'Limited Liability'—Transnational Communities and Cross-Border Governance," in *Research in the Sociology of Organizations*, vol. 33, *Communities and Organizations*, ed. Christopher Marquis, Michael Lounsbury, and Royston Greenwood (Bingley, UK: Emerald Group, 2011), 73–85.

6. Djelic and Quack, "Transnational Communities," 13.

7. Charles le Maistre, "Standardisation and Its Assistance to the Engineering Industries," *Electrician* 78 (October 13, 1916): 38–40. Reprinted in *Engineering*, the *Engineer*, the *Mechanical*

Engineer, the *Horseless Age: Automobile Engineering Digest, Science and Industry*, and the Victorian Institute of Refrigeration *Annual Proceedings*.

8. P. G. Agnew, "International Standardization," *American Machinist* 37, no. 17 (October 26, 1922): 634.

9. Djelic and Quack, "Transnational Communities," 18.

10. For the concept of community of practice, see Jean Lave and Etienne Wenger, *Situated Learning: Legitimate Peripheral Participation* (Cambridge: Cambridge University Press, 1991). Djelic and Quack ("Transnational Communities," 21) apply this term to standardizers.

11. As early as 1923, Germans used a professional designation whose English translation is "standardization engineer," but they applied that name to consulting engineers who performed time and motion studies within firms following Fredrick Taylor's method, not to men who served on technical committees (Brady, *Industrial Standardization*, 75). After World War II a group of primarily US engineers formed the Standards Engineering Society (SES) for the professional development of men and women employed by companies to sit on technical committees and to assess conformity with voluntary standards ("About SES," SES, accessed June 29, 2017, http://www.ses-standards.org/?page=A2).

12. A. A. Stevenson, "The American Engineering Standards Committee and National Standardization," *Proceedings of the Second National Standardization Conference Held in Connection with the Twenty-Fourth Annual Conference of the American Mining Congress*, Chicago, October 17–22, 1921, 121.

13. Stevenson, "American Engineering Standards Committee," 127–28, 271. DIN was "the first to publish standards in looseleaf form. . . . The idea is that the sheets shall be issued directly to designers, draughtsmen, and foremen, in much the same way as working drawings" (Norman F. Harriman, *Standards and Standardization* [New York: McGraw-Hill, 1928], 174–75).

14. This is a case where le Maistre's instincts about best practices proved wrong. In 1926 he recommended that, to save money, the IEC abandon this practice, which it had followed since publication of its first standard in 1913, and just publish a yearbook. The idea appears to have been dropped almost immediately. On the decision not to publish a yearbook, see IEC Committee of Action, *RM 34: Proces Verbaux de Reunions* (New York: IEC, April 21, 1926); and IEC, *RM 49* (London: IEC, September 23, 1929).

15. Brady, *Industrial Standardization*, 122–23. Information on the bodies for which Brady does not provide a founding date can be obtained by following the links to specific countries in "ISO: A Global Network of National Standards Bodies," ISO, accessed June 29, 2017, https://www.iso.org/members.html. Lino Camprubi, *Engineers and the Making of the Francoist Regime* (Cambridge, MA: MIT Press, 2014), 147, reports that the Spanish Normalization Association was founded in 1924, citing a 1935 book of DIN standards translated to Spanish; we have used this date.

16. Brady, *Industrial Standardization*, 122.

17. Alain Durand, *AFNOR: 80 Années d'Histoire* (Paris: AFNOR Éditions, 2008), 30.

18. Most of the governments of the industrialized countries had established such metrological bodies at the turn of the century, just as the United States did in 1901, as discussed in the introduction. Britain's National Physical Laboratory, discussed in chapter 2, was established to perform many of the same functions performed by the US and French government bodies.

19. Durand, *AFNOR*, 30.

20. "Obituary Notices: Charles Delacour le Maistre," *Journal of the Institution of Electrical Engineers* 1953, no. 9 (September 1953): 308.

21. Standards New Zealand, *Standards Council Annual Report for the Twelve Months Ended 30 June 2012* (Wellington: Standards New Zealand), 2.

22. Le Maistre to Weir, February 23, 1932, Papers of Viscount Weir, University Archives, Glasgow University, collection 96, box 15.

23. Standards New Zealand, *Annual Report*, 2, gives July 7, 1932, as the date of foundation and asserts that le Maistre's crucial visit was at the beginning of the same month. Le Maistre's letter to Weir of February 23 indicates that the visit began much earlier, suggesting that le Maistre was not as much the miracle worker as the official account suggests.

24. "Russian Affairs," *Electrical Review* 92, no. 2373 (May 18, 1923): 761–62, provides an account of le Maistre's visit, including his accompaniment of "Senator Wheeler of Montana U.S.A." (762). Burton K. Wheeler's memoir, *Yankee from the West: The Candid, Turbulent Life of the Yankee-Born U.S. Senator from Montana* (Garden City, NY: Doubleday, 1962), 199–203, provides an account of his trip without mentioning le Maistre. The IEC Records, Box CA-1 20.11.1924-15.11.1926, includes le Maistre's correspondence with Soviet colleagues about participating in international meetings; it begins in 1925 with discussions of whether the United States, where the next major conference would take place in April 1926, would be willing to grant visas to the Soviet delegation. The photograph at the front of IEC, *Publication 40: Report of the General Conference held in Bellagio, September 1927* (London: IEC, [ca. 1927]), IEC Records, includes a delegation from the USSR. This may be the first international standardization conference that Soviet engineers attended.

25. BESA, *Unofficial Conference of the Secretaries of the National Standardising Bodies Convened under the Instructions of the Main Committee of the British Engineering Standards Association / Report of Meetings of Secretaries at the Offices of the Association* (London: BESA, September 1921), marked "Private and Confidential," archives of the Institute for Civil Engineering, ICE BSI 3.1-33 (1), 5. France, Italy, and Sweden were unable to attend. Czechoslovakia was in the process of forming such an organization. Germany was not listed in the minutes, yet elsewhere Agnew, the secretary of the AESC, reported extensively on what he learned about German standard setting at the meeting (P. G. Agnew, "Industrial Standardization in Europe," *Proceedings of the Second National Standardization Conference Held in Connection with the Twenty-Fourth Annual Conference of the American Mining Congress*, Chicago, October 17–22, 1921, 124–27.).

26. "Second Unofficial Conference of the Secretaries of the National Standardizing Bodies at Baden and Zürich Switzerland July 3rd–6th 1923," Papers of Viscount Weir, University Archives, Glasgow University, collection 96, box 15; Minute #1446 of AESC Executive Committee Meeting, November 12, 1925, ANSI Records.

27. This process may be seen as typical of what Djelic and Quack ("The Power of 'Limited Liability,'" 89–92) call a process of "bottom-up" development of a transnational community: relatively informal "lateral interaction" among people with common goals leads to "mutual learning" followed by periodic "ritualized gatherings" at which the inevitable conflicts of the originally separate groups are worked out and some "broad-based" international aggregation is created. But this process is equally consistent with le Maistre exerting a subtle and strategic leadership without seeming to lead.

28. Undated typescript explaining the background of the photo of le Maistre landing in the United States in April 1926, IEC Records, Box CA-1 20.11.1924-15.11.1926.

29. Le Maistre to Clayton Sharp in New York, February 23, 1926, IEC Records, Box CA-1 20.11.1924-15.11.1926.

30. Mansfield Merriman, "The Work of the International Association for Testing Materials (IATM)" (chairman's address, 2nd Annual Meeting of the American Section, August 15–16, 1899), in *Proceedings of the ASTM* 1, no. 4 (1899): 24.

31. This graph is a Google Ngram of the phrase "standardization movement" (case insensitive) in English-language books scanned by Google Books. Although the selection of books

is not random, there is no reason to think it is skewed in any way that would affect the relative frequency of the phrase.

32. Comfort A. Adams, "National Standards Movement—Its Evolution and Future," *National Standards in a Modern Economy*, ed. Dickson Reck (New York: Harper & Brothers, 1956), 23–24.

33. Quoted in "The Engineer as a Citizen," Items of Interest, *Proceedings of the American Society of Civil Engineers* 45, no. 4 (April 1919): 420.

34. Quoted in "The Engineer as a Citizen." Tanks, a new weapon introduced in World War I, were often depicted in metaphorical cartoons and posters during the war and immediately afterward, such as this one in *Punch*, March 3, 1920, of Winston Churchill rolling over a parliamentary opponent, https://farm8.static.flickr.com/7372/12629410303_bb8cd82eca_b.jpg.

35. Summary of Adams's speech in "Duties of an Engineer in Government Affairs," *Electrical World* 78, no. 13 (March 29, 1919): 648.

36. Winton Higgins, *Engine of Change: Standards Australia Since 1922* (Blackheath, AU: Brandl & Schlesinger, 2005): 39–40.

37. On the antislavery, women's rights, and anti-foot-binding movements, see Margaret E. Keck and Kathryn Sikkink, *Activists beyond Borders: Advocacy Networks in International Politics* (Ithaca, NY: Cornell University Press, 1998), 39–78. On the Red Cross and other early humanitarian movements, see Michael A. Barnett, *Empire of Humanity: A History of Humanitarianism* (Ithaca, NY: Cornell University Press, 2011), 49–95.

38. Cecelia Lynch, *Beyond Appeasement: Interpreting Interwar Peace Movements in World Politics* (Ithaca, NY: Cornell University Press, 1999), 2, 50–57; Craig N. Murphy, *International Organization and Industrial Change: Global Governance since 1850* (New York: Oxford University Press, 1994), 151, 167, 28; Iwao Frederick Ayusawa, "International Labor Legislation," *Columbia University Studies in History, Economics, and Public Law* 91, no. 2 (1920): 15–58, 255. Daniel T. Rodgers, *Atlantic Crossings: Social Politics in the Progressive Age* (Cambridge, MA: Harvard University Press, 1998) emphasizes the shared world of progressive politics on both sides of the North Atlantic from the 1890s through World War II. The engineering movements were both, in one way, larger and, in another, smaller than the world of the movements that are his focus—larger in that the profession of engineering also united people in Latin America, Japan, China, and all of the European colonies, and smaller in that it was limited to a particular professional elite and its allies. Its members, for example, might conceive of themselves as *for* working people, but they certainly were not *of* them.

39. Frank Trentmann, *Free Trade Nation: Commerce, Consumption, and Civil Society in Modern Britain* (Oxford: Oxford University Press, 2008).

40. Charles le Maistre, "Standardization: Its Fundamental Importance to the Prosperity of Our Trade" (Paper read before the North East Coast Institution of Engineers and Shipbuilders on March 24, 1922), reprinted by the order of the council, p. 1, Papers of Viscount Weir, University Archives, Glasgow University, box 15, collection 96.

41. Minute #1356 of AESC Executive Committee Meeting, June 19, 1922, ANSI Records.

42. Minutes of the AESC meeting, August 15, 1919, ANSI Records.

43. "Russian Affairs," 71.

44. Minutes of the AESC, August 15, 1919.

45. IEC, *Publication 30: Report of the Berlin Meeting Held September 1913* (London: IEC, June 1914), 54–59, IEC Records.

46. André Lange, "Charles le Maistre: His Work, the IEC," in *The 1st Charles le Maistre Memorial Lecture* (Geneva: International Electrotechnical Commission, 1955), 3.

47. Conrad Noel, *The Labour Party: What It Is and What It Wants* (London: T. Fisher Unwin, 1906), 176; "Cobden Club," *Liberal Yearbook, Second Year* (London: Liberal Publication

Department, 1906), 14; see the address on the cover of H. G. Wells, *A Reasonable Man's Peace* (London: International Free Trade League, 1917).

48. Jacques Bardoux, *L'Angleterre radical: Essai de psychologie sociale 1906–1913* (Paris: Librairie Félix Alcan, 1913), 27–28.

49. "Police Raids (Enemy Propaganda)," Parl. Deb. H.C., November 26, 1917, vol. 99, cols. 1628–30, accessed June 29, 2017, http://hansard.millbanksystems.com/commons/1917/nov/26/police-raids-enemy-propaganda.

50. Thorstein Veblen, *The Theory of Business Enterprise* (New York: Charles Scribner's Sons, 1904), 23.

51. Veblen, *The Theory of Business Enterprise*, 36.

52. Thorstein Veblen, *The Engineers and the Price System* (New York: B. W. Huebsch, 1921).

53. IEC, *Publication 33: Fourth Plenary Meeting* (London: IEC, October 1919), 8–10, IEC Records.

54. IEC, *Publication 33*, 46–50.

55. Douglas F. Dowd, "Against Decadence: The Work of Robert A. Brady (1901–63)," *Journal of Economic Issues* 28, no. 4 (December 1994): 1031–61.

56. Brady, *Industrial Standardization*, 14–15.

57. Robert A. Brady, *The Rationalization Movement in German Industry: A Study in the Evolution of Economic Planning* (Berkeley: University of California Press, 1933), 4, citing Thorstein Veblen, *Imperial Germany and the Industrial Revolution* (New York: Macmillan, 1915).

58. Brady, *Rationalization Movement*, 12.

59. Brady, *Rationalization Movement*, 12–13.

60. Jeffrey Allan Johnson, "Chemical Engineering and Rationalization in Germany 1919–33," in *Neighbors and Territories: The Evolving Identity of Chemistry, Proceedings of the 6th International Conference on the History of Chemistry*, ed. José Ramón Bertomeu-Sánchez, Duncan Thorburn Burns, and Brigitte Van Tiggelen (Leuven: Mémosciences, 2008), 481.

61. Thomas Wölker, *Entstehung und Entwicklung des Deutschen Normenausschusses 1917 bis 1925* (Berlin: Beuth Verlag, 1992), 244–59. "Ein Wahl Grenzacher: Waldemar Hellmich Erfinder der DIN Norm," *Musée Sentimental de Grenzach-Whylen* (2011), accessed June 30, 2017, http://www.zeitzeugengw.de/ZeitungenMusent/zeitungHellmich.pdf.

62. Waldemar Hellmich, "Zehn Jahre deutsche Normung," *Zeitschrift des Vereines deutscher Ingenieure* 71 (1927): 1526, translated and quoted in Frank Dittmann, "Aspects of the Early History of Cybernetics in Germany," *Transactions of the Newcomen Society* 71, no. 1 (1999): 50.

63. Waldemar Hellmich and Ernst Huhn, *Was will Taylor? Die arbeitssparende Betriebsführung und kritische Bemerkungen über das "Taylorsystem"* (Berlin: VDI, 1919).

64. See "Systematic Bibliography of Works and Articles Recently Published in German on Scientific Management," in Paul Deviant, *Scientific Management in Europe*, International Labour Office Studies and Reports, Series B: Economic Conditions, no. 17 (Geneva: ILO, 1927), 172–210.

65. See John D. McCarthy and Mayer N. Zald, "Resource Mobilization and Social Movements: A Partial Theory," *American Journal of Sociology* 82, no. 6 (May 1977): 1212–41; JoAnne Yates and Craig N. Murphy, "Charles le Maistre: Entrepreneur in International Standardization," *Entreprise et Histoires* 51 (2008): 10–27; Hans Gerhard De Greer, *Rationaliseringsrörelsen i Sverige: Effektivitetsidéer och socialt ansvar under mellankrigstiden* (Stockholm: Studieförb, 1978); Russell, *Open Standards*; Ellis W. Hawley, "Herbert Hoover, the Commerce Secretariat,

and the Vision of an 'Associative State,' 1921–1928," *Journal of American History* 61, no. 1 (June, 1974): 116–40; Lee Vinsel, "Virtue via Association: The National Bureau of Standards, Automobiles, and Political Economy" (paper presented at the Business History Conference Annual Meeting, Frankfurt, March 15, 2014); and Consumer Union, "A Fifteenth Anniversary Report from Consumers Union: Consumer Problems in a Period Of International Tension," *Proceedings of Conference in Cooperation with Vassar Institute for Family and Community Living*, July 27–29, 1951.

66. Sidney Tarrow, "States and Opportunities: The Political Structuring of Social Movements," in *Comparative Perspectives on Social Movement: Political Opportunities, Mobilizing Structures and Cultural Framings*, ed. Doug McAdam, John D. McCarthy, and Mayer N. Zald (Cambridge: Cambridge University Press, 1996) is an important milestone in the analysis of political opportunity structures and social movements.

67. Trentmann, *Free Trade Nation*, 259; George H. Nash, *The Life of Herbert Hoover: The Humanitarian, 1914–1917* (New York: W. W. Norton, 1988).

68. Trentmann, *Free Trade Nation*, 260.

69. Murphy, *International Organization*, 160–61.

70. Ellis Hawley, "Three Facets of Hooverian Associationalism: Lumber, Aviation, and Movies, 1921–1930," in *Regulation in Perspective: Historical Essays*, ed. Thomas K. McCraw and Morton Keller (Boston: Harvard University Graduate School of Business Administration, 1981), 99, 104.

71. See Brady, *Rationalization Movement*; Mauro F. Guillén, *Models of Management: Work, Authority, and Organization in a Comparative Perspective* (Chicago: University of Chicago Press, 1994); De Greer, *Rationaliseringsrörelsen i Sverige.*

72. The current popularity of the term originated in the management school field of corporate strategy. Klaus Schwab and Hein Kroos, *Moderne Unternehmensführung im Maschinenbau* (Frankfurt: Maschinenbau-Verlag, 1971); and R. Edward Freeman *Strategic Management: A Stakeholder Approach* (Boston: Pitman, 1984) were important early works. As we have seen, early German standardizers did use *Interesessengruppen*, which was the term originally used by Schwab and Kroos (see chap. 9). The term is now often translated as "stakeholders."

73. Charles le Maistre, "Summary of the Work of the British Engineering Standards Association," *Annals of the American Academy of Political and Social Science* 82 (March 1919): 247–248.

74. BESA, *Unofficial Conference*, 19.

75. Stevenson, "American Engineering Standards Committee," 120–21.

76. Kaare Heidelberg, "Die ISA (International Federation of Standardizing Associations) 1926–1939," *DIN-Mitteilungen: Zentralorgan der Deutschen Normung* 56, no. 3 (1977): 135; and Thomas Wölker, "Der Wettlauf um die Verbreitung nationaler Normen in Ausland nach dem Ersten Weltkrieg und die Gründung der ISA aus der Sicht deutscher Quellen," *Vierteljahrschrift für Sozial- und Wirtschaftsgeschichte* 80, no. 4 (1993): 495.

77. Le Maistre to Guido Semenza, March 24, 1924, IEC Records, Box CA-1 20.11.1924-15.11.1926.

78. IEC Committee of Action, *RM 7: Proces Verbaux de Reunions* (IEC, April 28, 1924), IEC Records.

79. IEC Committee of Action, *RM 8: Proces Verbaux de Reunions* (IEC, July 17, 1924), IEC Records.

80. Le Maistre to Karl Strecker, January 1, 1925, IEC Records, Box CA-1 20.11.1924-15.11.1926.

81. M. Kloss to Charles le Maistre, March 8, 1925, document marked "Translation of letter from Dr. Kloss," IEC Records, Box CA-1 20.11.1924-15.11.1926.

82. Nalle Sturén, interview by Maria Nassén, January 13, 2008.

83. Le Maistre, "Summary of the Work," 252.

84. Ian Stewart, *Standardization Association of Australia Monthly Information Sheet*, April 1977, 4, quoted in Higgins, *Engine of Change*, 144; see 126 for background on Stewart.

85. Higgins (*Engine of Change*, 141) notes that Stewart's argument about the "legitimacy and superiority of decisions hammered out in uninhibited discussions between equals who represent a variety of interests" is often associated with Jürgen Habermas.

86. Ivar Herlitz, "The IEC, Yesterday, Today, and Tomorrow," in *Eighth Charles le Maistre Memorial Lecture* (Geneva: Central Office of the IEC, 1962), 17.

87. Jürgen Habermas, *The Inclusion of the Other: Studies in Political Theory* (Cambridge, MA: MIT Press, 1998); Jane Mansbridge et al., "The Place of Self-Interest and the Role of Power in Deliberative Democracy," *Journal of Political Philosophy* 18, no. 1 (2010): 64–100.

88. Jürg Steiner, *The Foundations of Deliberative Democracy: Empirical Research and Normative Implications* (Cambridge: Cambridge University Press, 2012), 268–71.

89. These are things that sociologist Jill Kielcolt has described as typical of social movements whose members develop a strong, lasting identification with the cause. K. Jill Kiecolt, "Self-Change in Social Movements," in *Self, Identity, and Social Movements*, ed. Sheldon Stryker et al. (Minneapolis: University of Minnesota Press, 2000): 125–26.

90. Both game theorists and students of deliberative decision making identify those situations as ones in which difficult-to-solve "cooperation" problems become much-easier-to-solve "coordination" problems; when properly channeled, uncertainty about one's own interests can make people "nice." Randall L. Calvert, "Leadership and Its Basis in Problems of Social Coordination," *International Political Science Review* 13, no. 1 (1992): 7–24; Thomas Risse, "'Let's Argue': Communicative Action in World Politics," *International Organization* 54, no. 1 (2000): 1–39.

91. Charles le Maistre, "The Effect of Standardisation on Engineering Progress" (paper read before the Royal Society of Arts, February 4, 1931), 3, Papers of Viscount Weir, University Archives, Glasgow University, box 15, collection 96.

92. IEC, "Meeting of the Committee on Screw Caps and Holders, Geneva, November 25, 1922," in *RM 4: Proces Verbaux de Reunions* (IEC), IEC Records.

93. IEC, "Meeting of the Committee on Lamp Sockets, Koninklijk Instituut van Ingenieurs, The Hague, April 17, 1925," in *RM 18: Proces Verbaux de Reunions* (IEC), IEC Records.

94. Percy Good, "Electrical Standardization, 1929–1930," *Journal of the Institution of Electrical Engineers* 69, no. 411 (March 1931): 404–13; US Department of Commerce, "International Electrotechnical Commission," in *International Standards Yearbook* (Washington, DC: US Department of Commerce, Bureau of Standards Miscellaneous Publication, 1932), 77.

95. Mansbridge et al., "Deliberative Democracy," 74.

96. Although the standardization movement was committed to a kind of technocracy, it should not be confused with the short-lived and ultimately relatively inconsequential US "technocracy movement," the brainchild of Columbia University industrial engineer Walter Rautenstrauch and his flamboyant but untrained colleague, Howard Scott, who championed a caricature of Veblen's views, in the early 1930s. See Donald R. Stabile, "Veblen and the Political Economy of the Engineer," *American Journal of Economics and Sociology* 45, no. 1 (January 1986): 51; and David Adair, "The Technocrats, 1919–1967: A Case Study of Conflict and Change in a Social Movement" (MA Thesis, Department of Political Science, Sociology, and Anthropology, Simon Fraser University, January 1970), 32–61.

97. The founding meetings of AESC focused on the lack of uniformity in the rules that they followed: "At the present time many bodies are engaged in the formulation of standards.

There is no uniformity in the rules for such procedure in the different organizations; in some cases the committees engaged in the work are not fully representative; and in a considerable proportion of cases they do not consult all the allied interests." Minutes of AESC Meeting, October 19, 1918, ANSI Records. The major exception from a strict hierarchy in the United States was, of course, ASTM, which did follow these rules.

98. Harriman, *Standards and Standardization*, 177.

99. Brady, *Industrial Standardization*, 100–123.

100. For example, ASTM's official history reports as the society's first "major contribution" to the World War II effort its 1942 publication of "more than 1,000 standard specifications available to industry and government. Since more than half of these were purchase specifications, they could be written directly into tens of thousands of government contracts for war-essential materials." ASTM, *ASTM 1898–1998: A Century of Progress* (West Conshohocken, PA: ASTM International, 1998), 41.

101. Brady, *Rationalization Movement*, 12–13.

102. Harriman, *Standards and Standardization*, 178–79.

103. Harriman, *Standards and Standardization*, 75.

104. SIS, *Standardiseringen i Sverige, 1922–1992* (Stockholm: SIS, 1993), 29.

105. SIS, *Standardiseringen i Sverige*, 30.

106. Timothy W. Luke, *Ideology and Soviet Industrialization* (Westport, CT: Greenwood Press, 1985), 123–24; Brady, *Industrial Standardization*,122.

107. Herbert Hinnenthal, "The 'Reichskuratorium für Wirtschaftlichkeit' (RKW)," *Commercial Standards Monthly* 7, no. 5 (November 1930): 155–56. Mary Nolan states, "The term 'rationalization movement' was coined by Herbert Hinnenthal, the first business manager of the RKF, and the contemporary American scholar Robert Brady used it in the title of his book. The term captures not only the comprehensiveness of the efforts undertaken but also the ideological and emotional commitment of so many to the theory and practice of rationalization." Mary Nolan, *Visions of Modernity: American Business and the Modernization of Germany* (New York: Oxford University Press, 1994), 274n10.

108. J. Ronald Shearer, "The Reichskuratorium für Wirtschaftlichkeit: Fordism and Organized Capitalism in Germany, 1918–1945," *Business History Review* 71, no. 4 (Winter 1997): 569–602.

109. Olle Sturén, "Toward Global Acceptance of International Standards" (speech given at the National Bureau of Standards in Washington, DC, June 1972), 2, Sturén Papers.

110. Oskar E. Wikander quoted in "Introducing Industrial Standards," *Comments on the Argentine Trade* 2, no. 4 (November 1922): 21.

111. Wölker, "Der Wettlauf um die Verbreitung nationaler Normen," 490.

112. [Charles le Maistre], "Memorandum in regard to the Work of the British Engineering Standards Association in Furtherance of British Export Trade" [1925], ICE Holdings of the BSI, formerly in the Science Museum, Part 3, Envelope 5.

113. Charles le Maistre, "Industrial Standardisation and Simplification: Report of Lecture Delivered at the Twenty-Sixth Conference for Works Directors, Managers, Foremen etc." (held at Balliol College, Oxford, April 19–23, 1928), quoted by Robert C. McWilliam, "The Evolution of British Standards" (PhD thesis, University of Reading, 2002), 204.

114. Russell, *Open Standards*, 87–88.

115. Minute #1844 of AESC Executive Committee Meeting, July 21, 1927, ANSI Records.

116. "ASA Growth Curves," *Industrial Standardization* 14, no. 12 (December 1943): 316.

117. Durand, *AFNOR*, 26.

118. Durand, *AFNOR*, 26.

119. Strecker to Charles le Maistre, December 16, 1926, Swiss Federal Archives, E2001D #1000/1553#286, Huber-Ruf, Alfred, Ing., Bern, 1940–1945 (Dossier), Topic "Organization, Standards, Rationalization."

120. Huber-Ruf's resume covering 1906–1941 is attached to a letter from Alfred Huber-Ruf to "Direktor Lusser," December 11, 1941, Swiss Federal Archives, E8190(A)1981/#196, Eidgenössisches Elecktishes Amt. Huber-Ruf, Alfred, Ing.

121. Semenza to Charles le Maistre, n.d., the sequence in the box suggests early November 1926, IEC Records, Box CA-1 20.11.1924-15.11.1926.

122. Le Maistre to Guido Semenza, November 8, 1926, IEC Records, Box CA-1 20.11.1924-15.11.1926.

123. The most extensive record of an AESC Executive Committee debate about the options for improving the working of "the organization or movement as a whole" is in Minute #1793 of AESC Executive Committee Meeting, May 19, 1927, ANSI Records.

124. P. G. Agnew, "Development of the ASA," *Industrial Standardization* 14, no. 12 (December 1943): 328.

125. Minute #1657 of AESC Executive Committee Meeting, October 14, 1926, ANSI Records.

126. Minute #1657 of AESC Executive Committee Meeting.

127. Minute #2045 of AESC Executive Committee Meeting, October 12, 1928, ANSI Records. Huber-Ruf's full-time employer and its address are on his resume (attached to letter from Alfred Huber-Ruf to "Direktor Lusser," December 11, 1941, Swiss Federal Archives). The distance was calculated using Google Maps. The records on Huber-Ruf in the Swiss Federal Archives also make it clear that, at least from 1939 onward, he ran the ISA from his home.

128. Minute #2168 of the ASA Board of Directors, October 16, 1929, ANSI Records (note: after the change in name and organization of the AESC, the minutes of the old AESC Executive Committee continue in sequence as the minutes of the ASA Board of Directors).

129. McWilliam, "The Evolution of British Standards," 209. In 1937, Great Britain attended the Paris ISA meeting as a member, but Australia, Canada, and New Zealand did not. "ASA Represents American Industry at International Meetings," *Industrial Standardization and Commercial Standards Monthly* 8, no. 9 (September 1937): 241. Although not members until then, they had previously participated in specific committees.

130. Durand, *AFNOR*, 30; Wölker, "Der Wettlauf um die Verbreitung nationaler Normen in Ausland."

131. Le Maistre to Clayton Sharp, November 17, 1927, IEC Records, Box CA-1 20.11.1924-15.11.1926.

132. The depth of the displeasure of German standardizers, including Hellmich, with what they considered to be the inch countries' efforts to protect their markets is demonstrated in Wölker, "Der Wettlauf um die Verbreitung nationaler Normen in Ausland."

133. Minute #1657 of AESC Executive Committee Meeting.

134. Minute #1717 of AESC Executive Committee Meeting, February 3, 1927, ANSI Records.

135. "USNC Standardization Work Reorganized," *ASA Bulletin* 3, no. 6 (June 1932): 199.

136. "Overview of ITU's History," ITU, accessed June 12, 2017, http://www.itu.int/en/history/Pages/ITUsHistory.aspx. CCIF and CCIT were merged into the CCITT in 1956, and today are the ITU-T. Today's equivalent of the CCIR is the ITU-R, or ITU Radiocommunication Sector.

137. Since the 1990s, with the rise of electronic mail handling, the Universal Postal Union has maintained a similar UPU Standards Board (Universal Postal Union, "General Information on UPU Standards," Bern, June 28, 2017).

138. IEEE EMC Society, *50 Years of Electromagnetic Compatibility: The IEEE Electromagnetic Compatibility Society and Its Technologies, 1957–2007* (Piscataway, NJ: IEEE EMC Society, 2007).

139. "International Radiotelegraph Conference (Berlin, 1906)," ITU, accessed August 14, 2017, http://www.itu.int/en/history/Pages/RadioConferences.aspx?conf=4.36.

140. "International Radiotelegraph Convention of Washington, 1927," ITU, accessed August 14, 2017, http://search.itu.int/history/HistoryDigitalCollectionDocLibrary/5.20.61.en.100.pdf.

141. International Radiotelegraph Convention of Washington, 1927, *General Regulations Annexed to the International Radiotelegraph Convention and General and Supplementary Regulations, Washington, November 25, 1927* (London: His Majesty's Stationery Office, 1928), accessed August 14, 2017, http://www.itu.int/dms_pub/itu-s/oth/02/01/S02010000144002PDFE.pdf.

142. John Braithwaite and Peter Drahos, *Global Business Regulation* (Cambridge: Cambridge University Press, 2000), 329.

143. International Radiotelegraph Convention, *General Regulations*, 111–12.

144. George Valensi, *The First Five Years of the International Advisory Committee for Long-Distance Telephone Communications* (Berlin: Verlag Europäischer Fernsprechdienst, 1929), 5–6.

145. Ernest K. Smith, "The History of the ITU, with Particular Attention to the CCIT and CCIR, and the Latter's Relation to URSI," *Radio Science* 11, no. 6 (June 1976): 499–500; see also "International Radiotelegraph Conference (Madrid, 1932)," ITU, accessed August 14, 2017, http://www.itu.int/en/history/Pages/RadioConferences.aspx?conf=4.41.

146. Vol. 1 of *World Engineering Congress Tokyo 1929 Proceedings: General Reports* (Tokyo: Kogakki, 1931). The data are from the front matter, vii, while the quotation is from "Sectional Meetings, Section XII Scientific Management," 47.

147. *WEC Proceedings*, 1:1.

148. Stanislav Špacek, "History of and Proposition for the Foundation of a World Engineers' Federation," in *WEC Proceedings*, 2:4.

149. *WEC Proceedings*, 2:2.

150. Tadashiro Inouye, "The Engineer as a Factor in International Relations," in *WEC Proceedings*, 2:17.

151. Inouye, "The Engineer as a Factor in International Relations," 2:18–19.

152. A. Huber-Ruf, "ISA: International Federation of the National Standardising Associations," in *WEC Proceedings*, 2:26.

153. Huber-Ruf wrote, "The question of giving preference to one particular system should by no means be raised in connection with standardisation work" ("ISA," 2:26).

154. Charles E. Skinner, "Standardization," in *WEC Proceedings*, 2:31.

155. F. A. E. Neuhaus, "Die Normung in Deutschland," and S. Konishi, "On Engineering Standardization in Japan," in *WEC Proceedings*, 2:45–78.

156. John Hays Hammond, "International Cooperation of Engineers," in *Abstracts of Papers to Be Read at World Engineering Congress* (Tokyo: Kogakki, 1929), 491.

157. Axel Enström, "The Engineer's Profession," *WEC Proceedings*, 2:123.

158. Johannes-Geert Hagmann, "Ambassadors of the 'Fifth Estate': The American Venture in the World Engineering Congress 1925–1929" (paper presented at the annual meeting of the Society for the History of Technology, Singapore, June 22–26, 2016).

159. *Commercial Standards Monthly* 7, no. 9 (March 1931).

160. I. Guttmann, "Short Courses in Standardization for Soviet Russia," *ASA Bulletin* 3, no. 6 (June 1932): 187–88.

161. José Luciano Dias, *História da Normalização Brasileira* (São Paulo: ABNT, 2011), 47–53. Information on the other Latin American bodies appears in "Anniversary Messages from Other Countries," *Industrial Standardization* 14, no. 12 (December 1943): 333.

Chapter 4 · *Decline and Revival of the Movement, the 1930s to the 1950s*

1. Listed at the back of *ISA Bulletin*, no. 29 (November 1940).

2. Willy Kuert, "The Founding of ISO," in *Friendship Among Equals: Recollections from ISO's First Fifty Years*, ed. Jack Latimer (Geneva: ISO, 1997), 15.

3. *ISA Bulletin No. 6: Conversion Tables: Inches-Millimeters* (August 1934).

4. *ISA Bulletin No. 7: Paper Sizes* (August 1934). On the US view, see John Gaillard, "A System of Paper Sizes as Developed in Europe," *Industrial Standardization* 3, no. 7 (July 1932): 201–8.

5. Howard Coonley, "The International Standards Movement," in *National Standards in a Modern Economy*, ed. Dickson Reck (New York: Harper, 1956), 39. The standard is first reported in *ISA Bulletin No. 16: Sound Film 16mm* (May 1938); Harvey Fletcher, "International Agreement Determines Standard Noise Measurement Units," *Industrial Standardization and Commercial Standards Monthly* 9, no. 1 (January 1938): 18–20. W. H. Martin, "Decibel—the Name for the Transmission Unit," *Journal of the AIEE* 48, no. 3 (March 1929): 223.

6. *ISA Bulletin 30: Symbols for Magnitudes and Units* (November 1940). See "nano" on the Sizes Inc. website, accessed July 12, 2017, http://sizes.com/units/nano.htm.

7. IEC Committee of Action, *RM 177* (Paris: IEC June 28, 1939).

8. Charles le Maistre, "The Effect of Standardisation on Engineering Progress" (paper read before the Royal Society of Arts, February 4, 1931), 3–4, Papers of Viscount Weir, University Archives, Glasgow University, box 15, collection 96.

9. Le Maistre, "The Effect of Standardisation," 3–4.

10. Le Maistre, "The Effect of Standardisation," 3–4.

11. Hans Gerhard De Greer, *Rationaliseringsrörelsen i Sverige: Effektivitetsidéer och socialt ansvar under mellankrigstiden* (Stockholm: Studieförb, 1978), 360.

12. SIS, *Standardiseringen i Sverige, 1922–1992* (Stockholm: SIS, 1993), 36.

13. "ASA Growth Curves," *Industrial Standardization* 14, no. 12 (December 1943): 316.

14. P. G. Agnew, "Standards in Our Social Order," *Industrial Standardization* 11, no. 6 (June 1940): 146.

15. The most comprehensive study that has made this case with reference to the United States is Robert J. Gordon, *The Rise and Fall of American Growth: The US Standard of Living since the Civil War* (Princeton, NJ: Princeton University Press, 2016). Gordon cites the role of electrical standardization in the generalization of the nineteenth-century electrical inventions to almost every American home in the 1920s through the 1940s (121–22) and tells specific standardization stories in other industries throughout his book, ending with a discussion in his conclusion about the "prosaic but very important" role of standardization in "the Great Leap Forward" in the American standard of living from the 1920s through the 1950s (561–62).

16. "ASA Growth Curves," 316.

17. "ASA Growth Curves," 316.

18. In July 1943 (vol. 14, no. 7) the title became simply *Industrial Standardization* and reference to federal cooperation was dropped from the masthead.

19. Alain Durand, *AFNOR: 80 Années d'Histoire* (Paris: AFNOR Éditions, 2008), 18, 76–79.

20. Robert C. McWilliam, "The Evolution of British Standards" (PhD thesis, University of Reading, 2002), 109–11, 204.

21. McWilliam, "Evolution of British Standards," 109–11.

22. Robert A. Brady, *Business as a System of Power* (New York: Columbia University Press, 1943), 153, 240, quoted in McWilliam, "The Evolution of British Standards," 110, 204.

23. McWilliam, "The Evolution of British Standards," 203–5.

24. Debate about the role of the American tariff in the severity and length of the global depression continues. Douglas A. Irwin, *Peddling Protection: Smoot Hawley and the Great Depression* (Princeton, NJ: Princeton University Press, 2011) concludes that it triggered retaliation and contributed to the remarkable decline in world trade in the 1930s and to discrimination against US goods that lasted for decades.

25. Charles le Maistre, "Empire Trade Requires Uniform Standards" (address to the Empire Club of Canada, Toronto, April 29, 1932), accessed July 5, 2017, http://speeches.empireclub.org/61016/data?n=2.

26. Charles le Maistre, *Director's Report on His Visit to the Argentine (1936)* (London: British Standards Institution), CE(OC)2700, 1937, ICE Holdings of the BSI, 389.6. Percy Good, *Standardisation and Certification: Deputy Director's Report on his visit to Australia and New Zealand (Canada and USA)* (London: British Standards Institution), CF(OC) 2067, March 1939, ICE Holdings of the BSI.

27. McWilliam, "The Evolution of British Standards," 205.

28. Winton Higgins, *Engine of Change: Standards Australia Since 1922* (Blackheath, AU: Brandl & Schlesinger, 2005), 345.

29. IEC Committee of Action, *RM 176* (Torquay: IEC, June 29, 1938).

30. The list of ISA standards, memorabilia from Huber-Ruf's visit to Mussolini, and his 1940 commendation from DIN are in the Swiss Federal Archives, in the Huber-Ruf—Organization, Standards, Rationalization collection. They were included by Huber-Ruf in letters written in late 1941 and early 1942 to heads of Swiss government agencies about the importance of standardization and his own significance in the international movement.

31. Alexander L. Bieri, *Traditionally Ahead of Our Time* (Basel: Roche Historical Archive, 2016), 27. Hellmich retained the title of curator of DIN and acted as a distant *éminence grise* for many on the DIN staff in Berlin (Peter Anthony, archivist of DIN, email to C. Murphy, June 15, 2016).

32. Bieri, *Traditionally Ahead of Our Time*, 26.

33. Durand, *AFNOR*, 75–79.

34. Arild Sæle, Erik Sundt, and Kay E. Fjørtoft, "Overview of Standardisation Processes and Standardisation Organisations" (Norwegian Maritime Technology Research Institute, Trondheim, February 1, 2002), 11.

35. Minute #3378 of ASA Board of Directors meeting, March 26, 1941, records in the Headquarters of the American National Standards Institute [subsequently referred to as ANSI Records], New York. It is unclear whether Huber-Ruf received any payment. In December 1941, Huber-Ruf's support from the ISA ended, and he sought employment in the Swiss Federal Electricity Commission. Letters exchanged between Huber-Ruf and Direkter Lusser, December 1–December 20, 1941, Swiss Federal Archives, E8190(A)1981/#196, Eidgenössisches Elecktishes Amt. Huber-Ruf, Alfred, Ing.

36. Minute #3384 of ASA Standards Council Meeting, April 10, 1941, ANSI Records.

37. Minute #3410 of ASA Standards Council Meeting, April 10, 1941, ANSI Records.

38. Minute #3438 of ASA Standards Council Meeting, September 18, 1941, ANSI Records.

39. Minutes #3438 and #3448 of ASA Standards Council Meeting, September 18, 1941, ANSI Records.

40. P. G. Agnew, "Legal Aspects of Standardization and Simplification: A Discussion from the Point of the Lay Worker," *Industrial Standardization* 12, no. 10 (October 1941): 260.

41. H. S. Osborne, "Events of the Year," *Industrial Standardization* 14, no. 12 (December 1943): 337.

42. "ASA Growth Curves," 316.

43. Charles le Maistre, "Wartime Standardization" (address to the North East Coast Institution of Engineers and Shipbuilders at Newcastle-upon-Tyne, January 8, 1943), ICE Holdings of the BSI, 3.3, 2.

44. Le Maistre, "Wartime Standardization."

45. P. G. Agnew, "War-Time Methods of the ASA," *Industrial Standardization* 14, no. 12 (December 1943): 347.

46. Ralph E. Flanders, "How Big Is an Inch?," *Atlantic Monthly*, January 1951, 48.

47. Minute #3709, item 9, of the ASA Standards Council Meeting, September 14, 1944, ANSI Records.

48. "New Foreign Standards Now in ASA Library," *Industrial Standardization* 14, no. 12 (December 1943): 355.

49. "War-Jobs 1943," *Industrial Standardization* 14, no. 12 (December 1943): 348–50. Percy Good's comment appears on "BS/ARP Secret Specification for the Lighting Perimeters of Internment and Prisoners of War Camps," n.d., ICE Holdings of the BSI, 3.4 (3).

50. Flanders, "How Big Is an Inch?," 44.

51. Flanders, "How Big Is an Inch?," 44–45.

52. "UNSCC," *Economist* 148 (March 3, 1945): 286.

53. Minute #3410 of ASA Standards Council Meeting, April 10, 1941.

54. Minute #3674 of ASA Standards Council Meeting, May18, 1944, ANSI Records.

55. Minute #3674 of ASA Standards Council Meeting, May18, 1944.

56. Minute #3707 of ASA Standards Council Meeting, September 14, 1944, ANSI Records.

57. Kuert, "The Founding of ISO," 16. The final list of members comes from Coonley, "International Standards Movement," 39.

58. Charles le Maistre, form letter inviting standards organizations to join UNSCC, n.d. [content indicates it was written in winter 1944–45 or spring 1945], Early records of the ISO and its predecessors, ISO Headquarters, Geneva. Technically, the Soviet Union had not joined the organization at its first meeting, but the UNSCC constitution included representatives of the Soviet Union's national standards body on its Executive Committee (along with representatives of those in Britain, Canada, and the United States), and the constitution explained that members of that committee need not necessarily be from standards bodies that were members of the UNSCC. "UNSCC," 287.

59. Minute #3758 of ASA Standards Council Meeting, December 8, 1945, ANSI Records.

60. Le Maistre, form letter, reports that he had resigned from his chairmanship of the Executive Committee of BSI in April 1944. The letterhead provides his new address. "Obituary: Charles le Maistre, C.B.E.," *Engineer* 196, no. 5086 (July 17, 1953): 81, suggests he resigned as secretary of BSI in 1942.

61. "UNSCC," 286–87.

62. ASA, for example, agreed to work on radio interference, shellac, and the testing of textiles (Minute #3795 of ASA Board of Directors Meeting, May 25, 1945, ANSI Records), and during the period in which the UNSCC Executive Committee handed out these assignments (July–October 1944) it also considered setting standards for flat-bottomed rails, airfield lighting, gas cylinders, building materials and equipment, food containers, and terms used relative to the plastics industry. UNSCC, "November 1944 Report on Progress," Early records of the ISO and its predecessors, ISO Headquarters, Geneva.

63. On Wollner's work at the Treasury, see H. J. Wollner et al., "Isolation of a Physiologically Active Tetrahydrocannabinol from Cannabis Sativa Resin," *Journal of the American Chemical Society* 64, no. 1 (1943): 26–29.

64. Minute #3758 of ASA Standards Council Meeting, December 8, 1945.

65. International Business Conference, foreword to *Final Reports of the International Business Conference, Westchester Country Club, Rye, N.Y., November 10–18, 1944* (New York: International Business Conference, 1944), 1.

66. Luther H. Hodges, "Hints for Headline Readers: We Aren't Waiting This Time to Plan for Peace after War Has Been Won," *Rotarian* (April 1945): 11.

67. Craig N. Murphy, *International Organization and Industrial Change: Global Governance since 1850* (New York: Oxford University Press, 1994), 160–61.

68. Sections 2, 5, and 7 of International Business Conference, *Final Reports of the International Business Conference.*

69. On Ryan's role in his company and BSI, see "New Members [of the Royal Commission of Awards to Inventors]," *Glasgow Herald*, December 22, 1952, 8.

70. Le Maistre, form letter, 2. "I.C.I." refers to the International Commission on Illumination, an organization established in 1913 to settle technical questions about the measurement of light from gas lamps. It became operational only in the 1920s and extended its interest to electric lighting, as well. By 1935, it worked by studying various questions, such as the definition of colors, through technical committees. Even though the reports of these committees were treated as de facto international standards, the organization did not consider itself a standard-setting body until the 1980s. See Martina Paul, "Nearly 100 Years of Service—CIE's Contribution to International Standardization," *ISO Focus* (May 2009): 17–19.

71. "UNSCC Meeting of 8.9.10 & 11 October 1945 in New York, List of Participants," handwritten document, handwriting appears to be that of Charles le Maistre. Early records of the ISO and its predecessors, ISO Headquarters, Geneva. Minute #3851 of ASA Standards Council Meeting, December 7, 1945, ANSI Records, reports the presence of New Zealand, Norway, and the Soviet Union, which are not included on the handwritten document. The Swiss are reported as speaking on October 8 in UNSCC, "Proceedings of the New York Meeting," October 8–11, 1945. Early records of ISO and its predecessors, ISO Headquarters, Geneva.

72. Minute #3808 of ASA Standards Council Meeting, September 27, 1945, ANSI Records; and UNSCC, "Proceedings."

73. UNSCC, "Proceedings"; and Minute #3851 of ASA Standards Council Meeting, December 7, 1945, which reports the entertainments.

74. UNSCC, "Proceedings," 9.

75. UNSCC, "Proceedings," 9.

76. "Founding Member States," United Nations, accessed July 6, 2017, http://www.un.org/depts/dhl/unms/founders.shtml.

77. UNSCC, "Proceedings," 19–33, 60–74.

78. UNSCC, "Proceedings," 74.

79. UNSCC, "Proceedings," 210–22, 283.

80. UNSCC, "Proceedings," 107–18.

81. UNSCC, "Proceedings," 47–52.

82. Minute #3851 of ASA Standards Council Meeting, December 7, 1945.

83. Minute #3896 of ASA Standards Council Meeting, April 25, 1946, ANSI Records.

84. The quotation is from Minute #3896 of ASA Standards Council Meeting, April 25, 1946. On Huber-Ruf's employment from 1941 to 1949, see Swiss Federal Archives, E8190(A)1981/#196, Eidgenössisches Elecktishes Amt. Huber-Ruf, Alfred, Ing.

85. Kuert, "The Founding of ISO," 17. Huber-Ruf's employment records in the Swiss Federal Archives (see note 84) suggest that illness may not have been the only problem. In December 1945 he applied for three days' leave in summer 1946 to take care of ISA business. There is no record that that request was granted.

86. JoAnne Yates and Craig N. Murphy, "From Setting National Standards to Coordinating International Standards: The Formation of the ISO," *Business and Economic History* 4 (2006): 20.

87. IEC Council, *RM 179* (Paris: IEC, July 1, 1946), 7.

88. IEC Council, *RM 179*, 7.

89. UNSCC, *Report of Conference of the United Nations Standards Co-ordinating Committee together with Delegates from Certain Other National Standards Bodies* (London, October 14–26, 1946), 9–10.

90. UNSCC, *Report*, 35.

91. UNSCC, *Report*, 6.

92. Minute #3990 of ASA Standards Council Meeting, November 21, 1946, ANSI Records.

93. Minute #3990 of ASA Standards Council Meeting, November 21, 1946.

94. Huber-Ruf's employment records (see note 84) contain a December 17, 1946, request to work at home one day because of ISA business, the last time the ISA is mentioned in these records. The records contain three more years of unsuccessful attempts by Huber-Ruf to obtain a pension from the Swiss government, including letters from attorneys, none of which mention funds from the ISA. Huber-Ruf's "wartime" employment ended in August 1947, but he continued to be employed on occasional temporary government contracts until December 1949, when the records end.

95. IEC Committee of Action, *RM 183* (Brussels: IEC, October 28, 1947), 6, IEC Records.

96. Roger Maréchal, "We Had Some Good Times," in *Friendship among Equals: Recollections from ISO's First Fifty Years*, compiled by Jack Lattimer (Geneva: ISO, 1997), 30.

97. Maréchal, "We Had Some Good Times," 25–26.

98. UN Economic and Social Council, "List of Non-governmental Organizations in Consultative Status with the Economic and Social Council as of 1 September 2014," E/2014/INF/5, 4–7.

99. Lino Camprubi, *Engineers and the Making of the Francoist Regime* (Cambridge, MA: MIT Press, 2014), 147.

100. IEC, *RM 183*, 2.

101. IEC, *RM 183*, 3.

102. Percy Good to Dr. Frank, Deutscher Normenauschuss, March 13, 1947, in *Packet Prepared for Mr. Cooke for a Trip to Germany in 1947*, ICE BSI 33.3 (2), ICE Holdings of the BSI. The previous letter is from Percy Good to E. G. Lewin, Esq., March 13, 1947, same packet.

103. Letter marked "confidential" from E. G. Lewin, H. Q. Control Commission for Germany (British Element) to J. O. Cooke, Esq., British Standards Institution," March 22, 1947, in *Packet Prepared for Mr. Cooke for a Trip to Germany in 1947*.

104. Lewin to Cooke, March 22, 1947.

105. Konrad H. Jarausch, *The Unfree Professions: German Lawyers, Teachers, and Engineers, 1900–1950* (New York: Oxford University Press, 1990), 209–10.

106. Thomas P. Hughes, "Elmer Sperry and Adrian Leverkühn: A Comparison of Creative Styles," in *Springs of Scientific Creativity: Essays on Founders of Modern Science*, ed. Rutherford Aris, Howard Ted Davis, and Roger H. Stuewer (Minneapolis: University of Minnesota Press, 1983), 201.

107. Jarausch, *The Unfree Professions*, 209–10.

108. Jarausch, *The Unfree Professions*, 210–11.

109. Letters reporting votes in Box: APPLICATIONS FOR MEMBERSHIP (by country), File: "Germany DIN 1951," IEC Records.

110. Richard Vieweg, "Measuring—Standardizing—Producing," in *Fourth Charles le Maistre Memorial Lecture* [Stockholm, July 10, 1958] (Geneva: IEC, 1958), 5.

111. "Obituary: Charles le Maistre, C.B.E.," 81. Photographs of Lea Gate House can be found in the Royal Institute of British Architects image library, accessed August 31, 2018, https://www.architecture.com/image-library/ribapix.html?keywords=lea%20gate%20house%20bramley.

112. Vieweg, "Measuring—Standardizing—Producing," 5.

113. J. F. Stanley to Mr. Ruppert, April 11, 1957, in Box: APPLICATIONS FOR MEMBERSHIP (by country), File: "People's Republic of China 1957," IEC Records.

114. Count of letters in Box: APPLICATIONS FOR MEMBERSHIP (by country), File: "People's Republic of China 1957," IEC Records.

115. ISO, undated and unsigned "Draft of Letter to UN Secretary-General," to Mr. Trygve Lie [served February 2, 1946 through February 1, 1951]; and "Statement by the International Organization of Standardization Regarding Coordinating of the Activities of the United Nations Organs and Agencies in the Sphere of Standardization," 30, Early records of the ISO and its predecessors, ISO Headquarters, Geneva.

116. IEC Committee of Action, *RM 243* (Estoril: IEC, July 10, 1951), 12.

117. ISO, "Draft of Letter"; and "Statement," 26, Early records of the ISO and its predecessors, ISO Headquarters, Geneva.; Johan Schot and Frank Schipper, "Experts and European Transport Integration, 1945–1958," *Journal of European Public Policy* 18, no. 2 (2011): 274–93.

118. Craig N. Murphy, *The United Nations Development Programme: A Better Way?* (Cambridge: Cambridge University Press, 2006), 88–102.

119. Olle Sturén, typescript, with handwritten corrections, speech to the ISO Council, 1986, 1, Sturén Papers.

120. Mohammed Hayath, "What the IEC Means to the Developing Countries," in *Seventh Charles le Maistre Memorial Lecture* [Interlaken, June 19, 1961] (Geneva: IEC, 1961), 20.

121. Hayath, "What the IEC Means to the Developing Countries," 14–15.

Chapter 5 · Standards for a Global Market, the 1960s to the 1980s

1. Olle Sturén, typescript, with handwritten corrections, speech to the ISO Council, 1986, 1, Sturén Papers.

2. Sturén, typescript; and Nalle Sturén, interview by Maria Nassén, January 13, 2008.

3. Lars and Lolo Sturén [Olle Sturén's sons], interview by Maria Nassén, August 20, 2007.

4. A. Scott Henderson, *Housing and the Democratic Ideal: The Life and Thought of Charles Abrams* (New York: Columbia University Press, 2000), 181.

5. The photograph, clipped from the paper, is in the folder titled "ISO," Sturén Papers. The same folder contains copies of the *Bulletin*. On the purpose of the *Bulletin*, Lars and Lolo Sturén, interview by Maria Nassén.

6. "Final Edition," *Bulletin of the IEC* [General Meeting], July 18, 1958, Stockholm, "ISO" folder, Sturén Papers.

7. Vince Grey, "ISO—a New Time, a New Start, the Transport Success Story," in *100 Year Commemoration for International Standardization: Addresses Presented* (Geneva: ISO, 1986), 49; 1961 in "Notebook Listing Travel," 1953-[1987], Sturén Papers.

8. Bob Bemer, "A History of Source Concepts for the Internet/Web [ca. 2002]," accessed September 4, 2018, https://web.archive.org/web/20161002194504/http://www.bobbemer.com/CONCEPTS.HTM.

9. Lolo Sturén, interview by Maria Nassén.

10. Olle Sturén, "Standardization and Variety Reduction as a Contribution to a Free European Market," SIS Report, May 16, 1958, Archives of the Institution for Civil Engineers (ICE BSI 3.3, Part 1).

11. International Committee for Scientific Management (CIOS), *Report of the European Management Conference, Berlin* (Geneva: CIOS, 1958), 159–72, "ISO" folder, Sturén Papers.

12. Roger Maréchal, "We Had Some Good Times," in *Friendship among Equals: Recollections from ISO's First Fifty Years*, compiled by Jack Lattimer (Geneva: ISO, 1997), 31.

13. Olle Sturén, "Toward Global Acceptance" (speech to National Bureau of Standards in Washington, DC, June 1972), 2–3, Sturén Papers.

14. "Statement by the Netherlands Delegation," "ISO" folder, Sturén Papers.

15. Henry St. Leger (ISO general secretary) to Olle Sturén, Sveriges Standardiseringskommission, Stockholm, February 16, 1965, Sturén Papers.

16. Olle Sturén to Mr. Henry St. Leger, general secretary ISO, Geneva, February 23, 1965, "ISO" folder, Sturén Papers.

17. Olle Sturén to J. M. Madsen and F. F. van Rhijn, March 29, 1965; Olle Sturén [note for files], NEDCO: arbetsuppgifter [duties] April 13, 1965; "Elaboration of Statement of Netherlands Delegation," April 12, 1965, ISO/NEDCO (Netherlands-1), 1; Olle Sturén to Mr. V. Clermont, Mr. F. Hadass, Mr. J. M. Madsen, Mr. F. F. van Rhijn, Mr. J. Wodzicki, and "One representative from the USSR," April 22, 1965, all in "ISO" folder, Sturén Papers.

18. "Report of the Committee for the Study of Netherlands' Statement Concerning ISO Liaisons and Activities," July 1965, Sturén Papers.

19. Olle Sturén to H. A. R. Binney (BSI), September 30, 1965, marked "private and confidential," "ISO" folder, Sturén Papers.

20. "Exchange of Letters about St. Leger Leaving ISO" file, Sturén Papers.

21. Maréchal, "We Had Some Good Times," 31.

22. ISO assistant general secretaries R. Maréchal and W. Rambal to the president of ISO Sir Jehangir Ghandy, Jamshedpur (India), January 7, 1966, "Exchange of Letters . . ." file, Sturén Papers.

23. Maréchal and Rambal to Ghandy.

24. Maréchal and Rambal to Ghandy.

25. Alain Durand, *AFNOR: 80 Années d'Histoire* (Paris: AFNOR Éditions, 2008), 120.

26. Maréchal, "We Had Some Good Times," 31.

27. Typescript marked "1969," section headed "The Secretary-General, Olle Sturén, made the following speech," 14–15, "ISO" file, Sturén Papers.

28. Sturén, "Toward Global Acceptance," 2–4.

29. Sturén, "Toward Global Acceptance," 5.

30. Sturén, "Toward Global Acceptance," 8.

31. Sturén, "Toward Global Acceptance," 6–7.

32. "Reference to Standards in Legislation," [1978], Old Fellows History File, Box CA-2, IEC Records. On the complexity of the network of organizations setting standards for electronic components in 1970, see E. H. Hayes, "Some International Aspects of Reliability and Quality Control," *Microelectronics Reliability* 9, no. 2 (1970): 137–43. Although the IEC source argues that the US government was particularly concerned that the European Organization for Quality Control might establish standards that would be, in effect, a nontariff barrier to trade, Hayes points out that the organization was, "in reality, more international than the name implies. The United States, Japan, and USSR are represented in the EOQC" (138).

33. Olle Sturén, "The Scope of the ISO" (address given at the annual meeting of the Standards Council of Canada, Ottawa, June 1977), 1–2, Sturén Papers; Olle Sturén, "Responding to the Challenge of the GATT Standards Code" (address given at the American National Standards Institute Evaluation Update Meeting, Washington, DC, March 18, 1980), Sturén Papers.

34. ISO and IEC, "ISO/IEC Code of Principles on 'Reference to Standards,'" in *ISO/EEC Guide 15-1977 (E)* [approved by ISO Council in September 1973 and by IEC Council in January 1974], accessed August 31, 2018, https://www.iso.org/files/live/sites/isoorg/files/archive/pdf/en/iso_iec_guide_15_1977.pdf.

35. Sturén, "Notebook Listing Travel."

36. Nalle Sturén, interview by Maria Nassén.

37. Sturén, "Scope of the ISO," 2.

38. Olle Sturén, "International Standardization: Why?" (address given to the Japan Standards Association, Tokyo, April 1978), 6, Sturén Papers.

39. Olle Sturén, "The Geography of the ISO" (address given before the Standards Council of the Standards Association of New Zealand, Wellington, March 1977), 3–4, Sturén papers.

40. Sturén, "Scope of the ISO," 2–3.

41. Sturén, "Scope of the ISO," 3.

42. Sturén, "International Standardization: Why?," 2.

43. Sturén, "International Standardization: Why?,"6.

44. E.g., Vince Grey, "Setting Standards: A Phenomenal Success Story," in Lattimer, *Friendship among Equals*, 33–42.

45. See, for example, Tineke M. Egyedi, "The Standardised Container: Gateway Technologies in Cargo Transport," *Homo Oeconomicus* 17, no. 3 (2000): 231–62; Lawrence Busch, *Standards: Recipes for Reality* (Cambridge, MA: MIT Press, 2011).

46. Quoted in Eric Rath, *Container Systems* (New York: John Wiley & Sons, 1973), 7.

47. Rath, *Container Systems*, 3.

48. Rath, *Container Systems*, 404–5.

49. Rath, *Container Systems*, 4.

50. Continental European transport uses smaller "swap bodies" filled with pallets for railroad and trucking (but not ocean) transportation, for example (Egyedi, "The Standardised Container," 253–57).

51. See "History," Bureau International des Containers, accessed July 14, 2017, https://www.bic-code.org/about-us/history/. In 1948, when the organization started up again after a wartime hiatus, it was renamed the Bureau International des Containers et du Transport Intermodal (BIC), and in 1970 it would devise the BIC-CODE system for marking containers.

52. Marc Levinson, *The Box: How the Shipping Container Made the World Smaller and the World Economy Bigger* (Princeton, NJ: Princeton University Press, 2006), 29–32.

53. Hans van Ham and Joan Rijsenbrij, *Development of Containerization: Success through Vision, Drive and Technology* (Amsterdam: IOS Press, 2012), 8. See also Francis G. Ebel, "Evolution of the Concept and Adoption of the Marine and Intermodal Container," in *Case Studies in Maritime Innovation* (prepared for the Maritime Transportation Research Board, Commission on Sociotechnical Systems, National Research Council, Washington, DC: National Academy of Sciences, May 1978), 5–6.

54. For example, the Department of Defense (as well as the Department of Commerce) sponsored a series of studies in the National Academy of Sciences (Ebel, "Marine and Intermodal Container," 20). Rath (*Container Systems*, 19) notes that these studies influenced work on standards. He also notes interest in and involvement with standardization by the US Department of Agriculture and the Department of Transportation, as well as the Interstate Commerce Commission, Civil Aeronautics Board, and the Federal Maritime Commission.

55. Levinson, *The Box*, 16–35. Cost estimates are on pp. 9–10, 34. Further discussion of the problem of port congestion may be found in Rath, *Container Systems*, 7; and in Ebel, "Marine and Intermodal Container," 3–6.

56. Van Ham and Rijsenbrij, *Development of Containerization*, 14. The details in this paragraph come from pp. 14–15.

57. See also Levinson, *The Box*, 49.

58. The following account of McLean and Sea-Land is based on Levinson, *The Box*, 36–59, 67–75.

59. Levinson, *The Box*, 49–50, with quote from p. 50.

60. According to Rath (*Container Systems*, 31), McLean chose these dimensions based on the truck bodies in the eastern part of the United States.

61. Levinson, *The Box*, 59–67.

62. Van Ham and Rijsenbrij, *Development of Containerization*, 28–29.

63. Van Ham and Rijsenbrij, *Development of Containerization*, 33–35.

64. Grey, "Setting Standards," 40. For Muller's affiliation, see Rath, *Container Systems*, 37. The Aluminum Company of America was known as Alcoa for most of the twentieth century, though it did not officially adopt Alcoa as its name until 1999.

65. Levinson, *The Box*, 128–31.

66. Grey, "Setting Standards," 40; Van Ham and Rijsenbrij, *Development of Containerization*, 45.

67. Levinson, *The Box*, 128–32. Marad stayed involved, however, and Levinson believes that the move toward standardization was bolstered by government interests and the specific standards agreed on were, in part, dependent on government approval (Levinson, email to J. Yates, June 12, 2014).

68. Email from Marc Levinson to J. Yates, based on his records collected while writing *The Box*, June 12, 2014; see also *The Box*, 132–37.

69. Egyedi, "The Standardized Container," 239–40; 244–45; 247–48.

70. Egyedi, "The Standardized Container," 239.

71. Levinson, *The Box*, 134–37; also email from Marc Levinson to J. Yates, June 12, 2014, and November 7, 2017.

72. Egyedi, "The Standardized Container," 239–40; Herman D. Tabak, *Cargo Containers: Their Stowage, Handling and Movement* (Cambridge, MD: Cornell Maritime Press, 1970), 2.

73. Quoted by Van Ham and Rijsenbrij, *Development of Containerization*, 44. They note that similar views were expressed by Matson's Leslie Harlander and various others involved in the process.

74. Egyedi, "The Standardized Container," 240. In 1971, according to Van Ham and Rijsenbrij (*Development of Containerization*, 42), they succeeded in getting the 24-foot and 35-foot sizes added into the ANSI standard (by then, ASA had become ANSI), which prevented further problems with legislation but did not affect the ISO standard or the overall movement toward the international standard.

75. Rath, *Container Systems*, 31.

76. Van Ham and Rijsenbrij, *Development of Containerization*, 24–26.

77. Rath, *Container Systems*, 31–32.

78. Grey, "ISO—a New Time, a New Start," 46–52.

79. Egyedi, "The Standardized Container," 240. Thirty-foot containers were not in the MH-5 Task Force proposal at the time they proposed the lengths to ISO TC104. They were added by the time that TC104 negotiated the Series 1 standards.

80. Egyedi, "The Standardized Container," 240–42; and Levinson, *The Box*, 138. There are minor differences between these two accounts, including the year of the Series 1 ISO standard (1964 or 1968), but both note that the Europeans ended up giving up the extra space of the slightly wider containers to assure the 40-foot length.

81. From ISO/TC104 (Sec. 196), 337, July 10, 1972, cited by Egyedi, "The Standardized Container," 242.

82. Grey, "Setting Standards," 42.

83. Levinson, *The Box*, 139.

84. Egyedi "The Standardized Container," 243.

85. Van Ham and Rijsenbrij, *Development of Containerization*, 44.

86. Egyedi ("The Standardized Container," 247) notes, "TC104 adhered to the basic rule of avoiding proprietary and patented solutions in order not to restrict the use of standards."

87. Levinson, *The Box*, 139.

88. Van Ham and Rijsenbrij, *Development of Containerization*, 45.

89. Levinson, *The Box*, 140.

90. Egyedi, "The Standardized Container," 242–43, quote from 243. On Strick Trailers, see also Levinson, *The Box*, 139, 142; and Strick's own history on its website, accessed July 14, 2017, http://www.stricktrailers.com/history.aspx.

91. Levinson, *The Box*, 141.

92. Levinson, *The Box*, 142.

93. Levinson, *The Box*, 142.

94. Van Ham and Rijsenbrij, *Development of Containerization*, 46.

95. Levinson, *The Box*, 142–44; Van Ham and Rijsenbrij, *Development of Containerization*, 46.

96. Tabak, *Cargo Containers*, 4.

97. Levinson, *The Box*, 148.

98. Ebel, "Marine and Intermodal Container," 18–19.

99. Levinson, *The Box*, 163–65; Ebel, "Marine and Intermodal Container," 19.

100. Ebel, "Marine and Intermodal Container," 10–11.

101. Van Ham and Rijsenbrij, *Development of Containerization*, 49. Unfortunately, the authors do not give a year for this event or their source.

102. Toby Poston, "Thinking inside the Box," *BBC News*, April 25, 2006, accessed July 24, 2017, http://news.bbc.co.uk/2/hi/business/4943382.stm.

103. Scott Baier and Jeffrey Bergstrand, "The Growth of World Trade: Tariffs, Transport Costs, and Income Similarity," *Journal of International Economics* 53, no. 1 (2001): 1–27; and David Hummels, "Transportation Costs and International Trade in the Second Era of Globalization," *Journal of Economic Perspectives* 21, no. 3 (2007): 131–54.

104. Daniel M. Bernhofen, Zouheir El-Sahli, and Richard Kneller, "Estimating the Effects of the Container Revolution on World Trade," *Journal of International Economics* 98 (January 2016): 26–50.

105. Egyedi, "The Standardized Container."

106. Egyedi, "The Standardized Container," 252–53.

107. In April 1974 the UN Conference on Trade and Development reached an agreement to regulate the small number of companies from the industrialized world that controlled transoceanic shipping. Many officials from the developing world feared that the wholesale shift of international shipping from the traditional liners to container ships would undermine these regulations. See Lawrence Juda, "World Shipping, UNCTAD, and the New International Economic Order," *International Organization* 35, no. 3 (1981): 493–516.

108. Van Ham and Rijsenbrij, *Development of Containerization*, 51.

109. Van Ham and Rijsenbrij, *Development of Containerization*, 51.

110. Van Ham and Rijsenbrij, *Development of Containerization*, 52. The origin of the quote is not given, and an online search for it in both English and Dutch does not show it before 2012, the year the book was published.

111. Martin Rowbotham, "The Relationship between Standardization and Regulation as They Affect Containerization," in *Proceedings of the Container Industry Conference, London, November/December 1977*, vol. 1 (London: Cargo Systems International, 1978), 229.

112. Burton Paulu, *Radio and Television Broadcasting on the European Continent* (Minneapolis: University of Minnesota, 1967), 33.

113. Donald G. Fink, "Television Broadcasting in the United States, 1927–1950," *Proceedings of the IRE* 39, no. 2 (February 1951): 117; Donald G. Fink, ed., *Color Television Standards: Selected Papers and Records of the National Television System Committee* (New York: McGraw-Hill, 1955), 1. Fink is identified on the title page of *Color Television Standards* as the vice chairman of NTSC, 1950–52.

114. Asa Briggs, *The History of Broadcasting in the United Kingdom*, vol. 4, *Sound and Vision* (Oxford: Oxford University Press, 1995), 448. For the French adoption of the German system and for the Soviet system, see Isabelle Gaillard, "The CCIR, the Standards and the TV Sets' Market in France (1948–1985)" (presentation at the Economic and Business History Association 11th Annual Conference, Geneva, September 14, 2007), 4, accessed July 14, 2017, http://www.ebha.org/ebha2007/pdf/Gaillard.pdf.

115. Briggs, *The History of Broadcasting*, 448.

116. "Report on the International Television Standards Conference," *Proceedings of the IRE* 38, no. 2 (February 1950): 116.

117. As quoted in Gaillard, "The CCIR," 5.

118. Erik B. Esping, "Study group XI of the CCIR," *IEEE Transactions on Broadcast and Television Receivers* 12, no. 2 (May 1966): 6.

119. Hugh R. Slotten, *Radio and Television Regulation: Broadcast Technology in the United States, 1920–1960* (Baltimore: Johns Hopkins University Press, 2000), 190.

120. Slotten, *Radio and Television Regulation*, 189–95. See also Fink, *Color Television Standards*, 5. In general, Slotten takes a critical perspective on events, focusing on various possible avenues of commercial advantage and influence on the FCC by RCA and its allies (GE, its broadcasting arm NBC, etc.), since RCA held the crucial patents underlying the monochrome standard. Fink sees the FCC as trying hard to serve the public interest and takes other players at closer to face value.

121. Slotten, *Radio and Television Regulation*, 195–203; Fink, *Color Television Standards*, 5–8.

122. Slotten, *Radio and Television Regulation*, 202–3; Fink, *Color Television Standards*, 8.

123. Slotten, *Radio and Television Regulation*, 203–4.

124. Slotten, *Radio and Television Regulation*, 215–19.

125. Slotten, *Radio and Television Regulation*, 214–16.

126. Slotten, *Radio and Television Regulation*, 216–26.

127. Fink, *Color Television Standards*, 17–19 (RCA system) and 13–14 (CTI system).

128. Fink, *Color Television Standards*, 9. Slotten (*Radio and Television Regulation*, 208–9) shows that one of the key FCC Commissioners, Robert Jones, questioned Fink's and others' objectivity and suspected them of bias toward RCA.

129. Fink, *Color Television Standards*, 10.

130. Slotten, *Radio and Television Regulation*, 226–28.

131. Slotten, *Radio and Television Regulation*, 228–29; Fink, *Color Television Standards*, 10.

132. Slotten, *Radio and Television Regulation*, 214–16, 229; Fink, *Color Television Standards*, 2–3, 21–37.

133. Fink, *Color Television Standards*, 40; Slotten, *Radio and Television Regulation*, 229.

134. Esping, "Study Group XI of the CCIR," 6–7.

135. Andreas Fickers, "The Techno-politics of Colour: Britain and the European Struggle for a Colour Television Standard," *Journal of British Cinema and Television* 7, no.1 (2010): 95–114.

136. Fickers, "Techno-politics of Colour," 95.

137. Fickers, "Techno-politics of Colour," 97; Rhonda J. Crane, *The Politics of International Standards: France and the Color TV War* (Norwood, NJ: Ablex, 1979), 13–14. Crane places the development of SECAM between 1958 and 1960, but Fickers claims that it was developed by 1956, which better fits the temporal unfolding of France's strategy as described by Crane.

138. Crane, *Politics of International Standards*, 38–40. She describes France's opposition to US technological and economic domination at this time, and its efforts to lead Europe in closing what it perceived as a technology gap with the United States. She also argues that, in

light of the opening of the European Economic Community, France could no longer depend on tariffs and import barriers to protect its industry, so it had to turn to non-tariff barriers such as standards (40–44). CFT was formed out of Compagnie générale de télégraphe sans fils for the purpose of developing and marketing SECAM equipment (47).

139. Crane, *Politics of International Standards*, 53–55.

140. Albert Abramson, *The History of Television, 1942 to 2000* (Jefferson, NC: McFarland, 2003), 100; Crane, *Politics of International Standards*, 57.

141. Paulu, *Radio and Television Broadcasting*, 35.

142. Crane, *Politics of International Standards*, 59.

143. Paulu, *Radio and Television Broadcasting*, 35; Crane, *Politics of International Standards*, 19.

144. Gerald Gross (ITU secretary general), *Report on the Activities of the International Telecommunication Union in 1964* (Geneva: ITU, 1965), 42, ITU Archives.

145. Fickers, "Techno-politics of Colour," 100–101; Crane, *Politics of International Standards*, 59–62. The following description of US delay and subsequent attempt to draw Soviet support is detailed in Crane, *Politics of International Standards*, 62–70.

146. Crane, *Politics of International Standards*, 72.

147. "Summary Record of the Opening Sessions (Thursday, 25 March, 1965 at 10 a.m.)" Doc. X/66-E, Doc. XI/69-E, March 29, 1965, ITU Archives.

148. Fickers, "Techno-politics of Colour," 102–3; Crane, *Politics of International Standards*, 72–76.

149. Paulu (*Radio and Television Broadcasting*, 35–36) reports the three-way vote, with 22 for SECAM, 11 for PAL, and 7 for NTSC; Crane (*Politics of International Standards*, 75) reports the two-way vote, with 21 for SECAM and 18 for the merged system.

150. Fickers, "Techno-politics of Colour," 103–5. For the British negative technical assessment of SECAM, Paulu (*Radio and Television Broadcasting*, 35) reports that "Britain's *Financial Times* (April 8, 1965) wrote: 'The television experts stated here time and time again that on objective technical grounds there was no doubt that NTSC was the best system although PAL was perhaps better suited to Europe. No one here doubts that SECAM is the worst of the three systems.'" Of course, the *Financial Times'* "technical" judgment may have been affected by the politics, as well.

151. Fickers, "Techno-politics of Colour," 105–8.

152. Esping, "Study Group XI of the CCIR," 6–7, quote from page 7.

153. George, H. Brown, "Comments on the NTSC System," *IEEE Transactions on Broadcast and Television Receivers* 12, no. 2 (May 1966): 83–85. Brown was an RCA engineer. Crane claims that it is technically possible to transmit the NTSC system on 625 or other line standards (*Politics of International Standards*, 17); thus, the rejection of NTSC reflects economic and political factors.

154. Crane, *Politics of International Standards*, 20.

155. Report by Sub-group XI-A-2, Annex to CCIR Doc.XI/1024-E, Study Group XI, "Report . . . Characteristics of Colour Television Systems," July 20, 1966, CCIR XIth Plenary Assembly, Oslo, 1966, 29–31, ITU Archives. Quote is from page 29.

156. Fickers, "Techno-politics of Colour," 108–10; Report by Sub-group XI-A-2, "Report . . . Characteristics of Colour Television Systems," 29–31.

157. Paulu, *Radio and Television Broadcasting*, 36.

158. Crane, *Politics of International Standards*, 77.

159. CCIR XIth Plenary Assembly, Oslo, 1966, Volume VI, 115, ITU Archives.

160. Fickers, "Techno-politics of Colour," 110.

161. See, for example, D. H. Pritchard and J. J. Gibson, "Color Television Part II: Worldwide Color Television Standards—Similarities and Differences," in *National Association of*

Broadcasters Engineering Handbook, ed. E. B. Crutchfield, 7th ed. (Washington, DC: National Association of Broadcasters, 1985), 51–52.

162. Sturén, "Responding to the Challenge," 2; Olle Sturén, "Collaboration in International Standardization between Industrialized and Developing Countries" (speech given at the German Institute for Standardization [DIN], Bonn, November 1981), 4, Sturén Papers.

163. Olle Sturén, "Developments in International Standardization and Their Relevance to Foreign Trade" (address given to the Canadian Export Association Annual Convention, Ottawa, October 19, 1981), 5. Sturén Papers.

164. Sturén, "Developments in International Standardization."

165. Sturén, "Developments in International Standardization," 4–5.

166. Sturén, "Developments in International Standardization," 6.

167. D. Linda Garcia, "Standards for Standard Setting: Contesting the Organizational Field," in *The Standards Edge: Dynamic Tension*, ed. Sherrie Bolin (Ann Arbor, MI: Sheridan Press, 2004), 8.

168. US Congress, Office of Technology Assessment (OTA), *Global Standards: Building Blocks for the Future* (Washington, DC: Office of Technology Assessment, 1992), 49, 13.

169. Garcia, "Standards for Standard Setting," 9.

170. Garcia, "Standards for Standard Setting," 8.

171. Mark S. Frankel, ed., "Professional Self-Regulation after the Hydrolevel Decision," *Perspectives on the Professions* 3, no. 3 (1983): 1–9.

172. Sturén, "Responding to the Challenge," 4.

173. OTA, *Global Standards*, 55–56. The circular can be found in the *US Federal Register* 45, no. 14 (January 21, 1980): 4326–29. With Circular A-119, the phrase "voluntary consensus standard setting," which had previously been used by ASTM to describe its work, became the accepted way in which US regulators referred to approved private standard setting.

174. OTA, *Global Standards*, 57.

175. Olivier Borraz, "Governing Standards: The Rise of Standardization Processes in France and in the EU," *Governance* 20, no. 1 (2007): 60–62.

176. Stephen Oksala, "The Changing Standards World: Government Did It, Even Though They Didn't Mean To," *Standards Engineering* 52, no. 6 (2000): 4.

177. Oksala, "The Changing Standards World," 4.

178. Oksala, "The Changing Standards World," 4.

179. Oksala, "The Changing Standards World," 6.

180. Olle Sturén, "A Non-political International Collaboration" (address to the Standards Institute of Israel January 23, 1986), 4, Sturén Papers.

181. Sturén, "Non-political International Collaboration," 2.

182. Sturén, "Non-political International Collaboration," 4.

183. Olle Sturén, "Music and Its Message for International Standardization," typescript, Sturén Papers, later revised and published in *100-Year Commemoration for International Standardization* (Geneva: ISO, 1986), 67–78.

184. Arturo F. Gonzales Jr., "Standardization: A Measure of Reason," *Rotarian* 131, no. 3 (September 1977): 46–47.

Chapter 6 • US Participation in International RFI/EMC Standardization, World War II to the 1980s

1. Susan J. Douglas, *Inventing American Broadcasting, 1899–1922* (Baltimore: Johns Hopkins University Press, 1987), 299–311.

2. Donald Heirman and Manfred Stecher, "A History of the Evolution of EMC Regulatory Bodies and Standards," *Proceedings of EMC Zurich 2005 Symposium* (Zurich: 2005): 83, 85.

3. Heirman and Stecher, "A History," 85; G. A. Jackson, "The Early History of Radio Interference," *Journal of the IERE* 57, no. 6 (1987): 244–50.

4. Heirman and Stecher, "A History," 92. These standards were published only in 1934.

5. Harold E. Dinger, "Radio Frequency Interference Measurements and Standards," *Proceeding of the Institute for Radio Engineering* 50, no. 5 (May 1962): 1313–14.

6. "Standards Committees Report Progress on Important Electrical Problems," *Industrial Standardization and Commercial Standards Monthly* 8, no. 3 (March 1937): 61 and 67. A list of 65 electrical committees under the auspices of ASA's Electrical Standards Committee includes C63, Radio-Electrical Coordination, administered by the Radio Manufacturers Association, and indicates that it held an initial meeting during 1936.

7. Nina Wormbs, "Standardising Early Broadcasting in Europe: A Form of Regulation," in *Bargaining Norms, Arguing Standards: Negotiating Technical Standards*, ed. Judith Schueler, Andreas Fickers, and Anique Hommels (The Hague: STT Netherlands Study Centre for Technology Trends, 2008), 115–18. See also F. L. H. M. Stumpers, "International Co-operation in the Suppression of Radio Interference—the Work of C.I.S.P.R.," *Proceedings, Institution of Radio and Electronics Engineers, Australia*, February 1971, 51–55. The original ten countries were Austria, Belgium, Czechoslovakia, France, Germany, Great Britain, the Netherlands, Norway, Spain, and Switzerland.

8. IEC Committee of Action, *RM 102* (IEC, January 25, 26, and 27, 1933), IEC Records; Stumpers, "Suppression of Radio Interference," 51; Heirman and Stecher, "A History," 85–86.

9. The numbers representing each organization were agreed on in the 1933 meeting; see IEC Committee of Action, *RM 106* (IEC, October 9, 1933), IEC Records.

10. IEC Committee of Action, *RM 115* (IEC, October 13, 1934), 7, IEC Records.

11. Heirman and Stecher, "A History," 86.

12. Stumpers, "Suppression of Radio Interference," 51–52. In 1966 CCIR officially recommended that ITU member administrations should follow CISPR recommendations to the extent possible, and that they should use the measurement methods and apparatus developed by CISPR in their national regulations on interference.

13. CISPR, *Report of the Meeting Held in London on 18th–20th November, 1946* (n.p.: Central Office of the IEC, 1946), 6, Leonard Thomas Papers, obtained from Daniel Hoolihan and to be deposited in Hagley Museum and Library [hereafter Thomas Papers]. See also Heirman and Stecher, "A History," 86. On the European character of the committee, see Stumpers, "Suppression of Radio Interference," 52. Occasionally observers from the United States and Japan attended. The woman in figure 6.1 was probably a translator or stenographer, as no women served in delegations until much later.

14. IEC Committee of Action, *RM 147* (IEC, June 23, 1937), IEC Records.

15. CISPR, *Report of the Meeting held in Paris on 3rd–4th July, 1939* (n.p.: Central Office of the IEC, 1939), Thomas Papers.

16. Heirman and Stecher, "A History," 84.

17. "Pioneers in EMC: Leonard W. Thomas," *ITEM 1990*, 362, found in Thomas Papers.

18. Minute #3771 of ASA Standards Council, meeting May 24, 1945, 13.

19. "Pioneers in EMC." According to Heirman and Stecher ("A History," 93), this standard was named "American War Standard for Radio Noise measuring instrumentation in the range 150 kHz to 20MHz."

20. CISPR, *Report of the Meeting Held in London*, 5.

21. Stumpers, "Suppression of Radio Interference," 52; Jackson, "History of Radio Interference," 245; Heirman and Stecher, "A History," 86.

22. CISPR, *Report of the Meeting held in London*, 30.

23. CISPR, *Report of Meeting held in Lucerne on 22th–25th* [*sic*] *October, 1947*, 8–10, 13–24, quote from 10 (Central Office of the IEC, n.d.), Thomas Papers.

24. CISPR, *Report of the Meeting held in London*, 9–13, quote from page 10.

25. CISPR, *Report of the Meeting held in London*, 18.

26. CISPR, *Report of the Meeting held in London*, 9. The link with CCIR would be particularly strong. In a 1971 paper, F. L. H. M. Stumpers, Dutch chairman of CISPR, 1967–73, stressed the close relations between CCIR and CISPR, mentioning two directors of CCIR who had been involved in CISPR's early work, the reciprocal sending of observers to the other's meetings, and CCIR's 1966 recommendation to follow CISPR recommendations (Stumpers, "Suppression of Radio Interference," 52).

27. CISPR, *Report of Meeting held in Lucerne*, 26, 8, 10.

28. CISPR, *Report of Meeting held in Lucerne*, 24; IEC Council, *RM 194* (IEC, October 13, 1948), IEC Records; Heirman and Stecher, "A History," 85–86.

29. Per Åkerlind, *Fifty Years of the CISPR: 1934–1984* (Paris: Ets Busson, 1984), 2. IEC Records. Åkerlind dates the decision to make CISPR a special committee of IEC to 1946, but Heirman and Stecher ("A History," 86) say it was formally constituted as such in 1950.

30. Åkerlind, *Fifty Years of the CISPR*, 5.

31. George Arthur Codding Jr., *The International Telecommunication Union: An Experiment in International Cooperation* (Leiden: Brill, 1952; repr., New York: Arno Press, 1972), 180–84, 203; Anthony R. Michaelis, *From Semaphore to Satellite* (Geneva: International Telecommunication Union, 1965), 178, 180. According to Michaelis (191), the two coordinated conferences also agreed to make the ITU "the specialized agency" of the UN in telecommunications. The original proposal was to make it "*a* specialized agency," but, to give it unique status in this domain, it insisted on being termed "*the* specialized agency of the UN in telecommunications."

32. That body was made up of 11 qualified radio experts from different countries, chosen to represent all geographical areas and to "serve not as representatives of their respective countries, or of regions, but as custodians of an international public trust" (Michaelis, *Semaphore to Satellite*, 250).

33. Atlantic City 1947 convention, as quoted in Codding, *International Telecommunication Union*, 273. For establishment of the specialized secretariat, see R. C. Kirby, "50 Years of International Radio Consultative Committee (CCIR)," *Telecommunication Journal* 45, no. 6 (1978): 6.

34. Kirby, "50 Years of CCIR," 5, shows the meeting years and locations up to 1978.

35. SG IV Terms of Reference, SG IV, CCIR, Documents of the Xth Plenary Assembly, Geneva, 1963, Vol. 4, 343. Duties of the former SG IV were delegated to a different SG.

36. E. Metzler, "Report by the Director, CCIR (Covering the Period between the IXth and the Xth Plenary Assembly)," October 11, 1962, [in preparation for the] 10th Plenary Assembly, CCIR, New Delhi, 1963, 1–3, Doc. 16-E, Box 26, Donald MacQuivey Papers, Silicon Valley Archives, Courtesy of Department of Special Collections and University Archives, Stanford University Libraries [hereafter MacQuivey Papers]. Also [Col. Lochard, France], "Report by the Chairman of Study Group 1: Transmitters," [in preparation for the] 10th Plenary Assembly, CCIR, New Delhi, 1963, Doc. 1-E, November 2, 1962, Box 26; P. David [France], "Report by the Chairman of Study Group II: Receivers," [in preparation for the] 10th Plenary Assembly, CCIR, New Delhi, 1963, Doc. 2-E, 22 October 1962, Box 26; and H. C. A. van Duuren [Netherlands], "Report by the Chairman of Study Group III: Fixed Service Systems," October 1962, 3, [in preparation for the] 10th Plenary Assembly, CCIR, New Delhi, 1963, Doc. 3-E, Box 26, all in MacQuivey Papers. This subsection draws primarily on the papers of Donald MacQuivey, who worked for Stanford Research Institute and was a member of the CCIR Advisory Committee.

37. Metzler, "Report by the Director," 8.

38. Ivar Herlitz, "The IEC, Yesterday, Today, and Tomorrow," in *Eighth Charles le Maistre Memorial Lecture* (Geneva: Central Office of the IEC, 1962), 15.

39. "Summary Report of the 48th Meeting of the US National CCIR Organization Executive Committee on January 25, 1962," Item 331, Box 10, MacQuivey Papers.

40. [Lochard], "Study Group 1: Transmitters," 11–12; David, "Study Group II: Receivers," 4.

41. CCIR Doc. 2191-D, February 5, 1963, nested within CISPR (Secretariat) 550, April 1963, found in CISPR files of the Ralph M. Showers Papers, a large collection of uncatalogued papers obtained directly from the Showers family [hereafter Showers Papers]. These papers have now been deposited with Hagley Museum and Library.

42. Metzler, "Report by the Director," 9–10; 16–21.

43. Ernst Metzler, "Curriculum vitae," in "Erzwungene elektrische Schwingungen an rotationssymmetrischen Leitern bei zonaler Anregung" (doctoral thesis, Eidgenössischen Technischen Hochschule in Zürich, 1943), 101.

44. This related and sequential set of genres comprises a genre system. Wanda J. Orlikowski and JoAnne Yates, "Genre Repertoire: Examining the Structuring of Communicative Practices in Organizations," *Administrative Science Quarterly* 39 (1994): 541–74.

45. Metzler, "Report by the Director," 20.

46. A similar format to the one described below had been followed in the resolutions of most global level intergovernmental conferences going back to the nineteenth century. See "Debate, Resolutions, and Voting: The Normal Procedure," in Johan Kaufmann, *Conference Diplomacy: An Introductory Analysis*, 3rd ed. (London: Macmillan, 1996), 17–22.

47. *Drafting Committee summary report of the 1st meeting*, Friday, January 18, 1963, Appendix 1 to Annex 1 to Doc. 293-E, Box 26, MacQuivey Papers.

48. Metzler's CCIR director's report commented that the "considerings," many of which he considered unnecessary, added to the bulk of documentation, and that at the Interim Meetings some attendees had suggested shortening CCIR documents "by deletion of the 'considerings.'" (Metzler, "Report by the Director," 21). He went on to summarize the CCIR decision: "Although the 'considerings' of a text should be kept to a minimum, they ought not to be abolished altogether, as certain types of 'considerings' are essential to a clear understanding of the text."

49. Johan Kaufmann suggests that the rhetorical purpose is the same when a similar format is used in intergovernmental conferences even when the topic in question is very clearly political. The format links a new action to a prior settled agreement on facts. Kaufmann, "Debate, Resolutions, and Voting."

50. David, "Study Group II: Receivers," 1.

51. Van Duuren, "Study Group III: Fixed Service Systems," 9.

52. David, "Study Group II: Receivers," 5–6.

53. Department of State Telecommunications Division, Executive Committee 252, "Terms of Reference for the United States National C.C.I.R. Organization," April 1, 1960, 1, Doc. 252, Box 26, MacQuivey Papers.

54. See, for example, the discussions of participation problems faced by US chairmen of US SG I and SG II in "Summary Report of the 47th Meeting of the US National CCIR Organization Executive Committee on December 14, 1961," 4–5, Item 322, Box 10, MacQuivey Papers.

55. Percentages of government representatives ranged from 64 to 81 percent and are computed from the summary reports of the US National CCIR Organization Executive Committee meetings in 1961–1963, as follows: August 1961, 45th meeting, item 290; December 1961, 47th meeting, item 322; January 1962, 48th meeting, item 331; March 1962, 49th meeting, item 350; October 1962, 51st meeting, item 372; November 1962, 52nd meeting, item 375; January 1963, 53rd meeting, item 378, all in Box 10, MacQuivey Papers. Summary reports of the 46th and the 50th meetings did not survive in this collection. Private companies (e.g., AT&T, West-

ern Union, Raytheon) and nonprofit associations (e.g., the Institute of Radio Engineers) were also represented during this period. The role of location is emphasized in the "Summary Report of the 48th Meeting of the US National CCIR Organization," where an alternate who lived in Washington was named to serve on the US National CCIR Organization Executive Committee with the note that he would be able to take a more active part than former representatives who lived elsewhere.

56. "Summary Report of the 47th Meeting of the US National CCIR," 2; when the United States offered to host the March 1962 Interim Meeting for SG IV and SG VIII in Washington, DC, it requested contributions from industry "to put on a good program, i.e., hostship activities, entertainment, field trips, etc." "Summary Report of the 45th Meeting of the US National CCIR Organization Executive Committee," 6.

57. "Summary Report of the 45th Meeting of the US National CCIR Organization," 5.

58. "Summary Report of the 45th Meeting of the US National CCIR Organization," 10.

59. "Summary Report of the 48th Meeting of the US National CCIR Organization," 3.

60. "Summary Report of the 45th Meeting of the US National CCIR Organization," 3.

61. Doc. 350, Summary Record of the US National CCIR Organization Executive Committee meeting on March 1, 1962, 3, Box 10; Doc. IV/58, described in Doc. IV/78-E, "Study Group IV, Summary Record of the Second Meeting," 2, Box 7; Doc. 115-E, "Active Earth Satellite 'Telstar': Preliminary Results at the United Kingdom Ground Station of Goonhilly Downs, Cornwall," October 4, 1962, Box 26, all in MacQuivey papers.

62. "Ralph M. Showers—Vita," ca. 1991; *Greystones 1935*, Haverford High School yearbook, 41; letters, 1940–43 re: Ralph M. Showers employment and draft deferment, all courtesy of Janet Showers Patterson. See also Don Heirman, "Completed Careers: Ralph Showers 1918–2013," *IEEE Electromagnetic Compatibility Magazine* 2, no. 4 (2013): 44–46.

63. Dan Hoolihan identified the other two men as John F. Chappell and John J. O'Neill, other members of the US EMC standards community (email from Dan Hoolihan to Janet Showers Patterson, July 24, 2017, forwarded to J. Yates, May 5, 2018). According to Hoolihan, Chappel was part of the US CISPR delegation in 1958 and 1965, but the photograph of Showers seems more likely to be 1958.

64. Ralph Showers to Bea[trice] Showers, The Hague, Sun. Eve., [1958], in "Traveling the World with BeaBea and DaDa: 1958–2005," a personal collection of letters, postcards, photographs, menus, etc., from their travels, unpublished document compiled by Janet Showers Patterson for Ralph Showers's grandchildren, courtesy of Janet Showers Patterson.

65. Minutes of Sectional Committee on Radio-Electrical Coordination C63 [hereafter C63], May 5, 1960, 112–13, Thomas Papers. Information on Aeronautical Radio, Inc., from Donald A. McKenzie, *Inventing Accuracy: A Historical Sociology of Nuclear Missile Guidance* (Cambridge, MA: MIT Press, 1993), 172n17.

66. Minutes of C63, February 10, 1960, 29–30, Thomas Papers.

67. ANSI C63 Steering Committee Minutes, January 30, 1970, Thomas Papers.

68. Pakala and Showers, as well as C63 members Leonard W. Thomas and Harold A. Gauper, are listed as Administrative Committee members in the minutes of the Administrative Committee meeting of the IRE Professional Group on Radio Frequency Interference, March 26, 1962, Thomas Papers. Only Showers represented IRE on C63 at this time.

69. Pakala joined the AIEE in 1924 and was a local officer by 1926 (see "Membership—Applications, Elections, Transfers, etc.," *Journal of the AIEE* 44, no. 1 (1925): 109; "Officers of the AIEE, 1926–27," *Journal of the AIEE* 46, no. 3 (1927): 317).

70. See announcement of Pakala's resignation as chairman and summary of his career, as well as announcement of Showers's acceptance of the chairmanship and summary of his career in minutes of C63, July 26, 1968, 228–29. For Showers's changing affiliation, see

minutes of C63, May 5, 1960, 113; March 28, 1963,148; and February 1, 1968, 212, all in Thomas Papers.

71. Minutes of the IRE Professional Groups Committee, February 6, 1962, and May 8, 1962, and of Administrative Committee meeting of the IRE Professional Group on Radio Frequency Interference, March 26, 1962, all in the Thomas Papers; Robert W. Fairweather to Herman Garlan, November 25, 1962, letter containing minutes of the IRE Professional Groups Committee meeting, November 20, 1962; and B. M. Oliver and Hendley Blackmon memo to chairmen, Professional Technical Groups and Technical Committees, March 15, 1963, on IEEE letterhead, all in the Thomas Papers.

72. See minutes of Administrative Committee meeting of the IEEE Professional Technical Group on RFI, March 26, 1963; and June 3, 1963, quote from page 7, both in the Thomas Papers. The change in name was publicly announced as occurring "in order to reflect the increased scope and also to more closely identify the work of the Group with the same terminology now being used in most government work." IEEE, *Professional Technical Group Radio Frequency Interference Newsletter*, no. 27 (April 1963), 1.

73. "IEEE Standards and Association History," accessed August 16, 2017, http://ethw.org/IEEE_Standards_Association_History; minutes of the Administrative Committee meeting of the IEEE EMC Group, June 8, 1964; and November 18, 1964, Thomas Papers.

74. Ralph Showers, who played major roles in CISPR and the IEC, actually had at least a minimal connection with the CCIR, serving as a member of CCIR US Study Group 1A, on the efficient use of the frequency spectrum and problems of frequency sharing, from 1978 to 1990. "Ralph M. Showers—Vita," ca. 1991.

75. The three standards are listed in Minutes of C63, December 6, 1960, 127–28, Thomas Papers. The finalization dates come from Ralph M. Showers, "Influence of IEC Work on National Electromagnetic Compatibility Standards," *ElectroMagnetic Compatibility 1993: 10th International Zurich Supplement to Symposium and Technical Exhibition on Electromagnetic Compatibility* (Zurich: EMC Proceedings Editor, March 1993), 7–8.

76. Minutes of C63, February 4, 1964, 159; February 4, 1965, 176; July 8, 1966, 191, Thomas Papers.

77. Minutes of C63, December 12, 1968, 242–43; July 24, 1969, 254–56; December 11, 1969, 264–67, Thomas Papers.

78. In 1963, for example, one of its member bodies, NEMA, proposed that one of its standards be made an American Standard (Minutes of C63, March 28, 1963, 154, Thomas Papers).

79. Minutes of C63, February 1, 1968, 213–16; July 26, 1968, 230–35; December 12, 1968, 242 and appendix A (dated January 1969), all in the Thomas Papers.

80. Ralph Showers to Bea[trice] Showers, The Hague, Midnight, Tuesday eve., [1958], and Showers to Bea, Amsterdam, Sun. Eve., [1958], in "Traveling the World with BeaBea and DaDa."

81. Minutes of C63, May 5, 1960, 117–18; December 6, 1960, 132–33, Thomas Papers.

82. Minutes of C63 Steering Committee, February 10, 1960, 29–30; May 5, 1960, 120–21, Thomas Papers.

83. Letter from Pakala to US representatives, October 17, 1960, reproduced in minutes of C63, December 6, 1960, 134, Thomas Papers; [Pakala] to US representatives on CISPR Working Groups and to all members of C63, memorandum, January 30, 1961, exhibit H to minutes of C63, December 6, 1960, Thomas Papers.

84. Minutes of C63, June 14, 1961, 142–45, Thomas Papers. By the first post-CISPR meeting of C63, a preliminary financial report on the CISPR Philadelphia meeting was provided, but the main CISPR office had not yet issued its report of the meeting. Minutes of C63, March 28, 1963, 150–52 and exhibit B, Thomas Papers.

85. A. H. Ball and W. Nethercot, *Radio Interference from Ignition Systems: Comparison of American, German and British Measuring Equipment, Techniques and Limits*, paper no. 3550 E (London: Institution of Electrical Engineers, May 1961), 273–78.

86. Committee Correspondence, B. H. Short to Mr. W. E. Pakala, subject: "Report of February 9 and 10, 1960, Meeting of the CISPR Working Group #4—Interference from Ignition Systems," Frankfurt, Germany, exhibit B to minutes of C63, May 5, 1960, 2, Thomas Papers.

87. Committee Correspondence, B. H. Short to Mr. A. C. Doty Jr., exhibit E to minutes of C63, December 6, 1960, Thomas Papers. SAE statement is from appendix #1 to this document, John M. Roop, SAE, to B. H. Short, September 20, 1960, 2.

88. Committee correspondence, B. H. Short to Mr. A. C. Doty Jr., exhibit E, 2.

89. Minutes of C63, December 6, 1960, 133 (quote), as well as exhibit G, W. E. Pakala to S. David Hoffman, January 13, 1961, Thomas Papers.

90. "Horn Not Enough—Motorists May Signal Others by Radio," *Broadcasting: The Weekly Newsmagazine of Radio*, August 19, 1946, 66.

91. *Anderson Daily Bulletin*, February 2, 1960, 4.

92. Ball and Nethercot, *Radio Interference*, 274.

93. Ball and Nethercot, *Radio Interference*, 277.

94. B. H. Short to Mr. A.C. Doty Jr., December 6, 1960, 2.

95. Minutes of C63, February 4, 1964, 160–61, Thomas Papers.

96. Minutes of C63, February 4, 1964, 163–64. All quotes in this paragraph are from this source.

97. Minutes of C63, June 19, 1964, 168, Thomas Papers.

98. Minutes of C63, February 4, 1965, 3, Thomas Papers.

99. Minutes of C63, February 4, 1965, 3.

100. Bauer, like Short, wrote very candidly about WG4, but he was also much more positive, as subsequent quotes will demonstrate.

101. Frederick Bauer to Ralph M. Showers, November 30, 1998, courtesy of Janet Showers Patterson.

102. Minutes of C63, July 8, 1966, exhibit B, 7, Thomas Papers.

103. Minutes of C63, February 4, 1965, 17 (quote), 178.

104. Minutes of C63, February 2, 1966, 184–18, Thomas Papers.

105. Minutes of C63, February 4, 1965, exhibit B, 4.

106. Frederick Bauer, report on meeting of WG4, CISPR, April 25–29, in minutes of C63, July 8, 1966, exhibit B, 2, Thomas Papers.

107. Bauer, report on meeting of WG4, CISPR, April 25–29, 10–11.

108. Report on meeting of WG4, CISPR, April 2–5, 1967, exhibit B in minutes of C63, June 19, 1967, 1, 3–5, 10–11, Thomas Papers.

109. Report on meeting of WG4, CISPR, April 2–5, 1967, exhibit B, 11.

110. Report on meeting of WG4, CISPR, April 2–5, 1967, exhibit B, 1.

111. For a similar transcending of the Cold War divide in the name of computer science, see Ksenia Tatarchenko, "'The Anatomy of an Encounter: Transnational Mediation and Discipline Building in Cold War Computer Science," in *Communities of Computing: Computer Science and Society in the ACM*, ed. Tom Misa (New York: ACM Books and Morgan and Claypool, 2016), 199–227.

112. Bauer's list of delegates at the 1966 WG4 meeting in Prague indicated that West Germany, Italy, the UK (three of its four delegates), and the United States were the only countries represented by private industry, while Austria, Czechoslovakia, Denmark, France, Netherlands, Norway, Sweden, the USSR, and Yugoslavia were represented by governmental delegates (Bauer, report on meeting of WG4, CISPR, April 25–29, 1966, in minutes of C63, July 8, 1966, exhibit B, Thomas Papers, emphasis in the original).

113. Bauer, report on meeting of WG4, CISPR, April 25–29, 1966, 11.

114. Report on meeting of WG4, CISPR, April 2–5, 1967, 8.

115. Report on meeting of WG4, CISPR, April 2–5, 1967, 8.

116. Minutes of C63 Sub-Committee 3, June 28, 1967, in exhibit C to minutes of C63, February 1, 1968, Thomas Papers.

117. S. A. Bennett, J. J. Egli, H. Garlan, G. C. Hermeling, and R. M. Showers, "Report of Delegates to the Plenary Assembly of the International Special Committee on Radio Interference (CISPR)," August 26–September 7, 1967, 5, in exhibit D to minutes of C63, February 1, 1968, Thomas Papers.

118. Minutes of C63, February 1, 1968, 219–20.

119. Compiled from a series of Showers's curriculum vitae, courtesy of Janet Showers Patterson.

120. Minutes of C63, June 23, 1978, 2, Thomas Papers. Moreover, the AIEE, one of the two societies that merged to form IEEE in 1963, had had its own Committee on Standardization since before 1900 (see chapter 1).

121. ANSI's reorganization was initiated, according to Showers, by OMB circular A-119, "Federal Participation in the Development and Use of Voluntary Standards," minutes of C63, August 21, 1981, 9, Thomas Papers. Fears of antitrust actions given the extremely expensive court decision against ASME in the Hydrolevel case were clearly also a factor; see Hedvah L. Shuchman, "Professional Associations and the Regulation of Standard-Setting," *Perspectives on the Professions* 3, no. 3 (1983): 6–10.

122. Listed firms are among many at the following meetings: minutes of C63, October 12, 1979; March 26, 1980; October 10, 1980, Thomas Papers.

123. See, for example, minutes of C63, December 9, 1994, 1–2, Showers Papers. Minutes for the previous meeting (December 2–3, 1993, 1–2, Showers Papers) list corporate attendees not representing associations as guests. Either the change happened during 1994, or the two different secretaries who prepared the minutes (Luigi Napoli in December 1993 and W. Kesselman in December 1994) had different interpretations of membership.

124. Sue Vogel (secretary, C63) to members of C63, memorandum, June 26, 1990, Showers Papers.

125. Minutes of special meeting of C63, January 16, 1979, 7, Thomas Papers.

126. From award citation for the Elihu Thomson Electrotechnology Medal from ANSI, October 12, 2011, as quoted in Heirman, "Completed Careers: Ralph Showers 1918–2013," 44.

127. Minutes of C63, April 6, 1983, 4, Thomas Papers. The meeting took place in the FCC offices in Washington.

128. Minutes of C63, "Policy on Immunity Standards" (adopted April 6, 1983), undesignated attachment to minutes of C63, April 6, 1983, 4, Thomas Papers.

129. Minutes of C63, October 12, 1979, 8; Ralph M. Showers, "Voluntary Standardization for Electromagnetic Compatibility," in *Proceedings of the 1978 Electromagnetic Interference Workshop*, ed. M. G. Arthur, NBS Special Publication 551 (Washington, DC: National Bureau of Standards, 1979), 12–19. The list referenced is on pages 17–18.

130. Minutes of C63, October 12, 1979, 8.

131. Minutes of C63, March 26, 1980, 3–5, plus attachments C-E, Thomas Papers. Quote from attachment B, "Scope, Purpose and Approach of an EMC Standard."

132. Minutes of C63, August 21, 1981, 3–4 (quotes from 3), and attachment C, Thomas Papers.

133. J. Yates, interviews of past chair of C63 Don Heirman (August 9, 2014) and current chair Dan Hoolihan (August 9, 2014, both transcripts deposited in Hagley Museum and Library, Wilmington, DE), highlight the ongoing shortage of active members. The update problem is explained in a personal email from Don Heirman to J. Yates, July 18, 2015.

134. Don Heirman to Ed Bronaugh and Al Smith, memorandum, February 7, 1983 (with a copy of ANSI C63.4-1981 attached); Don Heirman, marked-up copy of C63.4, March 30, 1983, and report of the voting on C63.4-1981; Ray Magnuson to committee, August 16, 1983, summary of comments on "Do Not Approve" ballots from that voting; "Notes on Meeting of Ad Hoc Committee on Open Area Test Sites (OATS) Amendment to C63.4," April 27, 1984; and Showers to Ad Hoc Committee on OATS Amendment to C63.4, May 15, 1984, with "a first attempt (Draft 11, May 1984) to accommodate many of the comments submitted during the voting on the OATS document," all in Showers Papers.

135. Siegfried Linkwitz and Ray Magnuson to Fred Huber, August 4, 1987, copied to the IEEE Board of Standards Review and the Ad Hoc Subcommittee working on incorporating OATS material into C63.4, Showers Papers. For the second flurry, see, among others, Ed Bronaugh to "Members Ad Hoc Committee to Work MP-4, MP-4 'A' and CBEMA Proposed MP-4 into C63.4," September 4, 1987, Showers Papers.

136. Fred Huber to Ralph Showers, September 21, 1988; and Showers to Huber (with results of his agreement to revise certain sections of C63.4), Draft 2, March 23, 1988, Showers Papers.

137. C63.4/D11, Revision of ANSI C63.4-1988, designated as unapproved draft, Feb 6, 1990, in Binder on C63.4/D11 balloting, Showers Papers. The vote tally is from an undated, handwritten note in the same binder, titled "ballot C63.4/D11 closed on 4/9/90."

138. Showers to Kristin Dittman, April 30, 1990; Glen Dash to Showers, May 1, 1990; "Further Corrections to C63.4D11," May 21, 1990; "Actions Taken with regard to Negative Ballots on C63.4D11," May 21, 1990; ballot on revised draft 11.3, June 14, 1990, from Art Wall; Showers to Vogel, June 25, 1990, all in Binder on C63.4/D11 balloting, Showers Papers.

139. Showers, "Actions Taken as a Result of the Reballotting of the Revision of C63.4," attachment to July 10, 1990, letter to Sue Vogel, in Binder on C63.4/D11 balloting, Showers Papers.

140. Vogel to members of C63, memorandum, subject: "Final draft-C63.4," August 5, 1990; Roger McConnell to Showers, comments on "Final Draft-C63.4," September 19, 1990; faxed letter from Heirman to Showers and others on McConnell's comments, October 2, 1990; John Hirvela to Showers, October 2, 1990; Bronaugh to Showers, Heirman, and Vogel, October 9, 1990; Heirman to Showers, October 10, 1990; Showers to Vogel, October 12, 1990; James B. Pate and Albert A. Smith Jr. to Showers, October 16, 1990; Robert McConnell to Showers, October 20, 1990, all in C63.4/D11 binder, Showers Papers.

141. "Chronology of C63.4-1991 Letter Ballot, 20 Feb 1991," binder on C63.4/D11 ballotting, Showers Papers. See also edited draft of article by Kristin Dittmann, "Accredited Standards Committee C63 Revises Its Standard on the Measurement of Radio Noise Emissions (C63.4)," for publication in the *IEEE Standards Bearer*, Showers Papers. At some point during the process, its range was broadened slightly, changing the name to "Methods of Measurement of Radio-Noise Emissions from Low-Voltage Electrical and Electronic Equipment in the Range of 9 kHz to 1 GHz (revision of ANSI C63.4-1988)," rather than from 10 kHz to 1 GHz (see ANSI Standard C63.4-1991, IEEE Explore Digital Library, accessed July 19, 2017, http://ieeexplore.ieee.org/stamp/stamp.jsp?tp=&arnumber=159236).

142. Janet Showers Patterson, Caroline Showers, and Virginia Showers White, memories of their father read at a memorial service in 2013, courtesy of Showers family.

143. As quoted in Heirman, "Completed Careers: Ralph Showers 1918–2013," 46.

144. For the story of this loss of the consumer electronics industry, see Alfred D. Chandler Jr., *Inventing the Electronic Century: The Epic Story of the Consumer Electronics and Computer Industries* (New York: Free Press, 2001).

145. "U.S. Industry's Stake in International Standards Making," special supplement, *EIA Weekly Report to the Electronic Industries* (no date indicated), exhibit C of minutes of C63, December 16, 1971, Thomas Papers.

146. Minutes of C63, December 16, 1971, 6, Thomas Papers.

147. USNC 1657, "Voting Policy of the USNC/IEC (Approved [by] the Executive Committee December 5, 1973)," attachment to minutes of C63, December 12, 1974, Thomas Papers.

148. Handwritten manuscript of Ralph Showers's acceptance speech when he was awarded the IEC Lord Kelvin Award in 1998, courtesy of Janet Showers Patterson.

149. Heirman, "Completed Careers: Ralph Showers 1918–2013," 44; IEC, "Granting of the IEC Lord Kelvin Award in Houston," news release, October 21, 1998, courtesy of Janet Showers Patterson.

150. Minutes of C63, December 12, 1974, Thomas Papers.

151. Dan Hoolihan, interview by J. Yates.

152. See, for example, letter, Showers to J. E. Bridges, Michael De Lucia, E. D. Knowles, and R. E. Sharp, January 23, 1973, Showers Papers, in which Showers invokes his position as technical advisor to request comments on a proposed CISPR document to help him establish a US position on it.

153. The deduction pages of his tax forms from 1971 to 2000 (courtesy of Janet Showers Patterson) show that he took deductions for large numbers of trips he paid for, both national and international. For example, in 1972, he itemized $1,111 of travel expenses, which is equivalent to $6715 in 2017 dollars. When he was chairman of CISPR, his travel increased considerably. In 1985, his last year in that position, he declared $8,936 in unreimbursed travel expenses, equivalent to $ 20,329 in 2017 dollars. (All conversions made with inflation calculator at https://www.officialdata.org/, accessed May 4, 2017.)

154. For his homesickness, see Ralph to Bea Showers, September 5, 1967, from Copenhagen, in "Traveling the World with BeaBea and DaDa." The table of contents of this document indicates each trip location, date, and who wrote letters/postcards. Starting in 1970, correspondence was mostly from Bea Showers.

155. Ralph Showers to John O'Neil, August 13, 1973, copy for Mrs. Showers, courtesy of Janet Showers Patterson. The contents of "Traveling the World with BeaBea and DaDa" reveals that she had accompanied him on four major international trips by then, at least three of which were CISPR meetings, either plenary or working group.

156. In a 1971 C63 meeting, Showers explained that the UK had proposed creating a purely IEC committee with the same domain; "The problem is that the CISPR is somewhat of an autonomous group competing with IEC committees for jurisdiction over radio interference matters." The compromise, he presumed, would likely be to bring CISPR processes more in line with IEC ones. Minutes of C63, December 16, 1971, 6, Thomas Papers.

157. Åkerlind, *Fifty Years of CISPR*, 12, 14.

158. Examples of the various parts of the process may be found in CISPR/A (Secretariat) and CISPR/A (Central Office / Bureau Central), obtained in scanned form directly from Donald Heirman, former CISPR chairman. These papers have since been made available in MSA 291, Donald N. Heirman Papers, Karnes Archives and Special Collections, Purdue University [hereafter Heirman Papers].

159. For the informal version, see, for example, CISPR/A (Central Office) 14, March 1980, "Report on the Voting under the Six Months Rule for the approval of Document CISPR/A (Central Office) 9," Central Office Documents 13–88, 1980–94; for the later IEC form, see, for example, CISPR/A (Central Office) 71, annex A, result of voting on DIS—Document CISPR/A (Central Office) 58, date of ballot November 30, 1991, in Central Office Documents 13–88, 1980–94, both in Heirman Papers.

160. See, for example, "Changes in CISPR/A (Central Office) 9 Suggested as a Result of the Voting under the Six Months Rule," appendix to IEC/CISPR/A (Central Office) 14, March 1980, "Report on the Voting under the Six Months' Rule for the Approval of Document CISPR/A (Central Office) 9," in Central Office Documents 13–88, 1980–94, Heirman Papers.

161. As indicated in a CV for Ralph Showers, ca. 1991, courtesy of Janet Showers Patterson.

162. Minutes of C63, June 23, 1978, 14; see also Åkerlind, *Fifty Years of the CISPR*, annex 2, 12.

163. V. A. Leonov to Showers, March 5, 1984, Showers Papers.

164. The first female chair of CISPR, Bettina Funk from Sweden, started her term in 2016 (IEC/CISPR, accessed August 18, 2017, http://www.iec.ch/emc/iec_emc/iec_emc_players_cispr.htm). When asked about the first female CISPR delegates, a former CISPR chair recalls two or three Russian women attending the meetings in Leningrad in 1989, in part because they spoke better English than many of their male colleagues, and no more than half a dozen women total since then (personal communication from Don Heirman to J. Yates, July 9–10, 2017).

165. For example, according to a condolence message from Katsuaki Hoshi, a Japanese delegate to CISPR and secretary of the CISPR / Japanese National Committee (posted in an internet memorial guestbook for Ralph Showers), September 17, 2013, Showers "introduced us [the Japanese delegates] into the CISPR society in 1973" and encouraged their participation.

166. Personal communication from Janet Showers Patterson, April 1, 2018.

167. David Dixon, posted September 16, 2013, and sent in a September 16, 2013, email to Janet Showers Patterson, courtesy Janet Showers Patterson.

168. His path was followed more recently by a younger colleague he mentored, Don Heirman, who was CISPR chairman, 2007–16. Don Heirman, interview by J. Yates; see also "Oral-History: Don Heirman," *Engineering and Technology History Wiki*, accessed August 16, 2017, http://ethw.org/Oral-History:Don_Heirman.

169. Hugh Denny, quoted in Heirman, "Completed Careers: Ralph Showers 1918–2013," 45.

170. Herb Mertel, as quoted in Heirman, "Completed Careers: Ralph Showers 1918–2013," 46. The next quote is from Bob Hofmann, quoted in ibid.

171. Fred Bauer to Showers, November 30, 1998, courtesy of Janet Showers Patterson.

Chapter 7 · Computer Networking Ushers in a New Era in Standard Setting, the 1980s to the 2000s

1. On the first meeting of TC97, see Andrew L. Russell, *Open Standards and the Digital Age: History, Ideology, and Networks* (Cambridge: Cambridge University Press, 2014), 147–48. For X3's founding date, see "SDO: InterNational Committee for Information Technology Standards," ANSI, accessed June 29, 2016, http://www.standardsportal.org/usa_en/sdo/incits.aspx.

2. For JTC1 and IEC technical committees, see the IEC's timeline, "Techline," IEC, accessed August 17, 2017, http://www.iec.ch/about/history/techline/swf/; and ISO's homepage for JTC1, "ISO/IEC JTC1—Information Technology," ISO, accessed August 17, 2017, http://www.iso.org/iso/home/standards_development/list_of_iso_technical_committees/jtc1_home.htm. In 1960, IEC also created a Technical Committee on Computers and Information Processing, IEC TC53, but it was dissolved in 1990.

3. Russell, *Open Standards*, 172.

4. Russell, *Open Standards*, 149; on ACM's formation, see JoAnne Yates, *Structuring the Information Age: Life Insurance and Information Technology in the 20th Century* (Baltimore: Johns Hopkins University Press, 2005), 296n3.

5. IFIP identifies itself as "the leading multinational, apolitical organization in Information & Communications Technologies and Sciences," representing IT societies and recognized by UNESCO. "About IFIP," IFIP website, accessed August 17, 2017, http://www.ifip.org/index.php?option=com_content&task=view&id=124&Itemid=439.

6. Russell, *Open Standards*, 147–261. We draw heavily on his account in this section, as well as on Janet Abbate, *Inventing the Internet* (Cambridge, MA: MIT Press, 1999). ICANN, the Internet Corporation for Assigned Names and Numbers, is often referred to as an instrument of internet governance, and it is consequently sometimes thought, incorrectly, to be a standardization body. Although it is a multi-stakeholder body, it does not set technical standards. It provides central oversight over the internet naming function but does not even administer that function directly. Internet governance expert Laura DeNardis speculates that the great focus on ICANN results from two factors: the domain names it oversees are more visible to internet users than the underlying technical protocols, and international controversy has surrounded the US role in forming ICANN (Laura DeNardis, *The Global War for Internet Governance* [New Haven, CT: Yale University Press, 2014], 22). For differentiation of the various bodies involved in internet governance and their functions, see Mark Raymond and Laura DeNardis, "Multistakeholderism: Anatomy of an Inchoate Global Institution," *International Theory* 7, no. 3 (2015): 572–616. They classify ICANN's function as controlling "critical internet resources" rather than as setting internet standards.

7. Abbate, *Inventing the Internet*, 8–41.

8. Abbate, *Inventing the Internet*, 73–74; see also Russell, *Open Standards*, 168–69.

9. Russell, *Open Standards*, 170–71. Cerf would soon end up in Washington working for ARPA. Also in 1972, ARPA added "Defense" to the beginning of its name and the letter D to its acronym, becoming DARPA (except for a short period in the 1990s when it reverted to ARPA), but for the sake of simplicity, we will call it ARPA throughout this chapter.

10. Russell, *Open Standards*, 172–73.

11. Russell, *Open Standards*, 173–75, with quote from 174. Reference to ISO TC97 is from a Cerf 1974 document quoted on 174–75.

12. Abbate, *Inventing the Internet*, 122, 127, 140.

13. Russell, *Open Standards*, 178–79, 183–87.

14. Abbate, *Inventing the Internet*, 160.

15. Russell, *Open Standards*, 187–88, 193–96. The INWG vote, which was reported only in spring 1976, after the CCITT Working Group had reached its decision on what recommendation to present to the plenary, showed a majority favoring the compromise but included many no votes and abstentions, including the abstention of Robert Kahn.

16. Russell, *Open Standards*, 202–13.

17. Nevertheless, ISO made sure that TC97 Subcommittee 16 established formal liaisons to coordinate and share this area of standards setting with CCITT; moreover, IBM intervened and slowed things down at various points through representation in the committee delegations from various countries (e.g., the UK, France) where it operated (Russell, *Open Standards*, 218–19).

18. For the seven layers, see ISO/IEC International Standard 7498-1, in *Information Technology—Open Systems—Interconnection—Basic Reference Model: The Basic Model*, 2nd ed. November 15, 1994, corrected and reprinted June 15, 1996 (Geneva: ISO and IEC, 1996), 34.

19. Russell, *Open Standards*, 241–58.

20. Carl F. Cargill, *Open Systems Standardization: A Business Approach* (Upper Saddle River, NJ: Prentice Hall PTR, 1997), 73–74. For Cargill's recent affiliation, see the Adobe website, accessed March 23, 2018, https://www.adobe.com/devnet/author_bios/carl-cargill.html.

21. Series II, X3T2 materials, Boxes 6–8, Freeman Papers, Haverford College Archives; the quotation is from "Meeting 36," Box 7, Folder 8.

22. Abbate, *Inventing the Internet*, 142–45; the paper is Fred C. Billingsley, "An Almost-Everything—Independent Data Transfer Method," March 23, 1987, X3T2 Documents 1985–1999, Freeman Papers.

23. Abbate, *Inventing the Internet*, 206–8; Russell, *Open Standards*, 239–41. For the date and history of the early meetings, see Scott Bradner, "The Internet Engineering Task Force," in *Open Sources: Voices from the Open Source Revolution*, ed. Chris DiBona and Sam Ockman (Sebastopol, CA: O'Reilly Media, 1999), 47–53.

24. Andrew L. Russell, "'Rough Consensus and Running Code' and the Internet-OSI Standards War," *IEEE Annals of the History of Computing* 28 (July-September 2006): 48–61. The David Clark quote is from his "A Cloudy Crystal Ball: Visions of the Future" (plenary presentation at 24th meeting of the Internet Engineering Task Force, Cambridge, MA, July 3–17, 1992), quoted in Russell, "'Rough Consensus and Running Code,'" 49, 55.

25. See Russell, *Open Standards*, 229–61, for an excellent discussion of the countervailing autocratic and democratic strands in the evolution of the internet. The IAB has since broadened its membership further.

26. Gary Malkin, "The Tao of IETF: A Guide for New Attendees of the Internet Engineering Task Force," RFC 1391, IETF (1993), accessed August 17, 2017, https://tools.ietf.org/rfc/rfc1391. For its present form (with slightly changed subtitle), see Paul Hoffman, ed., "The Tao of IETF: A Novice's Guide to the Internet Engineering Task Force," IETF (2012), accessed August 17, 2017, http://www.ietf.org/tao.html. This web version, which supersedes the RFC versions, is based on the 2006 version published as RFC 4677 and edited by Hoffman and Susan Harris. Its conversion to a webpage is described in RFC 6722, "Publishing the 'Tao of the IETF' as a Web Page," IETF (2012), accessed August 17, 2017, https://tools.ietf.org/html/rfc6722. The RFC was revised twice in 1994, then in 2001 and 2006, before being turned into a webpage in 2012.

27. Quote from 2012 webpage version, Hoffman, "The Tao of IETF," Sec. 2; it has only minor changes of wording from original 1993 version.

28. Quote from 2012 webpage version, Hoffman, "The Tao of IETF," Sec. 4.2; the wording is identical in earlier versions, but section numbers vary. In the original section from the 2001 version, additional text notes the following: "(And, if you think about it, how could you have 'voting' in a group that anyone can join, and when it's impossible to count the participants?)" Susan Harris, ed., "The Tao of IETF: A Novice's Guide to the Internet Engineering Task Force," RFC 3160, Sec. 3.2, IETF (2001), accessed August 17, 2017, https://tools.ietf.org/pdf/rfc3160).

29. For a thorough discussion of why and how humming is used in IETF, see P. Resnick, "On Consensus and Humming in the IETF," RFC 7282, IETF (2014), accessed August 17, 2017, https://tools.ietf.org/html/rfc7282.

30. See J. Galvin, "IAB and IESG Selection, Confirmation, and Recall Process: Operation of the Nominating and Recall Committees," in *BCP [Best Current Practices] 10*, RFC 3777, IETF (June 2004), accessed August 17, 2017, https://tools.ietf.org/html/rfc3777.

31. Timothy Simcoe found that "between 1992 and 2000 the median time from first draft to final specification grew from seven to fifteen months" ("Delay and De Jure Standardization: Exploring the Slowdown in Internet Standards Development," in *Standards and Public Policy*, ed. Shane Greenstein and Victor Stango [New York: Cambridge University Press, 2007], 260–95, quote from p. 262). He suggests that the growth of IETF attendees during this period of internet commercialization contributed to this increase. His measure starts at the point of the first published draft, thus omitting the time to get from the first idea to that point, which could also be significant.

32. Richard Barnes, interview by J. Yates, January 15, 2015.

33. DeNardis, *The Global War for Internet Governance*, 73.

34. O. Kolkman, S. Bradner, and S. Turner, "Characterization of Proposed Standards," RFC 7127, IETF (January 2014), accessed March 28, 2018, https://tools.ietf.org/html/rfc7127.

35. See, for example, archived IETF email list messages from the period right before the July 1992 meeting in ftp://ftp.ietf.org/ietf-mail-archive/ietf/ under 1992–07 (accessed July 7, 2016). See also Russell, *Open Standards*, 244–46, for reference to the flame wars and Russell, "'Rough Consensus and Running Code,'" 55, for a description of the meeting. This style of communication, as linguists of computer-mediated communication have observed since the mid-1990s, tends to attract male participants but to repel female participants. For the beginning of this literature, see Susan Herring, "Gender Differences in Computer-Mediated Communication: Bringing Familiar Baggage to the New Frontier" (keynote talk at panel entitled "Making the Net*Work*" at American Library Association annual convention, Miami, FL, June 27, 1994), accessed March 28, 2018, http://www.universalteacher.org.uk/lang/herring.txt.

36. Malkin, "The Tao of IETF."

37. All quotes in this paragraph from Russ Housley (IETF chair, 2007–13), interview by J. Yates, May 14, 2013, and May 16, 2013.

38. Russell, *Open Standards*, 217–19.

39. Andrew Russell (personal communication with J. Yates, January 10, 2018) notes that this lack of balance rules also avoided oversight, since in an ANSI committee a party could challenge a decision at a higher level on the basis of insufficient balance. As the internet has globalized, issues of global balance have also arisen, and they will be discussed later in the chapter.

40. Russ Housley, interview by J. Yates, May 14, 2013, and May 16, 2013.

41. Andrew Updegrove, "The Essential Guide to Standards," ConsortiumInfo.org, accessed August 17, 2014, http://www.consortiuminfo.org/essentialguide/. In a 2013 interview Updegrove said that he had set up 120 consortia, and that he was aware of only two other law firms that had set up "a fair number," but fewer than he had, and that most of the other consortia he was aware of were set up by lawyers who had just done one or two. Andrew Updegrove, interview by C. Murphy and J. Yates, February 27, 2013.

42. For a discussion of the various legal forms, see Updegrove, "Essential Guide," "II—What is a Consortium?"

43. Tineke M. Egyedi, "Consortium Problem Redefined: Negotiating 'Democracy' in the Actor Network on Standardization," *International Journal of IT Standards and Standardization* 1, no. 2 (July-December 2003): 23. For an extended contemporary discussion of the varieties of standards consortia, see Andrew Updegrove, "Dissecting the Consortium: A Uniquely Flexible Platform for Collaboration," *Standards Today* 9, no. 1 (January/February 2010): 1–17.

44. For example, see Updegrove, "Dissecting the Consortium," 6.

45. Andrew Updegrove, "1988—the Year of the Consortium," Gesmer Updegrove LLP, accessed August 17, 2017, http://www.gesmer.com/news/july-1989-issue-of-the-technology-law-bulletin-1988—the-year-of-the-consortium. Originally published in the July 1989 issue of the *Technology Law Bulletin*.

46. Biographical information on Cargill from the Adobe website, accessed March 23, 2018, https://www.adobe.com/devnet/author_bios/carl-cargill.html.

47. Carl F. Cargill, "Consortia and the Evolution of Information Technology Standardization," in *Standardisation and Innovation in Information Technology: SIIT'99 Proceeding: Aachen, September 15–17, 1999*, edited by K. Jakobs and R. Williams (Piscataway, NJ: IEEE), 37–42.

48. Andrew L. Russell, "Dot-org Entrepreneurship: Weaving a Web of Trust," *Entreprise et Histoire* 51 (June 2008): 48. He also notes additional legislation passed in subsequent years that supported standards consortia (49).

49. Cargill, "Consortia," 2–8.

50. Updegrove has argued, however, that consortia founded with strategic aims were less successful and long-lasting than nonstrategic consortia (Andrew Updegrove, "Forming, Funding, and Operating Standard-Setting Consortia," *IEEE Micro* 13, no. 6 (December 1993): 52–61).

51. Cargill, *Open Systems Standardization*, 222–25.

52. Cargill, *Open Systems Standardization*, 219–22.

53. See "About Us," Open Group, accessed August 17, 2017, http://www.opengroup.org/aboutus.

54. Cargill, "Consortia," 8.

55. Andrew Updegrove, "Changing Industries / Changing Consortia: 13A (A Case Study)," *Consortium Standards Bulletin* 2, no. 3 (February 2003), accessed August 17, 2017, http://www.consortiuminfo.org/bulletins/feb03.php#trends.

56. See "History of Ecma," Ecma International, accessed August 17, 2017, http://www.ecma-international.org/memento/history.htm. This paragraph draws on that history except as indicated.

57. ECMA bylaws, quoted from *ECMA Memento*, August 1962, and repeated every year through 1992. The *ECMA Memento* for 1993 shows revised bylaws with a different definition referring to hardware and software. *ECMA Mementos* from 1962 through 2016 are available for download at http://www.ecma-international.org/publications/memento_index.htm (accessed August 17, 2017).

58. Egyedi, "Consortium Problem Redefined," 23.

59. Egyedi, "Consortium Problem Redefined," 34.

60. Andrew Updegrove, "Openness and Legitimacy in Standards Development," *Standards Today* 11, no. 1 (November 2012), accessed September 4, 2018, https://www.consortiuminfo.org/bulletins/nov12.php#feature.

61. ISO/IEC JTC1 N535, August 31, 1989, "Directives for the Work of ISO/IEC Joint Technical Committee 1 (JTC1) on Information Technology," 25, accessed August 17, 2017, http://isotc.iso.org/livelink/livelink?func=ll&objId=6721404&objAction=browse. See also Tineke Egyedi, "Shaping Standardization: A Study of Standards Processes and Standards Policies in the Field of Telematic Services" (PhD thesis, Technische Universiteit Delft, 1996). Egyedi provides a compact summary of the steps for ISO/IEC in general (106–8, based on ISO/IEC Directives, Part 1, 1992, p. 16), and states that the Fast Track procedure was set up in 1987 for both ISO and IEC. She also traces the specifics of the evolution of the Fast Track process (108–11).

62. ISO/IEC JTC1 N535, 35–36; Tineke M. Egyedi, "Why Java Was—Not—Standardized Twice," *Computer Standards & Interfaces* 23 (2001): 253–65, see especially 257.

63. E.g., a message to the IEEE802 list, October 24, 2000, forwards an announcement that "ISO/IEC JTC1 has approved the Type A Liaison between the IEEE Computer Society and ISO/IEC JTC1/SC7 (Software and Systems Engineering)," accessed March 22, 2018, http://www.ieee802.org/secmail/msg00931.html.

64. John L. Hill, "Publicly Available Specification: A New Paradigm for Developing International Standards," *Computer* 28, no.10 (October 1995): 97–98. Egyedi ("Java," 257) dates the JTC1 PAS process as 1994/1999, implying that it became official (rather than just a trial) in 1999.

65. List of approved PAS submitters to JTC1 (past and current), accessed August 17, 2017, http://isotc.iso.org/livelink/livelink?func=ll&objId=8913248&objAction=browse&sort=name.

66. ISO/IEC Directives Part 1, Edition 12.0. 2016-05, IEC website, accessed August 18, 2017, http://www.iec.ch/members_experts/refdocs/. Indeed, a JTC1 supplement to the directives is still issued to highlight this and other JTC1-specific processes (see "ISO/IEC Directives,

Part 1: Consolidated JTC 1 Supplement 2015—Procedures specific to JTC 1, Based on ISO/IEC Directives Part 1 Eleventh Edition- 2014," IEC, accessed August 18, 2017, http://www.iec.ch/members_experts/refdocs/). It refers to earlier editions of this document, probably starting in the mid-1990s. In the broader ISO/IEC directives, a PAS is valid for a three-year period, and may be renewed for another, but must eventually lapse or follow the regular process to become an International Standard. The IEC itself currently defines a PAS as "a publication responding to an urgent market need," often reflecting "a consensus in an organization (e.g. manufacturers or commercial associations, industrial consortia, user group and professional and scientific societies) external to the IEC." It is "designed to bring the work of industry consortia into the realm of the IEC." "Publicly Available Specifications (PAS)," IEC, accessed July 13, 2016, http://www.iec.ch/standardsdev/publications/pas.htm.

67. Joseph Farrell and Garth Saloner, "Coordination through Committees and Markets," *RAND Journal of Economics* 19, no. 2 (1988): 235–52.

68. This discussion of the emergence of the World Wide Web and W3C is based on Tim Berners-Lee, *Weaving the Web: The Original Design and Ultimate Destiny of the World Wide Web by Its Inventor* (New York: Harper Collins, 1999); Russell, "Dot-org Entrepreneurship," 44–56; Andrew Russell, "Constructing Legitimacy: The W3C's Patent Policy," in *Opening Standards: The Global Politics of Interoperability*, ed. Laura DeNardis (Cambridge, MA: MIT Press, 2011), 159–76; and Abbate, *Inventing the Internet*, 212–18. For the transformation to URL in IETF standardization, see Tim Berners-Lee, Robert Cailliau, Ari Luotonen, Henrik Frystyk Nielsen, and Arthur Secret, "The World-Wide Web," *Communications of the ACM* 37, no. 8 (August 1994): 76–82.

69. Russell, "Dot-org Entrepreneurship," 45; JoAnne Yates and Craig N. Murphy, "Charles le Maistre: Entrepreneur in International Standardization," *Entreprise et Histoires* 51 (2008): 10–27.

70. Since then the European Research Consortium in Informatics and Mathematics has replaced INRIA as the European host, and a fourth host, Beihang University, was added in 2013. See "Facts about W3C," W3C, accessed August 18, 2017, https://www.w3.org/Consortium/facts.html.

71. Simpson Garfinkel, "The Web's Unelected Government," *Technology Review* 101, no. 6 (November/December 1998): 38–46.

72. The complete history of W3C's fee structure is available on the W3C website, at "W3C History," W3C, accessed August 18, 2017, https://www.w3.org/Consortium/fee-history.

73. See, for example, Garfinkel, "Unelected Government"; and Egyedi, "Consortium Problem Redefined," 29–30. Egyedi refers to his role as being that of a dictator, rather than a king.

74. "Memo from the Top," excerpting Tim Berners-Lee's response to critics' complaints that W3C is under his control, *Technology Review* 101, no. 6 (November/December 1998): 47.

75. Patents and patent policies were uncommon in standards before the 1980s. Rudi Bekkers and Andrew Updegrove, "A Study of IPR Policies and Practices of a Representative Group of Standards Setting Organizations Worldwide" (commissioned by the US National Academies of Science, Board of Science, Technology, and Economic Policy [STEP], Project on Intellectual Property Management in Standard-Setting Processes), National Academies of Science, September 17, 2012, accessed August 18, 2017, http://sites.nationalacademies.org/cs/groups/pgasite/documents/webpage/pga_072197.pdf. Bekkers and Updegrove note what they claim may be the first formal patent policy on standards in an ANSI Committee on Procedure recommendation, "that as a general proposition patented design or methods not be incorporated in standards. However, each case should be considered on its own merits and if a patentee

be willing to grant such rights as will avoid monopolistic tendencies, favorable consideration to the inclusion of such patented designs or methods in a standard might be given" (referenced by authors to minutes of ANSI Standards Council, Item 2564: "Relation of Patented Designs or Methods to Standards," November 30, 1932).

76. Except as indicated, this paragraph is drawn from Russell, "Constructing Legitimacy," 165–71. The role of the court case comes from an email from Andrew Updegrove to J. Yates, July 11, 2017. He also noted that RAND policies may require disclosure of patents, thus preventing the problems of submarine patents, which Russell ("Constructing Legitimacy," 165–71) includes as an advantage of a royalty-free policy for corporate members.

77. W3C, *W3C Patent Policy Framework: W3C Working Draft 16 August 2001*, W3C, August 16, 2007, accessed August 18, 2017, https://www.w3.org/TR/2001/WD-patent-policy-20010816/.

78. Tim Berners-Lee to Patent-Policy Comment mailing list, October 24, 2001, accessed August 18, 2017, https://lists.w3.org/Archives/Public/www-patentpolicy-comment/2001Oct/1642.html.

79. Mission and design principles come from "W3C Mission," W3C, accessed January 2, 2017, https://www.w3.org/Consortium/mission.

80. "About," OpenStand, accessed February 15, 2013, http://open-stand.org/about-us. By January 2, 2017, the OpenStand website had reworded the third principle as "collective empowerment" based on the items listed in the third bullet of the original statement quoted above.

81. See, for example, the Center for Democracy and Technology's paper, "The Importance of Voluntary Technical Standards for the Internet and Its Users," August 29, 2012 (the day of the OpenStand declaration), accessed August 11, 2017, https://cdt.org/insight/the-importance-of-voluntary-technical-standards/. This paper addresses what it perceives as the ITU threat very specifically, referencing OpenStand. The OpenStand declaration simply affirms the principles of open standards, with no mention of ITU. Indeed, when asked about this, Jeff Jaffe, W3C CEO, stated that although the OpenStand principles differ from those of the ITU, the purpose of the declaration was simply to affirm what is a best practice—not to distinguish it from any other approach. Jeff Jaffe, interview by J. Yates, November 1, 2016.

82. The OpenStand site defines balance very much in keeping with older principles: "Standards activities are not exclusively dominated by any particular person, company or interest group" ("Principles," OpenStand, accessed February 17, 2014, http://open-stand.org/principles/).

83. For a different perspective on openness in information and communication standards, see Russell, *Open Standards*.

84. W3C has three levels of confidentiality: public, members only, and team only; most working group sites and mailing lists are public ("W3C Process Document," W3C, accessed July 11, 2017, https://www.w3.org/2005/10/Process-20051014/comm.html). IETF committee lists are open to the public.

85. Jaffe, interview by J. Yates, November 1, 2016.

86. Indeed, in 2013, former IETF chair Housley noted that IETF was actively discussing diversity of various types, including international, because when new members had recently been chosen for the IESG (IETF's top governance structure), they were "all males from Western countries." Housley, interview by J. Yates, May 14, 2013, and May 16, 2013.

87. "Past Meetings," IETF, accessed March 24, 2018, https://www.ietf.org/how/meetings/past/.

88. "A Little History of the World Wide Web" and "Technical Plenary (TPAC) Meetings," W3C, accessed March 24, 2018, https://www.w3.org/History.html and https://www.w3.org/2002/09/TPOverview.html.

89. Andrew Edgecliffe-Johnson, "Lunch with the FT: Tim Berners-Lee," *Financial Times*, September 12, 2012, accessed September 13, 2018, https://www.ft.com/content/b022ff6c-f673-11e1-9fff-00144feabdc0#axzz2NRakIIKj.

90. "Sir Tim Berners-Lee Hopes Peace Will Be the Lasting Legacy of the World Wide Web," *Drum*, September 5, 2012, accessed September 5, 2017, http://www.thedrum.com/news/2012/09/14/sir-tim-berners-lee-hopes-peace-will-be-lasting-legacy-world-wide-web#wKcodIEx8g7x4jSg.99.

91. Trond Arne Undheim, "The Standardization Landscape in 2020," *Trond's Opening Standard* (Oracle blog), May 27, 2009, accessed May 22, 2017, https://web.archive.org/web/20090621101432/https://blogs.oracle.com/trond/.

Chapter 8 · Development of the W3C WebCrypto API Standard, 2012 to 2017

1. Although the email archives and meeting minutes for this group are public, the teleconference and face-to-face meetings themselves are not. Jeff Jaffe (W3C CEO) and working group chair Virginie Galindo gave J. Yates access to these meetings starting in late 2012. This access allowed Yates to follow the Web Cryptography Working Group (referred to as WebCrypto WG) from inside over a four-year period, reading all posts on the Working Group's mailing list (5,428 messages from its launch in April 2012 through January 2017, when WebCrypto API became a recommendation/standard), attending most regular (typically biweekly) meetings via audio teleconferencing and internet relay chat, attending two face-to-face meetings, and interviewing key members of the group twice.

2. The charter also specified an additional deliverable that was not on the recommendation track, a use cases and requirements document that would set out requirements for any features added to the spec beyond those listed in the charter, a common part of W3C standards development.

3. W3C, *Web Cryptography Working Group Charter* (W3C, September 20, 2016), accessed January 2, 2017, https://www.w3.org/2011/11/webcryptography-charter.html. This link is to the final version of the document, but only the milestone dates have changed since the original.

4. *World Wide Web Consortium Process Document* [hereafter *W3C Process Document*], September 1, 2015, accessed Sept. 10, 2018, https://www.w3.org/2015/Process-20150901/. This document also says, "The Director and CEO *may* delegate responsibility (generally to other individuals in the Team) for any of their roles described in this document." Yet a chance conversation between Yates and W3C director Berners-Lee, at MIT's 2016 doctoral hooding ceremony on June 2, 2016, revealed that Berners-Lee knew who chaired WebCrypto WG, suggesting that he was very much involved with W3C WGs.

5. According to WebCrypto WG member Mark Watson, W3C had a stronger editor than any other standardization body he had worked in, including ISO, IETF, ITU-T, and the European Telecommunications Standards Institute. Mark Watson, interview by J. Yates, April 9, 2013.

6. Watson, interview by J. Yates, April 9, 2013.

7. *W3C Process Document*, sec. 6, "W3C Technical Report Development Process," accessed January 4, 2017, https://www.w3.org/2015/Process-20150901/#Policies. This section and any quotes in it come from this document, unless otherwise indicated.

8. See *Web Cryptography Working Group Charter*, accessed September 10, 2018, https://www.w3.org/2011/11/webcryptography-charter.html. Its milestones include FPWD, last call, CR, PR, and recommendation. In the 2015 *W3C Process Document*, Last Call is not listed as a separate stage, and the text notes that "a Candidate Recommendation under this process corresponds to the 'Last Call Working Draft' discussed in the Patent Policy."

9. The history of the WebCrypto Charter in the W3C CVS log called for an end date of March 2014, two years after the charter was originally created, in March 2012.

10. Minutes of W3C WebCrypto, public meeting, May 7, 2012, accessed January 4, 2017, https://www.w3.org/2012/05/07-crypto-minutes.html. A two-month delay occurred between the creation of the charter and the launching of the WG, leaving one year and ten months on the original timeline.

11. Since the WebCrypto WG was launched, according to W3C CEO Jeff Jaffe, W3C has moved to a new model, adding Community Groups in which possible standards can be incubated and some preliminary consensus achieved before a Working Group is officially established (Jeff Jaffe, interview by J. Yates, November 1, 2016). Although that process will speed up time in working groups, it may not materially shorten the entire standards process from conception to recommendation.

12. Permissions to quote from WebCrypto WG participant interviews, which were obtained according to a procedure agreed to by MIT's institutional review board, allow us to refer to participants by real first names in the text and, when using approved quotations, by full names in the notes.

13. This background and the quote about her role come from Virginie Galindo, interview by J. Yates, April 25, 2013.

14. W3C and IETF working groups often have cochairs, with one driving process and one content (Karen O'Donoghue, interview by J. Yates, May 7, 2013).

15. For more on the workshop, see also "Call for Participation," W3C, accessed January 4, 2017, https://www.w3.org/2011/identity-ws/. For DomCrypt and its relation to WebCrypto, see the 2011 description of DomCrypt in the *Mozilla Wiki*, updated July 26, 2011, accessed September 11, 2018, https://wiki.mozilla.org/Privacy/Features/DOMCryptAPISpec/Latest.

16. WebCrypto, public meeting. Note that these teleconference minutes (and all other meeting minutes, teleconference or face to face, cited in this chapter) are created by a designated scribe for the meeting, who throughout the meeting typed a running summary of the discussion into internet relay chat (to which all attendees at the meeting were typically also connected and through which they could also comment inaudibly at any time during the meeting), creating a real-time running account. A link to these meeting minutes was sent out after the meeting, and the minutes were approved at the beginning of the subsequent meeting.

17. WebCrypto, public meeting.

18. David Dahl to the WebCrypto list, "Re: [WebCrypto WG] Agenda for next call on 14th of May," May 9, 2012. Unless otherwise indicated, all email from the WG list cited throughout the chapter is available in the WG's email archives at https://lists.w3.org/Archives/Public/public-webcrypto/, accessed January 6, 2017. In identifying email from the list, we cite date and subject line, but note that frequently there were multiple exchanges by the same individuals over the list on the same day and with the same subject line. Also, specific messages are often addressed to the author(s) of the previous message(s) as well as copied to the WebCrypto list as a whole, but for simplicity we indicate all messages only as to the list.

19. Ryan Sleevi to list, "Re: [WebCrypto WG] Agenda for next call on 14th of May," May 9, 2012.

20. David Dahl to list, "Re: [WebCrypto WG] Agenda for next call on 14th of May," May 10, 2012.

21. Minutes of W3C WebCrypto API WG, face-to-face meeting May 21, 2012, accessed January 6, 2017, https://www.w3.org/2012/05/21-crypto-minutes.html.

22. Raw survey data was shared by a link in David Dahl to list, "Fwd: Survey raw data," June 6, 2012; and also a summary in David Dahl to list, "Survey Summary (Google Docs Version)," June 6, 2012.

23. Minutes of W3C WebCrypto Working Group, face-to-face meeting June 11, 2012, accessed January 6, 2017, https://www.w3.org/2012/06/11-crypto-minutes.html.

24. Ryan Sleevi to list, "Strawman proposal for the low-level API," June 18, 2012. For David's comment on it, see minutes of W3C Web Cryptography WG teleconference, July 2, 2012, accessed January 6, 2017, https://www.w3.org/2012/07/02-crypto-minutes.html.

25. Minutes of Web Cryptography WG, face-to-face meeting, July 24, 2012, accessed January 6, 2017, https://www.w3.org/2012/07/24-crypto-minutes.html.

26. Other documents such as the use case document, for example, also had one or more editors at different points, but they were decidedly less influential in the process as a whole.

27. Data from "public-webcrypto@w3.org Mail Archives," accessed July 12, 2017, http://lists.w3.org/Archives/Public/public-webcrypto/.

28. Crypto Forum Research Group (CFRG) members to Zooko Wilcox-O'Hearn [now Zooko Wilcox], forwarded in Zooko Wilcox-O'Hearn to list, "feedback from CFRG," September 20, 2012.

29. Ryan Sleevi to list, "Re: Feedback from CFRG," September 20, 2012.

30. The complaint was posted to the WG's list for public comments, as was the response: Ryan Sleevi to the Public-webcrypto-comments list, September 19, 2012. Email from this public comments list is available in the WG's email archives, accessed September 11, 2018, at https://lists.w3.org/Archives/Public/public-webcrypto-comments/.

31. On the public-webcrypto-comments list, non–WG members commenting on the public working drafts repeatedly challenged some of the characteristics of the low-level API by calling for more safeguards to prevent what they often referred to as bad or insecure crypto on the part of developers. See, for example, thread with subject line "web crypto API: Side effects of a low-level API [1/6]," May, 24–25, 2013.

32. For example, in August 2012, the highest-traffic month (with 425 messages posted), Ryan wrote 120, or 28 percent of the total. Figures computed by sorting the public-webcrypto@w3.org Mail Archives by author for each month, counting the number from Ryan, and figuring the total by subtracting from the listed total the automated messages from the W3C Tracker and Bugzilla (used for discussion of debugging the spec). Only one month, October 2014, showed Ryan's participation as less than 10 percent.

33. See, for example, the thread with subject line "Re: Proposal for key wrap/unwrap (ISSUE-35), March 1–20, 2013.

34. Minutes of W3C Web Cryptography WG, November 27, 2012, accessed January 12, 2017, https://www.w3.org/2012/11/26-crypto-minutes.html.

35. Mark Watson to list, November, 28, 2012. Subsequent quotes are from further messages to this thread, subject line "Re: On Optionality," from Mike Jones, Ryan Sleevi, Mark Watson, and Virginie Galindo, all on the same date.

36. Ryan Sleevi to list, "Re: W3C Web Crypto WG—no conf call today, but voting instead," November 19, 2012.

37. Harry Halpin to list, "Re: W3C Web Crypto—no conf call today, but voting instead," November 19, 2012.

38. Minutes of Web Cryptography WG, face-to-face meeting, November 14–15, 2013, https://www.w3.org/2013/11/14-crypto-minutes.html and https://www.w3.org/2013/11/15-crypto-minutes.html, both accessed on January 12, 2017. This problem of time zones for phone conferences was never solved, despite the chair's multiple attempts to try alternative times and offers to hold additional phone meetings with the Koreans at other times.

39. This discussion took place on November 15, 2013. See Virginie Galindo to list, "W3C Web Crypto WG—Shenzhen F2F Take Away," November 27, 2013.

40. Minutes of Web Cryptography WG, November 14–15, 2013. This discussion took place on November 15 at the end of the meeting, and minutes do not indicate any response from Ryan to the offer of help.

41. Minutes of W3C Web Cryptography WG, December 2, 2013, accessed January 12, 2017, https://www.w3.org/2013/12/02-crypto-minutes.html.

42. Ryan Sleevi, interview by J. Yates, October 30, 2014.

43. Mark described the stylistic difference as follows: "I'm still not convinced by the approach that Ryan kind of imposed, in terms of having long procedural descriptions with a lot of duplicated procedures everywhere. So I would've done it differently, but he was very insistent on that from the beginning, so then we had to just keep jumping through those hoops." Mark Watson, interview by J. Yates, October 30, 2014. Following quotes are also from that source.

44. Sleevi, interview by Yates.

45. Minutes of W3C Web Cryptography WG, face-to-face meeting May 14, 2012, accessed January 13, 2017, https://www.w3.org/2012/05/14-crypto-minutes.html.

46. Minutes of W3C Web Cryptography WG, face-to-face meeting, October 30, 2014, accessed January 16, 2017, https://www.w3.org/2014/10/30-crypto-minutes.html. Minutes indicate that Chrome, Internet Explorer, and Firefox all had implementations, but no one knew about Apple Safari. In addition, Richard announced that he had built a polyfill implementation of the API (written in pure JavaScript on top of—rather than within—a browser) called PolyCrypt (Richard Barnes to list, "PolyCrypt," January 7, 2013).

47. See, for example, Virginie Galindo to list, "W3C Web Crypto WG—testing activity wiki," July 25, 2013. A W3C test expert explained its testing tools at the face-to-face meeting on October 30, 2014, but admitted that W3C still had trouble getting good tests because establishing and running them involved considerable work.

48. Minutes of Web Cryptography WG, September 14, 2015, W3C, accessed January 16, 2017, https://www.w3.org/2015/09/14-crypto-minutes.html. See, for example, thread starting with Harry Halpin to list, "WebCrypto edits on key material (Option 2)," January 15, 2016.

49. Ryan Sleevi to list, "Re: [W3C Web Crypto WG] how to progress?," January 21, 2016.

50. For example, see new members and discussion of testing in minutes of W3C Web Cryptography WG, March 7, 2016, accessed July 27, 2017, https://www.w3.org/2016/03/07-crypto-minutes.html; minutes of W3C Web Cryptography WG, April 4, 2016, accessed January 6, 2017, https://www.w3.org/2016/04/04-crypto-minutes.html; minutes of W3C Web Cryptography WG, June 6, 2016, W3C, accessed January 6, 2017, https://www.w3.org/2016/04/04-crypto-minutes.html.

51. In Virginie Galindo to list, "[W3C Web Crypto WG] CfC on moving the Web Crypto API to PR —> 24th of October," October 10, 2016, she announced that during the phone meeting that day, the group agreed to move to PR, pending the two-week waiting period for online objections. WebCrypto API's final approval as a Recommendation was announced in W3C staff contact Wendy Seltzer to list, "Fwd: The W3C Web Cryptography API is a W3C Recommendation," January 26, 2017.

52. WC3, *Web Cryptography Working Group Charter.* The three W3C working groups were the HTML Working Group, the Web Applications Working Group, and the Web Application Security Group. As W3C committees, these groups automatically had opportunities to comment on WebCrypto published working drafts.

53. Richard Barnes, interview by J. Yates, April 22, 2013. A search of the entire WebCrypto mailing list beginning in May 2012 showed only 25 messages (out of 5,433 total) that referred to ECMA, and the references were to ECMA standards for some aspect of JavaScript, not to any current contact with ECMA, the organization. In contrast, 104 messages referred to WHATWG, 307 to JOSE, and 639 to the IETF, and these often referred to current contacts.

54. W3C, "Dependencies and Liaisons."

55. Mike Jones (Michael B. Jones), primary editor of JOSE and active in WebCrypto WG, acted as de facto JOSE liaison to WebCrypto WG. Cochair Karen O'Donoghue was included as an invited expert and regularly attended teleconference meetings but was not active on the

list; cochair Jim Schaad was active in meetings and on the list, particularly during 2016. Other WebCrypto WG members such as Richard Barnes participated in both.

56. Ryan Sleevi to list, "Re: crypto-ISSUE-13: Relationship between the W3C Web Cryptography work product and the IETF JOSE WG [Web Cryptography API]," August 5, 2012.

57. Harry Halpin to list, "Re: crypto-ISSUE-13: Relationship between the W3C Web Cryptography work product and the IETF JOSE WG [Web Cryptography API]," August 8, 2012; and Ryan Sleevi to list, same subject, August 8, 2012.

58. For agreement, see for example David Dahl to list, "Re: crypto-ISSUE-13: Relationship between the W3C Web Cryptography work product and the IETF JOSE WG [Web Cryptography API]," August 8, 2012; see also Mike Jones to list, same subject, August 17, 2012.

59. Ryan Sleevi to list, "Proposed text to close ISSUE-13," August 29, 2012; Mike Jones to list, same subject, August 29, 2012.

60. Richard Barnes to list, "Re: IANA registry for WebCrypto?" January 18, 2013; Ryan Sleevi to list, same subject, January 18, 2013; Richard Barnes to list, same subject, January 18, 2013; the thread extends to January 24, 2013.

61. Barnes, interview by Yates.

62. Mike Jones to list, "FW: JOSE—19 drafts intended for Working Group Last Call," December 29, 2013. See also Ryan Sleevi to list and additional Mike Jones to list, same subject, all December 29, 2013.

63. See WHATWG—FAQ, accessed September 11, 2018, https://whatwg.org/faq.

64. Jaffe, interview by Yates. Ryan Sleevi also stated that this was the reason WHATWG was created (Sleevi, interview by Yates). For a look at the uneasy relationship as regards HTML5, see Stephen Shankland, "Growing Pains Afflict HTML5 Standardization," CNET, June 28, 2010, accessed March 12, 2017, https://www.cnet.com/news/growing-pains-afflict-html5-standardization/.

65. WHATWG—FAQ.

66. Barnes, interview by Yates.

67. Shankland, "Growing Pains." Hickson was at Opera when WHATWG was founded but soon after moved to Google. He is no longer the sole editor for WHATWG. The quote comes from "FAQ," *WHATWG Wiki*, accessed September 15, 2017, https://web.archive.org/web/20170715193408/https://wiki.whatwg.org/wiki/FAQ. This is no longer official policy. The language in WHATWG's current site talks about very rough consensus determined by the editor after listening to the community (WHATWG—FAQ).

68. Minutes of W3C Web Cryptography WG, June 3, 2013, accessed January 16, 2017, https://www.w3.org/2013/06/03-crypto-minutes.html. On Microsoft and WHATWG, see Garry Trinder, "You, Me and the W3C (aka Reinventing HTML)," in *Albatross!* (the personal blog of Chris Wilson), January 10, 2007, accessed July 12, 2017, https://blogs.msdn.microsoft.com/cwilso/2007/01/10/you-me-and-the-w3c-aka-reinventing-html/. Microsoft has changed its stance towards WHATWG more recently: "In 2017, Apple, Google, Microsoft, and Mozilla helped develop an IPR policy and governance structure for the WHATWG, together forming a Steering Group to oversee relevant policies," WHATWG—FAQ.

69. Ryan Sleevi to list, "Re: Registries and Interoperability," February 6, 2013.

70. Arun Ranganathan to list, "Re: Registries and Interoperability," February 7, 2013.

71. Barnes, interview by Yates.

72. Sleevi, interview by Yates. In the subsequent quote from the same interview, Ryan used the term "user agents" rather than "browsers"; for clarity I've substituted "browsers."

73. See, for example, Jeff Jaffe, "Decision by Consensus or by Informed Editor; Which Is Better?," W3C CEO's blog, October 7, 2014, accessed July 13, 2017, https://www.w3.org/blog/2014/10/decision-by-consensus-or-by-informed-editor-which-is-better/.

74. Virginie Galindo to list, "[W3C Web Crypto] CfC to make Key Discovery a Note," October 30, 2015; Virginie Galindo to list, same subject, March 23, 2016.

75. Mark said that his priorities were his company's priorities. He also noted that his company seemed to be unusual in WebCrypto WG, in that other very large application providers were not part of the WG—the big companies represented were mostly the browsers (Watson, interview by Yates, October 30, 2014). For W3C's priority of constituencies, see W3C, *HTML Design Principles: W3C Working Draft 26 November 2007* (W3C, November 27, 2007), accessed January 24, 2017, https://www.w3.org/TR/html-design-principles/#priority-of-constituencies: "In case of conflict, consider users over authors over implementors over specifiers over theoretical purity." The implementors are browser vendors, while the authors are website authors or application providers.

76. Based on the WebCrypto official list of participants in 2013 and 2017 that are no longer available online. Of the original 55 members, 9 were invited experts, and 46 represented 21 member organizations; of the 70 members listed at the end, 5 were invited experts, and 65 were from member organizations.

77. We tallied a list of attendees for each meeting from the minutes, as established by Zakim, the Semantic Web agent/bot that works in conjunction with internet relay chat to facilitate meetings. Before Zakim has learned the names and numbers of participants, it lists telephone numbers of those calling in, as well as separately listing names when someone has manually mapped a number to a name. Thus, the numbers and names that Zakim lists may include overlap, especially in the first year, leading to tallies that exceed the actual number present. During this period, 27 was the highest listed by Zakim. By late 2013, however, Zakim recognized most call-in numbers, and thus listed primarily or only names. From December 2013 onward, only 4 of 35 meetings exceeded single digits (with 13 the highest number), all during the contentious period around the Last Call milestone.

78. O'Donoghue, interview by Yates.

79. Don Heirman, interview by J. Yates, August 9, 2014.

80. Mark Watson, interview by Yates, April 9, 2013.

81. Vijay Bharadwaj, interview by J. Yates, May 1, 2013.

82. Galindo, interview by Yates.

83. Barnes, interview by J. Yates.

84. Bharadwaj, interview by Yates.

85. Russ Housley, interview by Yates, May 14, 2013, and May 16, 2013.

Chapter 9 · Voluntary Standards for Quality Management and Social Responsibility since the 1980s

1. Robin Kinross, "A4 and Before: Towards a Long History of Paper Standards" (Sixth Koninklijke Bibliotheek Lecture, Netherlands Institute for Advanced Study in the Humanities and Social Sciences, Wassenaar, 2009), 7.

2. Estimate of the size of the UN staff is from Craig N. Murphy, foreword to *The United Nations Development Programme and System*, by Stephen Browne (London: Routledge, 2011), xix.

3. John Seddon, *The Case against ISO 9000*, 2nd ed. (Dublin: Oak Tree Press, 2000), 1. Kristina Tamm Hallström writes, "A standard that was usually presented as having set the tone for the later development of standards for quality systems was the American MIL-Q-9858 that was established in 1959," *Organizing International Standardization: ISO and the IASC in Quest of Authority* (Cheltenham, UK: Edward Elgar, 2004), 53.

4. Tamm Hallström, *Organizing International Standardization*, 1.

5. Carl F. Cargill, *Open Systems Standardization: A Business Approach* (Upper Saddle River, NJ: Prentice Hall PTR, 1997), 216.

6. Quoted in Tamm Hallström, *Organizing International Standardization*, 54.

7. Olle Sturén, "Notebook Listing Travel, 1953-[1987]," Sturén Papers.

8. Lars and Lolo Sturén, interview by C. Murphy, November 20, 2007. On the history of Canada's standards network, see "History," Standards Council of Canada, accessed May 16, 2017, https://www.scc.ca/en/about-scc/history.

9. Lawrence D. Eicher, "International Standardization: Live or Let Die" (keynote address to the Canadian Forum on International Standardization, November 17, 1999), 6, ISO Records.

10. Masami Tanaka, "Address Under Agenda Item 1.1" (29th ISO General Assembly, Ottawa, September 13, 2006), ISO Records.

11. JoAnne Yates, *Control through Communication: The Rise of System in American Management* (Baltimore: Johns Hopkins University Press, 1989).

12. William M. Tsutsui, *Manufacturing Ideology: Scientific Management in Twentieth-Century Japan* (Princeton, NJ: Princeton University Press, 2001), 197–201.

13. ISO Central Secretariat, "What Is a Quality Management System?" in *ISO 9001: 2015* (Geneva: ISO, 2015), PowerPoint Presentation, accessed June 24, 2017, https://www.iso.org/files/live/sites/isoorg/files/standards/docs/en/iso_9001.pptx, slide 3.

14. "Report of the ISO Acting Secretary-General to the ISO General Assembly, Agenda Item 4," Stockholm, September 25, 2002, 9, ISO Records.

15. See, for example, John Braithwaite and Peter Drahos, *Global Business Regulation* (Cambridge: Cambridge University Press, 2000), 280.

16. Winton Higgins, *Engine of Change: Standards Australia Since 1922* (Blackheath, AU: Brandl & Schlesinger, 2005), 215.

17. "Our History," BSI, https://www.bsigroup.com/en-GB/about-bsi/our-history/; "History: Remaining True to Our Focus," CSA Group, http://www.csagroup.org/about-csa-group/history/ (both accessed May 16, 2017).

18. "History," ANAB, accessed May 16, 2017, http://www.anab.org/about-anab/history; Allison Marie Loconto, "Models of Assurance: Diversity and Standardization in Modes of Intermediation," *Annals of the American Academy of Political and Social Science* 671 (2017): 112–13.

19. "CB [certification body] Registry," ANAB, http://anabdirectory.remoteauditor.com; Perry Johnson Registrars, homepage, http://www.pjr.com; "Directory of Accredited Management Certification Bodies," Standards Council of Canada, https://www.scc.ca/en/accreditation/management-systems/directory-of-accredited-bodies-and-scopes, all accessed May 16, 2017.

20. "Accreditation by the ANSI-ASQ National Accreditation Board Granted to the Standards Institution of Israel, Quality and Certification Division," ANAB, accessed June 16, 2017, http://anab.jadianonline.com/Certificate.mvc?PKey=C264B28D-EABB-463C-AA49-9E05396FAC7E&useId=true&OrgId=e26e0b9e-9b4f-4434-8a00-05ea809a2d7a.

21. "Survey 2015," spreadsheet from ISO, accessed May 16, 2017, https://www.iso.org/the-iso-survey.html.

22. Higgins, *Engine of Change*, 215; "Survey 2015."

23. Denise Robitaille, "ISO 9000: Then and Now," *Quality Digest* 25, no. 11 (2006): 27; David Verboom, "The ISO 9001 Quality Approach: Useful for the Humanitarian Aid Sector?," ReliefWeb, January 23, 2002, www.reliefweb.int/rw/rwb.nsf/AllDocsByUNID/25f9cf5a7c0b4ab0c1256b4b00367719; Indian astrologer Rajendra Raaj Sudhanshu advertises his ISO 9001:2008 certification to his global clients on his website, http://www.sudhanshu.com/about_astrologer.htm, both accessed May 22, 2017.

24. "Survey 2015."

25. Eitan Naveh and Alfred Marcus, "Achieving Competitive Advantage through Implementing a Replicable Management Standard: Installing and Using ISO 9000," *Journal of Operations Management* 24, no. 1 (2005): 1–26; Pavel Castka, Daniel Prajogo, Amrik Sohal, and

Andy C. L. Yeung, "Understanding Firms' Selection of Their ISO 9000 Third-Party Certifiers," *International Journal of Production Economics* 162 (2015): 125–33.

26. Khalid Nadvi and Frank Wältring, "Making Sense of Global Standards," Institut für Entwicklung und Frieden der Gerhard-Mercator-Universität Duisburg, Heft 58 (2002); Isin Guler, Mauro F. Guillén, and John Muir Macpherson, "Global Competition, Institutions, and the Diffusion of Organizational Practices: The International Spread of ISO 9000 Quality Certificates," *Administrative Science Quarterly* 47, no. 2 (2002): 207–32; Nick Johnstone and Julien Labonne, "Why Do Manufacturing Facilities Introduce Environmental Management Systems? Improving and/or Signaling Performance," *Ecological Economics* 68, no. 3 (2009): 719–30; Xun Cao and Aseem Prakash, "Growing Exports by Signaling Product Quality: Trade Competition and the Cross-National Diffusion of ISO 9000 Quality Standards," *Journal of Policy Analysis and Management* 30, no. 1 (2011): 111–35; Cornelia Stortz, "Compliance with International Standards: The EDIFACT and ISO 9000 Standards in Japan" *Social Science Japan Journal* 10, no. 2 (2007): 217–41.

27. "ISO Flags, ISO Banners, ISO Logos," Standards Flags, accessed May 21, 2017, http://standardflags.com; information on the different standards from the relevant web pages at https://www.iso.org/. OHSAS 18001 was revamped and republished as ISO draft standard number 45001 in 2017, the organization's first labor standard.

28. ISO, *Recommendations from ISO TC 176 on Communicating and Marketing the ISO 9000:2000 Revisions* (Geneva: ISO, November 2000), 8.

29. Higgins, *Engine of Change*, 333.

30. Higgins, *Engine of Change*, 334.

31. Higgins, *Engine of Change*, 331.

32. James A. Thomas, "A Better Way of Doing Things," *ASTM Standardization News*, May 1999, 1.

33. ASTM, "Name Change Reflects Global Scope" (ASTM International news release #6261, December 11, 2001), accessed September 15, 2018, https://www.astm.org/HISTORY/astm_changes_name.pdf.

34. ASTM, "ASTM and ISO Additive Manufacturing Committees Approve Joint Standards under Partner Standards Developing Organization Agreement" (ASTM International News Release #9389, June 3, 2013), accessed September 15, 2018, https://www.astm.org/cms/drupal-7.51/newsroom/astm-and-iso-additive-manufacturing-committees-approve-joint-standards-under-partner. This standard was another instance of an anticipatory standard in an emerging computer-based industry, as discussed in chapter 7.

35. "Memorandum of Understanding between IULTCS and ISO on Cooperation in the Development of Standards Associated with the Testing of Tanned Leather and Tanning Products," Geneva, December 12, 2005, IULTCS, accessed August 1, 2017, http://www.iultcs.org/pdf/MoU_IULTCS.pdf; Duff Johnson, "It Just Works: PDF Turns 20!," PDF Association, June 15, 2013, accessed August 1, 2017, https://www.pdfa.org/it-just-works-pdf-turns-20/. In 2008, ISO took over maintenance of PDF (portable documents format) standards from their original corporate developer, Adobe Systems, working in cooperation with the PDF Association and the Association for Digital Document Standards.

36. Trond Arne Undheim, "The Messy Globalization of Standards," *Trond's Opening Standard* (Oracle blog), May 20, 2009, accessed May 22, 2017, https://web.archive.org/web/20090621101432/https://blogs.oracle.com/trond/.

37. Seddon, *Case against ISO 9000*, 142–43.

38. Deborah Cadbury, *Chocolate Wars: The 150-Year Rivalry between the World's Greatest Chocolate Makers* (New York: Public Affairs, 2010), quoted phrases from xii.

39. John G. Ruggie, "International Regimes, Transactions, and Change: Embedded Liberalism in the Postwar Economic Order," *International Organization* 36, no. 4 (1982): 379–415.

40. Paul M. Goldberg and Charles P. Kindleberger, "Toward a GATT for Investment: A Proposal for the Supervision of the International Corporation," *Law and Policy in International Business* 2 (1970): 295–325.

41. Tagi Sagafi-Nejad in collaboration with John H. Dunning, *The UN and Transnational Corporations: From Code of Conduct to Global Compact* (Bloomington: Indiana University Press, 2008), 23-33.

42. John Bolton, "Should We Take Global Governance Seriously?," *Chicago Journal of International Law* 1 (2000): 218 (first quote), 220–21 (second quote).

43. Sagafi-Nejad, *UN and Transnational Corporations*, 121. Boutros-Ghali moved the UNCTC from New York to Geneva, severely cut its staff, and made it a minor bureau within the UN Conference on Trade and Development.

44. Sagafi-Nejad, *UN and Transnational Corporations*, 71–73.

45. Craig N. Murphy and JoAnne Yates, *The International Organization for Standardization: Global Governance through Voluntary Consensus* (London: Routledge Press, 2009), 77–80.

46. "Survey 2015."

47. Murphy and Yates, *International Organization for Standardization*, 78–79; Aseem Prakash and Matthew Potoski, *The Voluntary Environmentalists: Green Clubs, ISO 14001, and Voluntary Environmental Regulations* (Cambridge: Cambridge University Press, 2006); Paul Langley, "Transparency in the Making of Global Environmental Governance," *Global Society* 15 (2001): 73–92; the Canadian study is Olivier Borial, "Corporate Greening through ISO 14001: A Rational Myth?," *Organization Science* 18, no. 1 (2007): 127–46, quotations from 127. "Standards Catalogue: ISO/TC 207 Environmental Management," ISO, accessed June 21, 2017, https://www.iso.org/committee/54808/x/catalogue/p/1/u/0/w/0/d/0.

48. "Survey 2015."

49. Kenneth W. Abbott and Duncan Snidal, "The Governance Triangle: Regulatory Standards Institutions and the Shadow of the State," in *The Politics of Global Regulation*, ed. Walter Mattli and Ngaire Woods (Princeton, NJ: Princeton University Press, 2009), 49–50. The estimate of the number of existing codes comes from Gare Smith and Dan Feldman, *Company Codes of Conduct and International Standards: An Analytical Comparison*, 2 vols. (Washington, DC: World Bank, 2003).

50. Suzanne Shanahan and Sanjeev Khagram, "Dynamics of Corporate Responsibility," in *Globalization and Organization: World Society and Organizational Change*, ed. Gili S. Drori, John M. Meyer, and Hokyu Wang (Oxford: Oxford University Press, 2006), 203, 222; Larry Catá Backer, "Creating Private Norms for Corporate Social Responsibility in Brazil," *Law at the End of the Day*, June 25, 2006, accessed July 9, 2017, lcbackerblog.blogspot.com/2006/06/creating-private-norms-for-corporate.html.

51. Declan Walsh and Steven Greenhouse, "Certified Safe, a Factory in Karachi Still Quickly Burned," *New York Times*, December 7, 2012, A1.

52. Alice Tepper Marlin, speaking in the film *Architect of Corporate Responsibility: The Story of Alice Tepper Marlin and the Founding of Social Accountability International* (Arlington, VA: Ashoka Global Academy for Social Entrepreneurship, in partnership with Skoll Foundation, 2006).

53. Tepper Marlin, speaking in *Architect of Corporate Responsibility*; the plan Tepper Marlin conceived for SA8000 was influenced by the work of her husband, John Tepper Marlin, who wrote a path-breaking 1973 *Journal of Accountancy* article suggesting how firms could measure, and accountants could verify, environmental pollution and in 1988 conducted the first comprehensive social audit, at Ben & Jerry's, the New England ice cream maker. Alice Tepper Marlin and John Tepper Marlin, "A Brief History of Social Reporting," *Business Respect*, no. 51 (2003), www.businessrespect.net/page.php?Story_ID=857.

54. Quotations from Tepper Marlin, speaking in *Architect of Corporate Responsibility*; background from Alice Tepper Marlin, interview by Honor McGee, August 18, 2009.

55. "Number of SA8000-Certified Organisations by Year," Social Accountability Accreditation Services, http://www.saasaccreditation.org/?q=node/110; "SA8000 Certified Organisations," Social Accountability Accreditation Services, http://www.saasaccreditation.org/certfacilitieslist, both accessed June 21, 2017.

56. *Setting the Standard for the Global Economy: Strategies from Alice Tepper Marlin, Founder of Social Accountability International* (Arlington, VA: Ashoka Global Academy for Social Entrepreneurship, in partnership with Skoll Foundation, 2006); Deborah Leipziger, ed., *SA8000: The First Decade: Implementation, Influence, and Impact* (Sheffield: Greenleaf, 2009); the study outlining how rigorous evaluative research might be done is included in that volume, Michael J. Hiscox, Claire Schwartz, and Michael W. Toffel, "Evaluating the Impact of SA8000 Certification," 147–65.

57. Keller Easterling, *Extrastatecraft: The Power of Infrastructure Space* (London: Verso, 2014), 197.

58. Jean Krasno, "Kofi Annan: From Ghana to the World Stage," in *Personality, Political Leadership, and Decision Making: A Global Perspective*, ed. Jean Krasno and Sean LaPides (Santa Barbara, CA: Praeger, 2015), 337–58; and Kofi Annan, "The Quiet Revolution," *Global Governance* 4 (1998): 123–38.

59. Kofi Annan quoted in George Kell and David Levin, "The Evolution of the Global Compact Network: An Historic Experiment in Learning and Action" (paper presented at the Academy of Management Annual Conference, Building Global Networks, Denver, August 11–14, 2002), accessed September 5, 2018, http://citeseerx.ist.psu.edu/viewdoc/download?doi=10.1.1.493.8153&rep=rep1&type=pdf, 6.

60. Annan quoted in Kell and Levin, "Global Compact, 6–7.

61. Biographical details from Roundtable: John G. Ruggie, Distinguished Scholar in International Political Economy, 40th Annual Convention of the International Studies Association, Washington, DC, February 19, 1999.

62. Kell and Levin, "Global Compact," 35.

63. Kell and Levin, "Global Compact," 35.

64. Transparency International, "International Corporations Decide to Add Anti-corruption Principle to UN Global Compact" (Transparency International press release, June 24, 2004).

65. Kell and Levin, "Global Compact," 39–40.

66. Kell and Levin, "Global Compact," 10.

67. Robert Beckett and Jan Jonker. "AccountAbility 1000: A New Social Standard for Building Sustainability," *Managerial Auditing Journal* 17 (2002): 36–42; "Global Sullivan Principles of Social Responsibility," CSRIdentity.com, accessed June 21, 2017, http://csridentity.com/globalsullivanprinciples/index.asp; "GRI's History," GRI, accessed June 22, 2017, https://www.globalreporting.org/information/about-gri/gri-history/Pages/GRI's%20history.aspx.

68. Tepper Marlin, interview by McGee.

69. The list of members of the Global Compact Council appears in Kell and Levin, "Global Compact," 36.

70. Tepper Marlin, interview by McGee.

71. "Our Participants," UN Global Compact, accessed June 21, 2017, https://www.unglobalcompact.org/what-is-gc/participants/.

72. S. Prakash Sethi and Donald H. Schepers, "United Nations Global Compact: The Promise-Performance Gap," *Journal of Business Ethics* 122 (2014): 193–208. On the Global Compact's policies regarding branding and use of its logo, see "UN Global Compact Logo

Policy," UN Global Compact, December 2015, accessed June 24, 2017, https://www.unglobal compact.org/docs/about_the_gc/logo_policy/Logo_Policy_EN.pdf.

73. "Human Rights Principles and Responsibilities for Transnational Corporations and Other Business Enterprises," UN Doc. E/CN.4/Sub.2/2002/XX, E/CN.4/Sub.2/2002/WG.2/WP.1 (February 2002 for discussion in July/August 2002) was the most important draft, University of Minnesota Human Rights Library, accessed August 18, 2017, http://hrlibrary.umn .edu/principlesW-OutCommentary5final.html.

74. John G. Ruggie, "Business and Human Rights: Treaty Road Not Traveled," *Ethical Corporation Newsdesk*, May 6, 2008, accessed August 9, 2017, https://sites.hks.harvard.edu /m-rcbg/news/ruggie/Pages%20from%20ECM%20May_FINAL_JohnRuggie_may%2010 .pdf.

75. International Chamber of Commerce and International Organisation of Employers, "The Sub-commission's Draft Norms," March 2004, https://www.humanrights.ch/upload /pdf/070706_ICC_IOE_subcomm.pdf; quotation from p. 10.

76. Thosapon Mengweha, "ISO 26000, a Social Responsibility Standard: Lesson Learned and Expectation to Drive the Pragmatic Sustainable Development Approach" (MA Thesis, Mälardalen University, Västerås, Sweden, May 23, 2007), 29.

77. "New Work Item Proposal: Social Responsibility," ISO26000.info, October 7, 2004, accessed June 23, 2017, http://iso26000.info/wp-content/uploads/2016/02/2004-10-07_-_New _work_item_proposal_-_Social_responsibility.pdf.

78. "History of ISO 26000," ISO26000.info, accessed June 23, 2017, http://iso26000.info /history/; Kristina Sandberg, interview by Maria Nassén, January 16, 2008; Dorothy Bowers, "Making Social Responsibility the Standard," *Quality Progress* 39 (April 2006): 35–38.

79. "Social Responsibility," ISO Working Group on Social Responsibility, accessed August 2, 2017, http://isotc.iso.org/livelink/livelink/fetch/2000/2122/830949/3934883/3935096 /home.html; "About the ISO26000.info Website," ISO26000.info, http://iso26000.info/about/; "Miljöpriset till svenskar som enade världen kring socialt ansvar," Sveriges Ingenjörer press release, April 18, 2016), https://www.sverigesingenjorer.se/Aktuellt-och-press/Nyhetsarkiv /Pressmeddelanden/Miljopris-till-svenskar-som-enade-varlden-kring-socialt-ansvar, all accessed June 24, 2017.

80. Rochelle Zaid, SAI senior director for standards and impacts, interview by Honor McGee, August 18, 2009; Tineke Egyedi and Sebastiano Toffaletti, "Standardising Social Responsibility: Analysing ISO Representation Issues from an SME Perspective," in *Proceedings of the 13th EURAS Workshop on Standardisation*, ed. Kai Jakobs (Aachen: Wissenschaftsverlag Mainz, 2008), 121–36; Kristina Sandberg, email to C. Murphy, May 19, 2008; Tepper Marlin, interview by McGee; Pavel Castka and Michaela A. Balzarova, "The Impact of ISO 9000 and ISO 14000 on Standardisation of Social Responsibility—an Inside Perspective," *International Journal of Production Economics* 113 (2008): 74–87; Halina Ward, "ISO 26000: Social Responsibility Talks Tread on Government Toes," *Ethical Corporation*, May 15, 2009, accessed June 24, 2017, http://www.ethicalcorp.com/content/iso-26000-social-responsibility-talks -tread-government-toes.

81. Adrian Henriques, *Standards for Change: ISO 26000 and Sustainable Development* (London: International Institute for Environment and Development, 2012), 21.

82. Oshani Perera, *How Material Is ISO Social Responsibility to Small and Medium-Sized Enterprises?* (Winnipeg: International Institute for Sustainable Development, 2008), 16; Hendriques, *Standards for Change.*

83. "ISO and Sustainability," iso26000.org, accessed July 3, 2017, http://iso26000.info /isosust/.

84. Zaid, interview by McGee.

85. Tepper Marlin, interview by McGee; Zaid, interview by McGee.

86. Tepper Marlin, interview by McGee.

87. Zaid, interview by McGee.

88. ISEAL Alliance, *Standard-Setting Code*, draft revision 5.2—March 30, 2014, side-by-side version, accessed August 9, 2017, available only to ISEAL members and subscribers, https://www.isealalliance.org/sites/default/files/Standard-Setting%20Code%20V5.2-Side-by-side_FINAL%201%20Apr2014.pdf.

89. Jason Potts et al., eds., *The State of Sustainability Initiatives Review 2014: Standards and the Green Economy* (Winnipeg: International Institute for Sustainable Development and International Institute for Environment and Development, 2014), 48; see also Loconto, "Models of Assurance," 112–32.

90. Potts et al., *State of Sustainability Initiatives Review 2014*, 31, 38.

91. Craig N. Murphy, "Globalizing Standardization: The International Organization for Standardization," *Comparativ—Zeitschrift für Globalgeschichte und vergleichende Gesellschaftsforschung* 23 (2013): 137–53.

92. Ethical Corporation Institute, *Guide to Industrial Initiatives in Corporate Social Responsibility* (London: Ethical Corporation Institute 2009).

93. Ruggie, "Business and Human Rights," 832.

94. John G. Ruggie, "Prepared Remarks" (Public Hearings on Business and Human Rights, European Parliament, Brussels, April 16, 2009), accessed July 5, 2017, https://business-humanrights.org/sites/default/files/reports-and-materials/Ruggie-remarks-to-European-Parliament-16-Apr-2009.pdf. Ruggie was influenced by the research conducted by Richard M. Locke and his colleagues, see Richard M. Locke, *The Promise and Limits of Private Power: Promoting Labor Standards in the Global Economy* (Cambridge: Cambridge University Press, 2013).

95. Walsh and Greenhouse, "Certified Safe." The fire-related updates are discussed in SAI, "Fire Safety a Key Focus in SA8000 Revision," March 11, 2013, accessed September 15, 2018, http://www.sa-intl.org/index.cfm?fuseaction=Page.ViewPage&PageID=1435#.WrvzvGaZO3g.

96. Scott Cooper emails to C. Murphy, January 16, February 19, and March 10, 2015.

97. Scott Cooper, "The International Labor Organization and the International Organization for Standardization," *Professional Safety* 63 (October 2018): 70–74.

98. Klaus Schwab and Hein Kroos, *Moderne Unternehmensführung im Maschinenbau* (Frankfurt: Maschinenbau-Verlag, 1971), 20; the translation appears in WEF, *A Partner in Shaping History: The First 40 Years 1971–2010* (Geneva: WEF, 2009), 7.

99. Richard Samans, Klaus Schwab, and Mark Malloch Brown, "Running the World, after the Crash," *Foreign Policy* 184 (2011): 80–83, gives an official description of the process; Harris Gleckman, "Multi-stakeholderism: A Corporate Push for a New Form of Global Governance," in *State of Power 2016: Democracy, Sovereignty, and Resistance* (Amsterdam: Transnational Institute, 2016), 91–106, provides a critical view.

100. WEF, *Everybody's Business: Strengthening International Cooperation in a More Interdependent World* (Geneva: WEF, 2010).

101. Gleckman, "Multi-stakeholderism," 97. Gleckman's exhaustive critique, "Readers' Guide: Global Redesign Initiative," can be found at https://www.umb.edu/gri, accessed August 21, 2017.

102. Braithwaite and Drahos, *Global Business Regulation*, 280, 579.

103. A balanced assessment of the impact of the new CSR standards appears in Tim Bartley, "Re-centering the State," in *Rules without Rights: Land, Labor, and Private Authority in the Global Economy* (Oxford: Oxford University Press, 2018), 258–83.

Conclusion

1. Mark Mazower, *Governing the World: The Rise and Fall of an Idea 1815 to the Present* (New York: Penguin Books, 2012), 201–15.

2. Swedish Standards Institute, *Money Doesn't Make the World Go Round; Standards Do* (Stockholm: SIS Forum AB, 2013), 5.

3. See, for example, David Singh Grewal *Network Power: The Social Dynamics of Globalization* (New Haven, CT: Yale University Press, 2008); Keller Easterling, *Extrastatecraft: The Power of Infrastructure Space* (London: Verso, 2014), 171–209; Jonathan Sterne, *MP3: The Meaning of a Format* (Durham, NC: Duke University Press, 2012); and Lawrence Busch, *Standards: Recipes for Reality* (Cambridge, MA: MIT Press, 2011). Although we are not aware of feminist critiques of standardization, this would also be a very promising approach.

4. Dan Hoolihan, interview by J. Yates; and Don Heirman, interview by J. Yates, both on August 9, 2014.

5. The head of standardization of a major software company at the closing session of the 22nd EURAS Conference, DIN headquarters, Berlin, June 30, 2017.

6. Wang Ping, interview by C. Murphy, April 1, 2018.

7. Russ Housley, interview by J. Yates, May 14, 2013, and May 16, 2013.

8. Wang Ping and Zheng Liang, "Beyond Government Control of China's Standardization System—History, Current Status, and Reform Suggestions," in *Megaregionalism 2.0: Trade and Innovation within Global Networks*, ed. Dieter Ernst (Singapore: World Scientific, 2018), 333–61.

9. Wang, interview by Murphy.

10. Hoolihan, interview by Yates.

11. Craig N. Murphy, "Globalizing Standardization: The International Organization for Standardization," *Comparativ—Zeitschrift für Globalgeschichte und vergleichende Gesellschaftsforschung* 23, nos. 4/5 (2014): 137–53.

12. Amy Cohen, "On Being Anti-imperial: Consensus Building, Anarchism, and ADR," *Law, Culture, and the Humanities* 9, no. 2 (2013): 243–60.

index